Walter Dittrich (Ed.)

Recent Developments in Particle and Field Theory

Topical Seminar, Tübingen 1977

With 86 Figures

Friedr. Vieweg & Sohn   Braunschweig/Wiesbaden

CIP-Kurztitelaufnahme der Deutschen Bibliothek

Recent developments in particle and field theory
Walter Dittrich (ed.). — Braunschweig, Wiesbaden:
Vieweg, 1979.

ISBN-13:978-3-528-08426-4   e-ISBN-13:978-3-322-83630-4
DOI: 10.1007/978-3-322-83630-4

NE: Dittrich, Walter [Hrsg.]

1979

ISBN-13:978-3-528-08426-4

Preface

This is a collection of lectures and seminars delivered at the "Symposium on Particles and Fields" held at Tübingen University, West Germany, from June 20th to July 1st, 1977, on the occasion of the University's 500th anniversary.

We were very fortunate that so many excellent colleagues from Europe and the U. S. contributed their ideas to this meeting, whose main purpose ist was to cover a brod spectrum of the various aspects in the current development of particles and fields, rahter than to focus on a single subject. It was interesting to see source-, quark-, and string-people side by side attacking the very same unsolved problems in particle physics. Exchange of ideas and techniques between the diverse representatives of field- and particle physics was the principal goal of the meeting.

Last not least, we are most grateful to the President of our University A. Theis, and his assistant, H. E. Lang, for the generous support that enabled us to make this symposium a profitable and pleasant experience for all of us.

The Editor
Walter Dittrich

This volume is dedicated to
the memory of
Benjamin Lee

Contents

Using Field Theory in Hadron Physics

Henry D. I. Abarbanel

Fermi National Accelerator Laboratory, Batavia, Illinois 60510

I. Introduction

This decade has seen a marvelous return to quantum field theory among theorists in high energy physics. The compelling beauty of non-Abelian gauge theories, the striking empirical support for spontaneous symmetry breaking in weak interaction theories, the deep connection between asymptotic freedom and the approximate scaling in inelastic lepton scattering on hadrons, the entrancing suggestion of quark confinement—all this and more has drawn our focus once again on quantum field theories. We have learned to view <u>field theory</u> as providing us with <u>fundamental degrees of freedom</u> rather than thinking that each new particle or resonance as requiring a new field for its description. Indeed the idea that some small set of degrees of freedom (quarks and gluons) provides the basis for all observed mesons and baryons seems to be a concrete realization of the ideas of "nuclear democracy" advocates of the last decade. [Increasing numbers of them have been seen with path integrals and Lagrangians lately.] It makes explicit the concept that all hadrons are composites; not of each other, though through unitarity all the hadrons can become the other hadrons within the restrictions of conservation of charge, baryon number, isospin, etc. It is a deeper way: they are all composites of quarks and gluons.

These lectures have no pretentions to cover all possible topics in the connection of field theory and hadron physics. Rather the goal is much more modest: I hope to touch on a number of tantalizing questions which will serve to some extent as an introduction for the student as well as the research person curious for more than a peek. Several "old" topics will be treated—the renormalization group and the infrared and ultraviolet limits of field theory, choosing Quantum Chromodynamics (QCD) from among all theories, various thoughts on spontaneous mass generation. Some newer ones are discussed here too—ideas on color confinement, instantons and the vacuum state in QCD, and related topics.

As general background material I recommend the article on "Gauge Theories" by E.S. Abers and B.W. Lee, Physics Reports 9C, 1 (1973); the lectures by S. Coleman at the 1975 Erice Summer School; the book by J.C. Taylor, Gauge Theories of Weak Interactions (Cambridge U. Press, 1976); and the review article by R. Jackiw, Rev. Mod. Phys. 49, 681 (1977).

II. The Renormalization Group and Some Consequences

Quantum field theories of relevance to particle physics all have divergences when one calculates in perturbation theory about the free theory characterized by propagators

$$\text{(Spin Numerator)}/(m^2 - p^2 - i\varepsilon) \tag{1}$$

Such theories are not defined by the classical Lagrange density; one has to give a prescription for making the theory finite in every order of perturbation theory before a calculational procedure of recognizable validity emerges. The generally accepted manner for doing this is to define the theory by giving the value of a few basic quantities (mass, coupling constant, ...) at some point in momentum space. So one takes the original theory defined by the classical Lagrangian and renormalizes, so the divergences are absorbed in scales of wave functions (or field operators) and other physically harmless locations. Since there is an enormous arbitrariness in how, precisely, one renormalizes, we can anticipate an invariance of physical quantities on changing the point in momentum space where that renormalization is done. The expression of that invariance is the renormalization group. The behavior of classes of quantum field theories under this group allows one to select those with controllable infrared or ultraviolet behavior and thus on the basis of the behavior of experiments which probe long or short wavelength phenomena to choose acceptable field theories.

To illustrate this in action let's look at a scalar field in D dimensions of space-time with Lagrangian density

$$\mathcal{L} = \tfrac{1}{2}(\partial_\mu \phi_0(x))^2 - \lambda_0(\phi_0(x))^N \qquad . \qquad (2)$$

When λ_0 is dimensionless, namely when $N = 2D/(D-2)$, all the divergences of quantities expanded in a perturbation series in λ_0 are logarithmic and the theory can be renormalized, i.e. made finite, by redefining the field

$$\phi(x) = Z^{-\frac{1}{2}}\phi_0(x) \qquad (3)$$

and coupling

$$\lambda = Z_\lambda Z^{N/2}\lambda_0 \qquad , \qquad (4)$$

where the dimensionless (infinite) factors Z and Z_λ are constructed so all Green functions in the theory are finite. These renormalization factors are defined by giving the value of certain Green functions at some point $p^2 = -\mu^2$, $\mu^2 > 0$, in momentum space. These Green functions are given by

$$G_0^{(n)}(p_1, \cdots p_n, \lambda_0)\,\delta^D(\sum_{j=1}^{n} p_j) =$$

$$\int dx_1 ... dx_n\, e^{-i\sum_{j=1}^{n} p_j \cdot x_j} \langle 0| T(\phi_0(x_1)...\phi_0(x_n))|0\rangle \qquad . \qquad (5)$$

To lowest order in λ_0 we have for $G^{(2)}$ and $G^{(N)}$

$$G^{(2)}(p^2, \lambda_0) = i/(p^2 + i\epsilon) \quad , \tag{6}$$

and

$$G^{(N)}(p_j, \lambda_0) = -i\lambda_0/(2\pi)^{D/2\,(N-2)} \quad . \tag{7}$$

We define renormalized Green functions by

$$G^{(n)}(p_1, \dots p_n, \lambda, \mu) = Z^{n/2} G_0^{(n)}(p_1, \dots p_n, \lambda_0) \quad , \tag{8}$$

and require (this is the <u>re</u>normalization)

$$\frac{\partial}{\partial p^2} i\, G^{(2)}(p^2, \lambda, \mu)^{-1} \bigg|_{p^2 = -\mu^2} = 1 = Z \frac{\partial}{\partial p^2} i\, G_0^{(2)}(p^2, \lambda_0)^{-1} \bigg|_{p^2 = -\mu^2} \quad , \tag{9}$$

and

$$G^{(N)}(p_1, \dots p_N, \lambda, \mu) \bigg|_{p^2 = -\mu^2} = \frac{-i\lambda}{(2\pi)^{D/2\,(N-2)}} \tag{10}$$

$$= Z^{N/2} G_0^{(N)}(p_j, \lambda_0) \bigg|_{p^2 = -\mu^2} \quad . \tag{11}$$

These determine Z and then Z_λ once one has calculated $G_0^{(2)}$ and $G_0^{(N)}$ to whatever accuracy desired using perturbation theory in λ_0 and some method of cutting of the divergence integrations. After rescaling by Z and Z_λ via (3) and (4) the cutoff is sent to infinity with ϕ, λ, and μ held fixed. What is remarkable, then, is that the resulting theory is then finite to all orders in λ.

But what are we to make of μ? We began with a Lagrangian with a field of dimension $(D-2)/2$ and a dimensionless coupling. Now we also seem to have a mass scale μ. Since μ is arbitrary, the consequences of the theory should be independent of it. We can insure that by noting that since $G_0^{(n)}(p_j, \lambda_0)$ never heard of μ

$$\mu \frac{\partial}{\partial \mu} \left(G_0^{(n)}(p_j, \lambda_0) \right) = 0 \qquad . \tag{12}$$

The appearance of μ will then be only apparent in this sense: only one real physical parameter enters this problem, namely the coupling λ. Since λ is defined by a Green function evaluated at $p_j^2 = -\mu^2$, λ may be traded off for μ. Physical masses, for example, must be of the form

$$M_{Physical} = \mu F(\lambda) \qquad , \tag{13}$$

but $F(\lambda)$ must depend on $\lambda(\mu)$ in such a way that

$$\frac{\partial}{\partial \mu} M_{Physical} = 0 \qquad . \tag{14}$$

From (12) and the definition of $G^{(n)}$ we learn

$$\left[\mu \frac{\partial}{\partial \mu} + \beta(\lambda) \frac{\partial}{\partial \lambda} - \frac{n}{2} \gamma(\lambda) \right] G^{(n)}(p_1, \dots p_n, \lambda, \mu) = 0 \qquad , \tag{15}$$

where

$$\beta(\lambda) = \mu \frac{\partial}{\partial \mu} \lambda \bigg|_{\lambda_0 \text{ fixed}} \qquad , \tag{16}$$

and

$$\gamma(\lambda) = \mu \frac{\partial}{\partial \mu} \log Z \bigg|_{\lambda_0 \text{ fixed}} \qquad . \tag{17}$$

This differential equation provides restrictions on the form $G^{(n)}$ can take, and, even more important, knowing the functions $\beta(\lambda)$ and $\gamma(\lambda)$ in perturbation theory can frequently give information on $G^{(n)}$ not possible to see in perturbation theory. We'll come to some examples of this.

Suppose the dimensions of $G^{(n)}$ are $\mathcal{D}$, then the solution of (15) is

$$G^{(n)}(\xi p_i, \lambda, \xi) = G^{(n)}(p_i, \tilde{\lambda}(-\log \xi), \mu) \times$$

$$\times \ \exp + \int_{-\log \xi}^{0} [\mathcal{D} + \frac{n}{2} \gamma(\tilde{\lambda}(t))] \, dt \tag{18}$$

where

$$\frac{d\tilde{\lambda}(t)}{dt} = -\beta(\tilde{\lambda}(t)) \ , \quad \tilde{\lambda}(0) = \lambda \qquad , \tag{19}$$

is the <u>running coupling constant</u>. It tells us that at a <u>scale</u> determined by ξ the strength of the non-linearity in $\mathcal{L}$ may vary. Clearly that is because we do not do perturbation theory in λ but really in the operator $\lambda \phi^N$ which does depend on x.

As ξ varies, the argument of $\tilde{\lambda}(-\log \xi)$ moves about and approaches zeroes of the function $\beta(\lambda)$. Clearly when λ is near λ_1 where $\beta(\lambda_1) = 0$, $\lambda(t) = 0$ and the effective coupling in $G^{(n)}$ is not λ, the renormalized coupling, but λ_1. Suppose near λ_1

$$\beta(\lambda) = \beta_1(\lambda - \lambda_1) \quad , \tag{20}$$

then

$$\tilde{\lambda}(-\log \xi) = \lambda_1 + \xi^{\beta_1}(\lambda - \lambda_1) \quad . \tag{21}$$

If $\beta_1 > 0$, then as $\xi \to 0$, $\tilde{\lambda}(-\log \xi) \to \lambda_1$. If $\beta_1 < 0$, then as $\xi \to \infty$, $\tilde{\lambda}(-\log \xi) \to \lambda_1$. Zeroes of β with positive slope govern the infrared behavior of the theory ($\xi p_i \to 0$); zeroes with negative slope, the ultraviolet behavior ($\xi p_i \to \infty$).

In QED in lowest order perturbation theory

$$\beta_{QED}(e) = +\frac{b}{2} e^3 \ , \ b > 0 \quad , \tag{22}$$

so the effective charge is

$$\tilde{e}^2(-\log \xi) = \frac{e^2}{1 - be^2 \log \xi} \quad , \tag{23}$$

so perturbation theory for $G^{(n)}(p_i, \ \tilde{e}(-\log \xi), \mu)$ is good for $\xi \to 0$ or really for $\xi \ll \exp 1/(be)^2$ with b order unity, and $e^2 \approx 1/137$. The infrared properties of QED are then calculable in perturbation theory; e.g. the electron or muon magnetic moment. The short distance behavior is unknown and the formula for $\tilde{e}^2(-\log \xi)$ clearly breaks down for ξ large.

In a pure Yang-Mills theory with the Lagrange density

$$\mathscr{L}_{YM} = -\frac{1}{4g^2} F_{\mu\nu}^a F_{\mu\nu}^a, \quad a = \text{gauge group index} \tag{24}$$

$$F_{\mu\nu}{}^{a} = \partial_{\mu}A_{\nu}{}^{a} - \partial_{\nu}A_{\mu}{}^{a} + f_{abc}A_{\mu}{}^{b}A_{\nu}{}^{c} \quad , \tag{25}$$

the function $\beta(g)$ for small g is

$$\beta(g) = -\frac{B}{2}g^{3} + O(g^{5}) \ , \ B > 0 \tag{26}$$

and

$$\tilde{g}(-\log \xi)^{2} = \frac{g^{2}}{1 + Bg^{2}\log \xi} \quad . \tag{27}$$

so the ultraviolet or short distance behavior of the theory is given by perturbation theory in the small coupling $\tilde{g}(-\log \xi)$. When we add fermions to the theory the total Lagrangian is

$$\mathscr{L} = \mathscr{L}_{YM} + \bar{\psi}[i \not{\partial} + T^{a}\not{A}^{a} - m_{0}]\psi \quad , \tag{28}$$

with T^{a} a representation matrix for the representation of the gauge group in which the fermions sit. If the gauge group is SU(3) and the fermions (quarks) are members of the fundamental triplet representation then B in (20) is positive if the number of such fermions (number of <u>flavors</u>) is ≤ 16. The phenomenon of $B > 0$ and predictable ultraviolet behavior is called <u>asymptotic freedom.</u>

Another consequence of the renormalization group follows from looking at the renormalized propagator $G^{(2)}(p^{2}, \lambda, \mu)$ when there is no mass term in the original Lagrangian. By dimensional reasoning

$$G^{(2)}(p^{2},\lambda,\mu) = \mu^{-2}F\left(\frac{p^{2}}{\mu^{2}}, \lambda\right) \quad , \tag{29}$$

where F is dimensionless. Now the renormalization group says

$$F\left(\frac{p^2}{\mu^2}, \lambda\right) = F\left(1, \tilde{\lambda}(-\log \frac{p^2}{\mu^2})\right)\left(\frac{\mu^2}{p^2}\right) \exp + \int_{-\log p^2/\mu^2}^{0} \gamma(\tilde{\lambda}(t))dt \quad . \tag{30}$$

Define $\chi(\lambda)$ by

$$\frac{d \log \chi(\lambda)}{d\lambda} = -\frac{1}{\beta(\lambda)} \quad , \tag{31}$$

then

$$\chi(\tilde{\lambda}(-\log p^2/\mu^2)) = \frac{p^2}{\mu^2}\chi(\lambda) \quad , \tag{32}$$

and

$$F(1, \tilde{\lambda}(-\log p^2/\mu^2)) = \mathscr{F}\left(\frac{p^2}{\mu^2}\chi(\lambda)\right) \quad , \tag{33}$$

and

$$G^{(2)}(p^2, \lambda, \mu) = \frac{1}{p^2}\mathscr{F}\left(\frac{p^2}{\mu^2}\chi(\lambda)\right) \exp \int_{-\log p^2/\mu^2}^{0} dt\, \gamma(\tilde{\lambda}(t)) \quad . \tag{34}$$

If $G^{(2)}$ has a pole at $p^2 = M^2(\lambda)$ away from $p^2 = 0$, it must take the form

$$M^2(\lambda) = \mu^2\chi(\lambda)^{-1} \times \text{constant}$$

$$= \mu^2 \exp - \int^{\lambda} \frac{dx}{\beta(x)} \tag{35}$$

Suppose

$$\beta(x) = \frac{b}{2} x^3 \qquad , \qquad (36)$$

then

$$M(\lambda)^2 = \mu^2 \exp 1/b\lambda^2 \qquad (37)$$

and we see that in a massless theory mass can be generated dynamically only outside perturbation theory; indeed $M(\lambda)^2$ is zero to every order of expansion in λ.

III. Selecting a Field Theory

With the renormalization group as a tool we can begin the selection of a field theory. {Of course, purists will wonder whether it's a sufficient tool. The wondering should cease, however; the answer is no, but it's about all we have of a non-perturbative nature in field theory.} Years of deep inelastic lepton scattering[1] (Figure 1) indicates that for $|q^2| \gtrsim 1$ $(GeV/c)^2$ the constituents (quarks or partons) of a proton carrying charge or isospin or hypercharge or charm are essentially point-like and free.

Boldly this is assumed to indicate the field theory of hadrons has asymptotic freedom which necessitates a non-Abelian gauge theory plus fermions carrying gauge charge plus ordinary Q, I, Y, C, ... Which gauge theory shall we choose? The answer is connected with hadron spectroscopy; that is, the character of bound states of the hadron degrees of freedom. Following the spirit of the old Fermi-Yang proposal that mesons are $N\bar{N}$ bound states one assumes that mesons are quark-antiquark bound states with three (or more) kinds of quarks carrying isospin, strangeness, charm, etc. for all the internal quantum numbers observed for mesons. Baryons are then made out of qqq states. The properties of the well-established

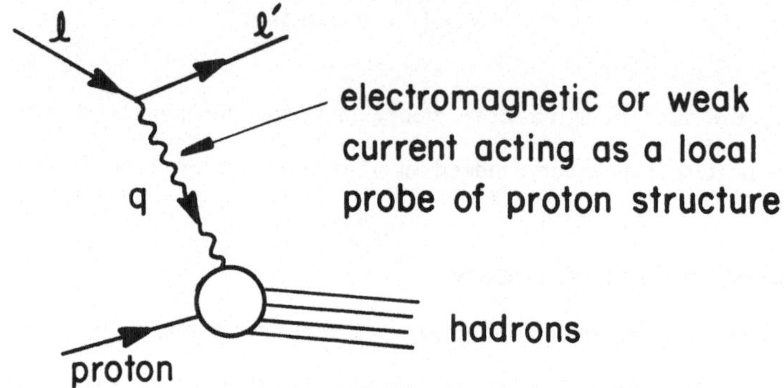

Fig. 1

	Q	I_3	Y	B	C
u	$2/3$	$+1/2$	$1/3$	$1/3$	0
d	$-1/3$	$-1/2$	$1/3$	$1/3$	0
s	$-1/3$	0	$-2/3$	$1/3$	0
c	$2/3$	0	$-2/3$	$1/3$	1

$$Q = I_3 + \frac{Y}{2} + C$$

Fig. 2

quarks are in Figure 2. More quarks will be called for by the discovery of states like the upsilon, if, indeed, it is made out of $q\bar{q}$ as we now presume the π, ρ , ψ , ... to be.

Some examples of the construction of observed states out of q's are as follows:

$$|\pi^+> = |u\bar{d}> \qquad |K^+> = |u\bar{s}> \qquad |\psi> = |c\bar{c}>$$

$$|p> = |uud> \qquad |\Lambda> = |uds> \qquad |\Omega^-> = |sss> \qquad ,$$

(38)

etc. The only flaw in this picture comes when one assumes the quarks in baryons are in relative L = 0 states. This requires $|qqq>$ to be <u>symmetric</u>. Since L = 0 is very likely to be the ground state, we require another label in which to anti-symmetrize the quark wave function. We want quarks to form triplets under the group of this additional degree of freedom, which is called <u>color</u>.[2] The natural color groups are SU(2) and SU(3). In either case a baryon will be

$$|B> = |\epsilon_{\alpha\beta\gamma}\, q_\alpha q_\beta q_\gamma > \quad \alpha, \beta, \gamma = 1, 2, 3 \quad .$$

(39)

Since no quantum number associated with color has been observed, we presume that baryons and mesons are singlets under color transformations. If we choose SU(2) for the color group, then singlets can be made as $|q_\alpha \bar{q}_\alpha>$, $|q_\alpha q_\beta q_\gamma \epsilon_{\alpha\beta\gamma}>$ as desired, but also $|q_\alpha q_\alpha>$, $|\bar{q}_\alpha q_\beta q_\gamma \epsilon_{\alpha\beta\gamma}>$ and other unwanted states are singlets. So choose SU(3). Then we can make $|q_\alpha \bar{q}_\alpha>$ and $|\epsilon_{\alpha\beta\gamma} q_\alpha q_\beta q_\gamma>$ into singlets using only two or three quarks. Of course $|q_\alpha \bar{q}_\alpha\, q_\beta \bar{q}_\beta>$ is a singlet, but it comes up as a problem (or virtue) for heavier bound states than (probably) yet observed.

Now we must have a binding agent to hold the quarks together in the bound states called hadrons. This glue must be free of the Q, Y, I, C,... _flavor_ quantum numbers listed above since it does not interact with the electromagnetic or weak current coupled to leptons. It does carry about 50% of the proton momentum, but not its quantum numbers. The choice made these days is to identify the glue with the gauge bosons of an SU(3) non-Abelian gauge theory, $A_\mu^a(x)$, a = 1,...8, and have them interact with the color of quarks.

So we have the following quarks and gluons:

$q_{\alpha f}(x)$ $\qquad$ $\alpha = 1,2,3$ $\qquad$ Local SU(3) gauge symmetry

$\qquad\qquad\qquad$ f=u,d,s,c,... $\qquad$ Global SU(N) flavor symmetry

$A_\mu^\alpha(x)$ $\qquad$ $\alpha = 1,...8$ $\qquad$ Local SU(3) gauge boson flavor singlet.

For these fields we write the Lagrangian density

$$\mathcal{L}_{QCD} = -\frac{1}{4g^2} F_{\mu\nu}^{\alpha}\, F_{\mu\nu}^{\alpha} + \bar{q}_{\alpha f}\left[i\,\slashed{\partial}\,\delta_{\alpha\alpha'}\delta_{ff'} + \left(\frac{\lambda^\beta}{2}\right)_{\alpha\alpha'} \gamma^\mu A_\mu^{\beta}\,\delta_{ff'} \right.$$

$$\left. - (M_0)_{ff'}\,\delta_{\alpha\alpha'} \right] q_{\alpha'f'} \qquad , \qquad\qquad (40)$$

this defines the theory now popularly known as Quantum Chromodynamics (QCD). In it λ^β are the usual SU(3) 3 x 3 representation matrices. M_0 is a matrix in flavor space that contains mass splittings; so we can accommodate in this term the phenomenological fact that $m_u \approx m_d < m_s < m_c$ corresponding to $m_\pi < m_K < m_\psi$. The renormalization group function $\beta(g)$ for this theory in lowest order is

$$\beta(g) = -\frac{g^3}{48\pi^2}\,(33 - 2F) \qquad\qquad (41)$$

when the flavor index $f = 1,...F$.

It is instructive to make a side by side list of comparisons between QCD and the more familiar QED:

QED	QCD
Photon $A_\mu(x)$	Gluon $A_\mu^{\,\alpha}(x)$

****

QED	QCD
"Matter" e, μ $\psi_e(x),\ \psi_\mu(x)$	"Matter" quarks $q_{\alpha f}(x)$ colors and flavors

****

QED	QCD
Spectrum of States: γ, e, μ-read from Lagrangian	Spectrum of States: Hadrons

****

QED	QCD
Bound states: positronium, H-atom	Bound states: π, K, ψ, p, Λ,...

****

QED	QCD
Interaction among bound states: molecular forces, Van der Waals force	Interaction among bound states: hadron scattering

****

QED	QCD
Infrared free; "easy" IR properties	Ultraviolet free; "easy" UV properties

****

QED	QCD
Photons have no charge	Gluons carry color charge

****

QED	QCD
Charge is observed. Charge is not confined	Charge (color) is not seen; no massless bosons; no quarks. Confinement

Unfortunately we are not going to provide the full solution of the theory defined by $\mathscr{L}_{QCD}$. Indeed, the properties of the ground state are still under study. Let's discuss it a bit, anyway. First of all, in $\mathscr{L}$ are massless vector bosons which carry color. That's bad, but not terrible since these bosons interact among themselves via the cubic and quartic terms in $(F_{\mu\nu})^2$, so the real vector boson could be massive, if it is even permitted as an asymptotic state. Second it has quarks of unknown mass carrying quantum numbers (color) probably never seen. Third, it is asymptotically free[3]; so its UV behavior is calculable and as far as one can tell in detailed comparison with inelastic muon scattering data, compatible with short distance behavior seen experimentally. Its infrared behavior is totally unknown and not inferable from $\mathscr{L}_{QCD}$ by just looking at it. This is just the opposite of QED where we read $\mathscr{L}_{QED}$ like a book; namely, there are photons and electrons and small, $0(\alpha)$, corrections.

So what evidence do we have that $\mathscr{L}_{QCD}$ is connected with the world of real hadrons? Certainly the charmonium spectroscopy of the states in the mass range ~3 to ~4 GeV/c^2 strongly indicates that for heavy quarks, c quarks, various qualitative features of QCD are in operation. In deep inelastic lepton scattering, in e^+e^- annihilation, in the qualitative features of hadron spectroscopy—in all these places we sense QCD at work.

A very nice set of calculations in this regard has been done over the past few years by the ITEP, Moscow, group of Shifman, Vainstein, Voloshin, Zakharov, Novikov and Okun.[4] They study the physics of currents formed from massive quarks. In particular the charmed quark. For example, they consider the current-current matrix element of heavy quarks

$$D_{\mu\nu}{}^{C}(q) = \int d^4x e^{-iq\cdot x} <0\,|T(J_{\mu}{}^{C}(x)J_{\mu}^{C}(0))\,|0> \quad , \qquad (42)$$

with

$$J_{\lambda}{}^{C}(x) = \bar{q}_{\alpha c}(x)\gamma_{\lambda}q_{\alpha c}(x) \qquad (43)$$

which is the $c\bar{c}$ contribution to the electromagnetic current.

In the q^2 plane for this matrix element one has for large $-q^2$ asymptotic freedom where one may calculate $D_{\mu\nu}{}^{C}$. Near $q^2 = 0$ not much is known. For positive q^2 there are known resonances: $\psi, \psi', \psi'',...$ to which $c\bar{c}$ couple. See Figure 3. In perturbation theory they are able to calculate graphs like the ones in Figure 4. The ITEP group argues that the distances in this perturbation theory which are essential in the calculation are $x \lesssim m_c^{-1}$. If $m_c \approx 1.5\ GeV/c^2$, as indicated by the ψ mass and the charmed meson masses, then they argue further that even at $q^2 = 0$, asymptotic freedom has set in for these heavy quark matrix elements. They view the world as in Figure 5.

Now using dispersion relations in q^2, they relate the values of $D_{\mu\nu}{}^{C}$ at $q^2 = 0$ they calculate from perturbation using $\mathscr{L}_{QCD}$ to q^2 in the resonance region. Given the masses of the resonances they calculate quite a few numbers of which $\Gamma(\psi \to e^+e^-) = 5\ keV$ is representative.

To study even charge conjugation states they look at forward light by light scattering via the electromagnetic coupling to charmed quarks. From these calculations using the ITEP Freedom assumption and dispersion relations again, they are able to calculate a variety of quantities related to the C-even X states.

This approach seems very attractive to me. It maintains gauge invariance and relativistic co-variance at every stage. It is necessary, of course, to use the dispersion relations with care and good sense so one doesn't demand information about too many q^2 derivatives of $D_{\mu\nu}{}^{C}$ at $q^2 = 0$, for then one is reconstructing $D_{\mu\nu}{}^{C}$ via a Taylor series, and somewhere that is bound to fail.

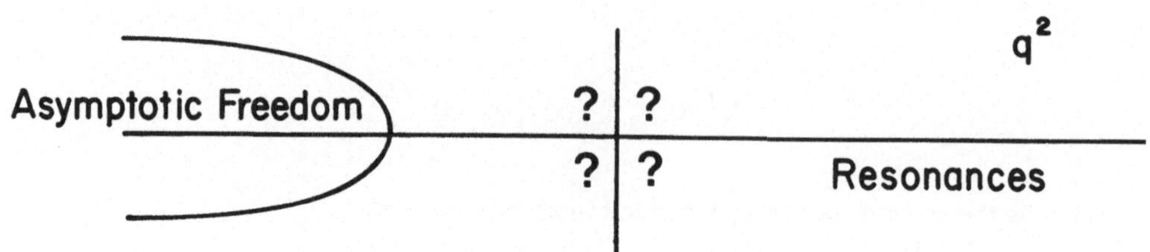

Fig. 3

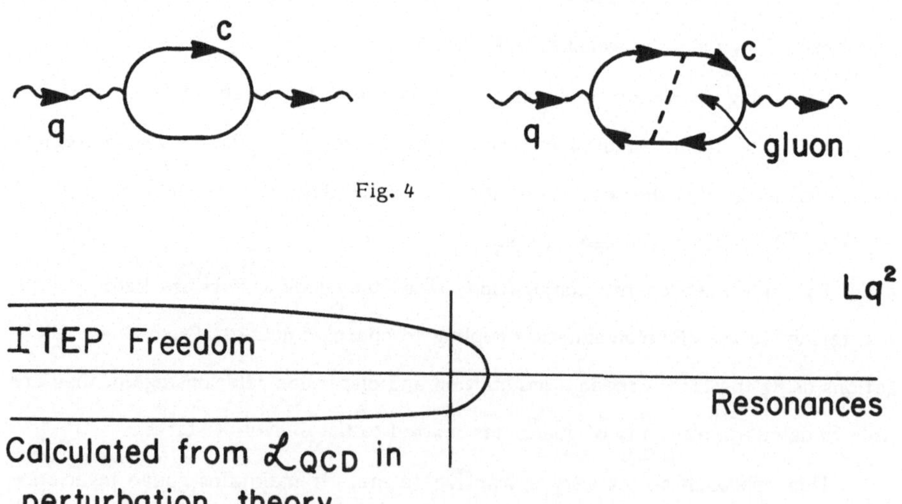

Fig. 4

Fig. 5

In any case a significant number of phenomena about heavy quarks are derivable in this fashion. A blind application to strange quarks works "OK," and to light quarks doesn't do well at all. That is as it should be from the distance arguments given before, but leaves a lot of work to be done on the QCD aspects of light quarks.

IV. Dynamical Mass Generation

The next topic I'll take up is the intriguing question of generation of mass in non-Abelian gauge theory by dynamical means. This is a situation where there appears no mass scale in the Lagrangian density (so in $\mathcal{L}_{QCD}$ above, $M_0 = 0$), but since a mass is introduced via the renormalization process, it is possible for the renormalized theory to have mass. We saw how this might operate by the study of the renormalization group in Section II; now we'll put a bit of flesh on these bones.

We really want to study QCD. So let's imagine $M_0 = 0$ in $\mathcal{L}_{QCD}$. It is then

$$\mathcal{L}_{QCD} = -\frac{1}{4g^2} F_{\mu\nu}{}^a F_{\mu\nu}{}^a + \bar{q} i \not{D} q \quad , \tag{44}$$

with

$$(D_\mu)_{ab} = \partial_\mu \delta_{ab} + i(T^C)_{ab} A_\mu{}^C \quad . \tag{45}$$

The symmetry in this is $SU(3)_{color} \otimes G$ where

$$G = U(F)_{vector} \otimes U(F)_{Axial} \tag{46}$$

and F is the number of flavors. Associated with the global symmetry G are the conserved currents

$$\mathcal{V}_\lambda = \sum_{f=1}^{F} \bar{q}_f \gamma_\lambda q_f \quad , \quad \mathcal{A}_\lambda = \sum_{f=1}^{F} \bar{q}_f \gamma_\lambda \gamma_5 q_f$$

$$\mathcal{V}_\lambda^\alpha = \sum_{f',f=1}^{F} \bar{q}_f \gamma_\lambda T_{ff'}{}^\alpha q_{f'} \quad , \quad \mathcal{A}_\lambda^\alpha = \sum_{f'',f=1}^{F} \bar{q}_f \gamma_\lambda \gamma_5 T_{ff'}{}^\alpha q_{f'} \tag{47}$$

with $\alpha = 1,\ldots F^2-1$. The T^α are $F \times F$ representation matrices of SU(F). These currents, except for $\mathcal{A}_\lambda$, remain conserved under interaction. The axial current has an "anomaly" coming from the short-distance structure of the theory, and its divergence is not zero, but

$$\partial^\lambda \mathcal{A}_\lambda \propto g^2 F_{\mu\nu}{}^\alpha \tilde{F}_{\mu\nu}{}^\alpha \quad , \tag{48}$$

where $\tilde{F}_{\mu\nu}{}^\alpha = \tfrac{1}{2} \epsilon_{\mu\nu\sigma\tau} F_{\sigma\tau}{}^\alpha$ is the dual gauge field.

Now the symmetry of the hadron <u>states</u> is not as large as G. It is only $U_{Vector}(1) \times SU(F)$ with F^2-1 pseudoscalar mesons which are more or less certainly massless Goldstone bosons. Even the SU(F) flavor symmetry is approximate, and the massless bosons pick up a mass possibly from the weak interactions.

To generate these Goldstone bosons we must break the chiral $(U_{Axial}(F) \otimes U_{Vector}(F))$ symmetry of the bare vacuum (as defined by $\mathcal{L}_{QCD}$). Because of the anomaly it is possible (below we'll even see it's likely) that there is no massless boson associated with the $U_{Axial}(1)$ symmetry. This is fine, actually, since there appears to be no such massless (read small mass in practice) object.

We can generate a mass scale in a Lagrangian theory by two presently acceptable methods: (1)let a scalar meson field, ϕ, coupled to quarks as $\phi\, q\bar{q}$ develop a vacuum expectation value $\langle\phi\rangle \neq 0$. This is a "spontaneous symmetry breaking" and is essentially kinematic in origin; that is, it is put in by hand. (2)We can have composite fields like $\bar{q}q$ itself develop a non-zero vacuum expectation value. This is dynamical and by the same renormalization group arguments used before must take the form

$$< \bar{q}(0)q(0) > = \mu^3 \exp - \int^g dx/\beta(x) \qquad . \tag{49}$$

How can we investigate this possibility in QCD? Let's look at the Ward identity satisfied by the proper quark-quark-axial vector current vertex $\Gamma_\lambda^{\ \alpha}$

$$q^\lambda \Gamma_\lambda^{\ \alpha}(p, p + q) = - \{\gamma_5 T^\alpha S^{-1}(p + q) + S^{-1}(p)T^\alpha \gamma_5\} \tag{50}$$

where $S(p)$ is the quark propagator and T^α is a flavor matrix. Let's look at this as $q_\lambda \to 0$. The right hand side approaches $\{\gamma_5 T^\alpha, S^{-1}(p)\}$. If this anti-commutator is non-zero, then $\Gamma_\lambda^{\ \alpha}$ must have the property

$$\Gamma_\lambda^{\ \alpha}(p, p + q) \underset{q \to 0}{\approx} F_\pi \mathscr{P}(p, p)\frac{q_\lambda}{q^2} T^\alpha \gamma_5 \qquad , \tag{51}$$

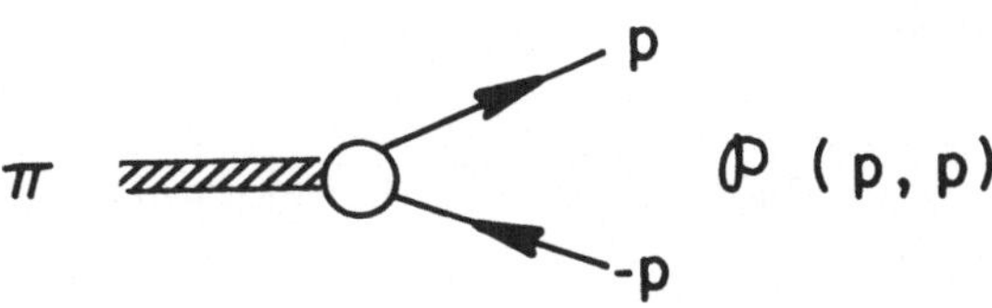

Fig. 6

where F_π is a constant and $\mathscr{P}$ is the wave function to find a $\bar{q}q$ pair in the pseudoscalar state called pion (Figure 6). In every order of perturbation theory $S^{-1}(p)$ is proportional to $A(p^2)\not{p}$, which anti-commutes with γ_5, so $\{\gamma_5 T^\alpha, S^{-1}(p)\}$ is zero. To generate a pion pole at $q^2 = 0$ we need a non-perturbative effect whereby

$$S^{-1}(p) = A(p^2)\not{p} - B(p^2) \qquad , \tag{52}$$

then the ward identity tells us

$$2B(p^2) = F_\pi \mathscr{P}(p, p) \quad .$$
(53)

So proving that a non-zero anomalous part to $S^{-1}(p)$ exists would be equivalent to showing there is a massless pion. This is not more than a statement of Goldstone's theorem, since $B(p^2)$ has dimensions of mass it introduces into the problem a scale thus breaking the chiral symmetry, thus giving rise to a massless boson.

To investigate the presence of B one looks at the Bethe-Salpeter equation for $\mathscr{P}$ and concludes

$$B = \int d^4p \, \frac{B}{p^2 A^2(p^2) - B^2(p^2)} \, K$$
(54)

where K is the irreducible kernel for $q\bar{q}$ scattering in QCD. Now Lane[5] has argued that only in an asymptotically free theory can we investigate this equation in some convincing way. Namely, we can look at the $p^2 \to \infty$ behavior of B and conclude that the equation <u>may</u> have a solution. Now since S^{-1} is a gauge dependent object, the conclusion still leaves one uneasy. Of course, the gauge dependence comes from the fact that

$$S(p) = \int d^4x e^{-ip\cdot x} <0 \, |T(\bar{q}(x)q(0))| 0 > \quad ,$$
(55)

and $\bar{q}(x)q(0)$ is gauge dependent except at $x = 0$. Since $p^2 \to \infty$ "takes" us to $x = 0$, the conclusion stated may be OK.

So we see that this standard Nambu-Jona-Lasinio approach is tricky in gauge theories, because gauge invariance (a crucial aspect of the theories) is treated cavalierly. What we need, for example, is a compact way to calculate quantities like $<0 \, |\bar{q}(0)q(0)| 0>$ directly. { You can see from (55) that it is zero if $B(p^2)$ is zero. } This is zero in every order of perturbation theory and evaluating it outside of perturbation theory has proven more or less intractable without going full circle to equations like (54) for B.

$$- \ 23 \ -$$

We can, however, get a feeling for the issues by looking at a model introduced by Ansel'm[6] in 1959 and analyzed in 1961 by Vaks and Larkin[7]. It exists in one space and one time dimension and has N fermions $\psi_1 \dots \psi_N$ coupled via

$$\mathscr{L} = \overline{\psi} i \not{\partial} \psi + g^2/2 \ (\overline{\psi}\psi)^2 \quad , \tag{56}$$

with

$$(\overline{\psi}\psi) = \sum_{i=1}^{N} \overline{\psi}_i \psi_i \quad . \tag{57}$$

The generating functional for this theory is

$$e^{iW[J]} = \int d\psi \, d\overline{\psi} \ e^{i \int [\mathscr{L} + \overline{J}\psi + \overline{\psi}J]} \tag{58}$$

which is equivalent to[8]

$$e^{iW[J]} = \int d\psi \, d\overline{\psi} \, d\sigma \ e^{i \int [\overline{\psi} i \not{\partial}\psi - \sigma^2/2 - g\sigma\overline{\psi}\psi + \overline{J}\psi + \overline{\psi}J]} \quad , \tag{59}$$

which can be seen directly by integrating over the auxiliary field σ. σ is not a dynamical variable, $\frac{\partial\sigma}{\partial t}$ does not appear in $\mathscr{L}$, but simply plays the role of separating the $(\overline{\psi}\psi)^2$ in the interaction and allowing us to deal with the more familiar form for $\mathscr{L}$ in (53). If $<\sigma> \neq 0$, then the fermion acquires a mass $m_F = g <\sigma>$; so σ is "substituting" for $\overline{\psi}\psi$.

To investigate whether $<\sigma> \neq 0$, let's set $\sigma(x) = v + \chi(x)$ and study the energy density of the resulting theory as v varies. If we make this substitution and introduce a source S for the χ ,field we have

$$e^{iW[S]} = \int d\psi \, d\overline{\psi} \, d\chi \ e^{i \int [\overline{\psi} i \not{\partial}\psi - \chi^2/2 - \chi v - gv\overline{\psi}\psi - g\chi\overline{\psi}\psi + S\chi]} \quad , \tag{60}$$

which becomes on integrating over ψ and $\bar{\psi}$

$$e^{iW[S]} = \int d\chi \exp i \int \left[-\frac{\chi^2}{2} - v\chi - i \, \mathrm{tr} \log [\, 1 - \frac{1}{i\partial - gv} g\chi \,] + S\chi \right] \,, \qquad (61)$$

$$= \int d\chi \, e^{i\int [\, \hat{\mathscr{L}} + S\chi \,]} \qquad\qquad (62)$$

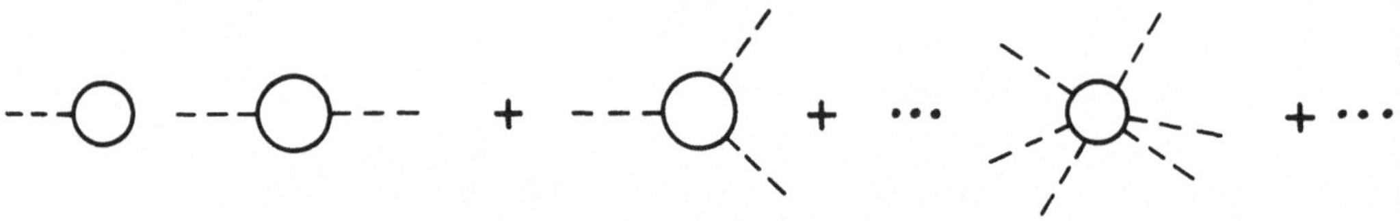

Fig. 7

This $\hat{\mathscr{L}}$ is non-local, non-Hermitean, and non-finite. It is non-local because it contains all the processes in Figure 7 where the dashed line represents a χ and the solid line a propagator $(\not{p} - gv)^{-1}$. It is non-Hermitean because it contains particle production. It is non-finite because the first two graphs in Figure 7 are infinite and are the wave function and mass renormalization of the original theory, which, by power counting, is just renormalizable at $D = 2$.

After renormalization, the vacuum, or ground state is defined by $\langle\chi\rangle = 0$ or equivalently $\frac{\partial \hat{\mathscr{L}}R}{\partial \chi}\Big|_{\chi=0} = 0$. In D space-time dimensions this means

$$v \left\{ 1 + \frac{2\lambda^2 \, \mathrm{tr}\, \underset{\sim}{1}\, \pi^{D/2} \Gamma(2 - \frac{D}{2})}{(2\pi)^D (2 - D)} \left[(D - 1) \int_0^1 dy [\, F(\lambda)^2 + y(1 - y)\,]^{\frac{D}{2} - 1} \right. \right.$$

$$\left. \left. - F(\lambda)^{D-2} \right] \right\} = S \qquad\qquad (63)$$

with the dimensionless coupling

$$\lambda = \sqrt{N}\, g(\mu)^{D/2 - 1} \qquad\qquad (64)$$

defined in terms of the normalization point μ, and

$$F(\lambda) = gv/\mu \quad . \tag{65}$$

In $D = 2 + \varepsilon$ dimensions we can study (63) as $S \to 0$. Requiring the vacuum to be stable which means $\partial^2(-\hat{\mathcal{L}}_R)/\partial \chi^2|_{\chi =0} > 0$, we find

$$S \to 0 \quad \left\{ \begin{array}{ll} v = 0 & \lambda^2 \leq \pi\varepsilon \\[2mm] v \neq 0 & \lambda^2 > \pi\varepsilon \end{array} \right. \quad . \tag{66}$$

At $D = 2$ exactly we can carry out the integral at $S = 0$ to find

$$\sqrt{1 + 4F(\lambda)^2} \; \log\left[\frac{\sqrt{1 +4F(\lambda)^2} - 1}{\sqrt{1 + 4F(\lambda)^2} + 1} \right] = - \frac{2\pi}{\lambda^2} \quad , \tag{67}$$

and discover $F(\lambda)$ behaves as shown in Figure 8. From the renormalization group we know

$$\frac{d \log F(\lambda)}{d\lambda} = - \frac{1}{2\beta(\lambda)} \quad , \tag{68}$$

so we deduce that $\beta(\lambda)$ behaves as in Figure 9. Near $\lambda = 0$ $\beta(\lambda)$ behaves as

$$\beta(\lambda) = - \frac{\lambda^3}{4\pi} + \lambda \, e^{-2\pi/\lambda^2} + \ldots \quad , \tag{69}$$

so the non-analytic structure in $\beta(\lambda)$, coming from the non-analytic behavior of $F(\lambda)$ (or m_F if you like), shows up in the non-perturbative requirement of vacuum stability. The infrared stable behavior near $\lambda^2 = \pi$ is a direct result of the non-analytic terms in (69) or more precisely (68).

The hope in treating this unrealistic model in detail[9] is to learn what must be the ingredients for a study of the more complicated vacuum of non-Abelian gauge theories. To date that has not been a realized hope.

F(λ)
$e^{-\pi/\lambda^2}$
$(\lambda/\pi)(\pi^2-\lambda^2)^{-1}$
λ^2
π

Fig. 8

β(λ)
π
λ^2

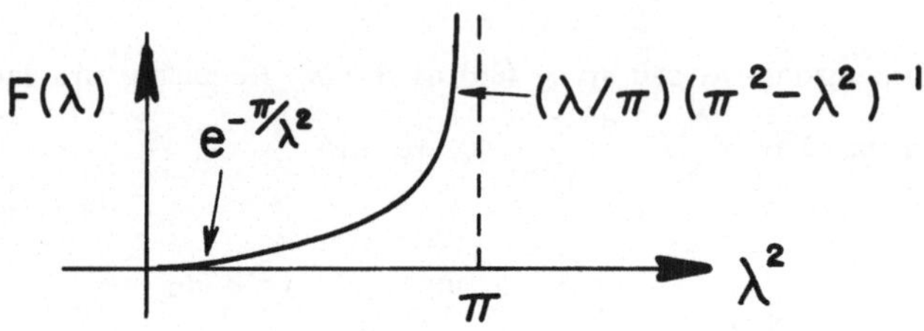

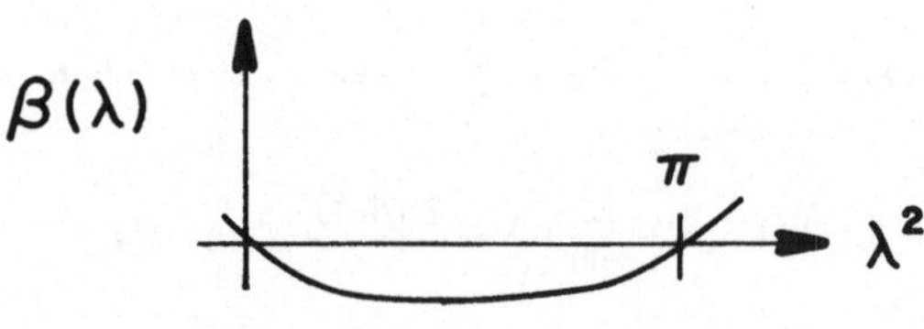

Fig. 9

V. Quark Confinement

We have talked at length in these lectures about quarks (with quantum numbers given in Figure 2) and the gauge bosons which are supposed to glue them together. To date no one has reported a convincing sighting of either a quark or a gluon. Perhaps we will find them. Even more intriguing: perhaps we will never find them. This, taken at its face, means $\mathcal{L}_{QCD}$ does not represent the spectrum of QCD but only the degrees of freedom: $A_\mu^\alpha(x)$ and $q_{\beta f}(x)$. The idea that these degrees of freedom can never be observed by themselves but only in the bound state combinations of color singlets we call hadrons goes by the label of color or quark confinement.

Clearly this question of observing color co-ordinates is an infrared or long distance problem: only if we can separate quarks can we study them individually. At short distances we may sense their existence by local probes of the weak current. With reference to the infrared behavior of QCD let's look at the effective coupling

$$\tilde{g}_{QCD}^{\,2}(p^2/\mu^2) = g^2/(1 + bg^2 \log p^2/\mu^2) \quad , \quad b > 0 \quad . \tag{70}$$

For $p^2 \to \infty$, it goes to zero; that's just asymptotic freedom. As $p^2 \to 0$, $\tilde{g}^2$ grows and in the approximation which leads to (70) blows up. So the long distance behavior of QCD is a strong coupling problem. In contrast, QED has $b < 0$ in (70) and the infrared behavior has $\tilde{g}_{QED}^{\,2} \to 0$. This ties in nicely with the fact that the spectrum of QED can be read directly from the Lagrangian.

In (70) there is the arbitrary renormalization scale μ which tells us when $\tilde{g}^2$ is small ($p^2 \gg \mu^2$) or large ($p^2 \lesssim \mu^2$). Presumably it is when $\tilde{g}^2$ is order of unity that strong coupling and binding effects come into play. So we can tentatively identify

$\mu \approx 1$ GeV as the hadronic scale and expect that when $p^2 > 1$ $(GeV)^2$, $\tilde{g}_{QCD}^2$ is small. Similarly when distances $> 1/\mu$, we can expect strong forces to come into play.

To investigate the forces between colors we develop now a picture originally due to Ken Wilson.[10] Picture two color charges at a fixed time separated by a distance R (Figure 10). As R grows, at some point the energy in the field between the charges becomes of order m_{hadron}. At that distance it becomes energetically favorable to put that energy into the creation of two color charges which then shield the color force between the original charges (Figure 11). Before this creation of pairs from the "vacuum", we can ask about the forces between the charges.

To do this in a relativistic fashion we need a space-time picture. So imagine that at t = 0 the pair is created and separates a distance R in a short time; then moves along for a long time ($\approx$T) at distance R and finally annihilate (Figure 12). What we have pictured here is a large charge loop of area $\approx$ RT located in the vacuum. Suppose we calculate the energy per unit volume of this loop

$$e^{-iE(R)T} = \langle e^{i \int J_{charge}^{\mu a}(x) A_{\mu}^{a}(x) d^4 x} \rangle \quad , \tag{71}$$

where the average is taken over configurations of the gauge field $A_{\mu}^{a}(x)$. For a single charge moving along the path $z_{\mu}(\tau)$ the current is

$$J_{charge}^{\mu}(x) = \int d\tau \, g \, \frac{dz^{\mu}(\tau)}{d\tau} \, \delta^4(x - z(\tau)) \quad , \tag{72}$$

so

$$\exp - iE(R)T = \langle \exp ig \int_{Loop} dz_{\mu} A^{\mu}(z(\tau)) \rangle \quad . \tag{73}$$

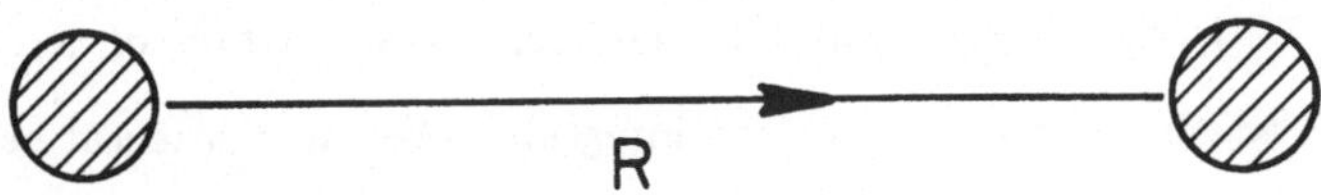

$R \ll \mu^{-1}$, $\tilde{g}^2$ small, charges free

$R \gtrsim \mu^{-1}$; $\tilde{g}^2$ large, strong binding(?)

Fig. 10

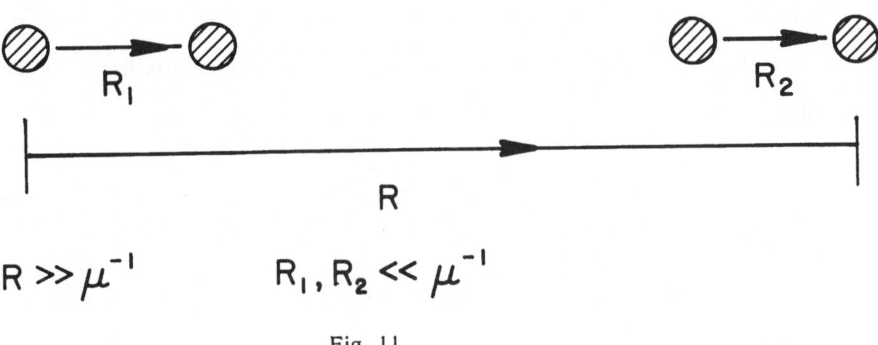

$R \gg \mu^{-1}$ $\qquad$ $R_1, R_2 \ll \mu^{-1}$

Fig. 11

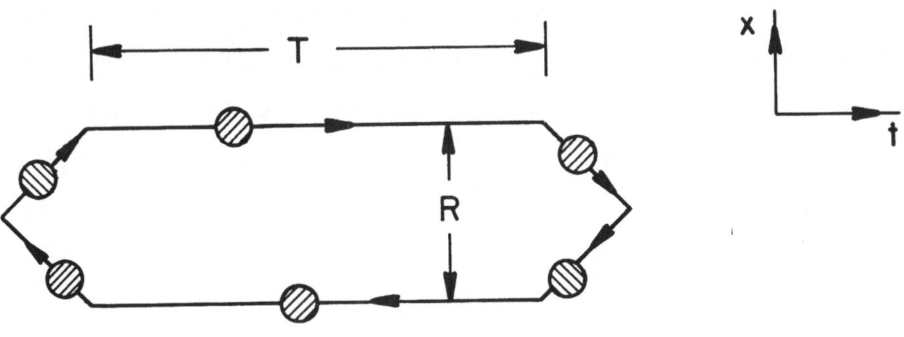

Fig. 12

To evaluate E(R), the energy stored in the loop as a function of R, we must evaluate the vacuum average of the loop integral. When that integral behaves as $\exp -iR^{\alpha}T$ with $\alpha > 0$, we can say the binding energy is confining since it rises with separation. In QED, $E(R) \xrightarrow[R \to \infty]{}$ finite number. In QCD it is conjectured that $E(R) \alpha R$ for large R; equivalently the loop integral behaves as $\exp -i$(Area of loop). Wilson has studied this loop integral in QED on a lattice and concludes that for e^2 large enough, the loop integral behaves as the area.

To address the question of confinement it seems we ought to look at the vacuum structure of QCD to be sure we have it in hand. Later we can come to the question of evaluating the energy of a loop in the vacuum.

To look at vacuua in quantum theory we must first go back to classical physics. Recall the elementary problem of the harmonic oscillator with hamiltonian

$$H = \frac{p^2}{2m} + \frac{m\omega^2 x^2}{2} \qquad . \tag{74}$$

The classical ground state has $x = 0$. The quantum ground state is characterized by the wave function

$$\psi_0 \sim e^{-x^2} \tag{75}$$

which represents fluctuations about the classical ground state at $x = 0$.

In QED our co-ordinate is the value of $A_\mu(x)$ at every point in space-time. The classical ground state has $F_{\mu\nu} = 0$ and the energy momentum tensor $\theta_{\mu\nu} = 0$. Indeed this allows us to conclude that $A_\mu = 0$ is a fine classical ground state. The quantum mechanical ground state is represented by a wave functional $\Psi_0[A_\mu(x)]$ something like

$$\Psi_0[A_\mu] = \exp - \int d^4x \, A_\mu(x) A^\mu(x) \tag{76}$$

In QCD we expect that in the classical ground state $F_{\mu\nu}^{\;a}(x) = 0$ again and also $\theta_{\mu\nu} = 0$. This does not mean, however, $A_\mu^{\;a} = 0$, and that's they key to the subtlety of the quantum vacuum in QCD. Let's go over to a matrix formulation of the theory by introducing a set of matrices T^α satisfying

$$[T^\alpha, T^\beta] = if_{\alpha\beta\gamma} T^\gamma \tag{77}$$

with $f_{\alpha\beta\gamma}$ the structure functions of our gauge group. Now let

$$F_{\mu\nu}(x) = -i \sum_\alpha T^\alpha F_{\mu\nu}^{\;\alpha}(x) \quad, \tag{78}$$

and

$$A_\mu(x) = -i \sum_\alpha T^\alpha A_\mu^{\;\alpha}(x) \quad, \tag{79}$$

so

$$F_{\mu\nu}(x) = \partial_\mu A_\nu(x) - \partial_\nu A_\mu(x) + [A_\mu, A_\mu] \quad . \tag{80}$$

If $F_{\mu\nu}(x) = 0$ it implies

$$A_\mu(x) = g^{-1}(x)\partial_\mu g(x); \quad g(x)^+ = g^{-1}(x) \quad, \tag{81}$$

and $g(x)$ is in the gauge group. So $A_\mu(x)$ is a pure gauge field. The representation of $A_\mu(x)$ in terms of $g(x)$ may not be unique. We must ask how many _different_ mappings there are from $(\vec{x}, t)$ to the gauge group. Each such inequivalent mapping will give a different classical vacuum with $F_{\mu\nu}(x) = 0$. Mathematicians have been kind enough to prove for us that for SU(N), (N $\geq$ 2), as the gauge group, there are an infinite number of inequivalent mappings labeled by an integer q = 0, ±1, ±2,... Essentially this represents the number of times (and "direction") that the $(\vec{x}, t)$ manifold is traced out on the SU(N) manifold.

A visualizable example is the mapping of the real line onto a circle. In the language above we consider elements g(x) in 0(2) and ask how many times the manifold of g(x), the circle, is covered as $-\infty < x < \infty$. Use the usual stereographic projection of the real line onto the unit circle (Figure 13) with θ the angle of the line joining the north pole to the real line. As x goes from $-\infty$ to $+\infty$ the angle θ runs from $-\pi/2$ to $+\pi/2$. Let Θ the angle on the 0(2) circle by $\Theta = \pm n(2\theta)$ then we cover the 0(2) manifold $\pm n$ times as x goes over its manifold.

From this observation on the classical vacuua in QCD which is due to Belavin, et al.,[11] we conclude there are an infinite number of inequivalent classical ground states. Which one of the $|\pm q >$, q = 0, 1, 2,... shall we choose for the quantum theory?

Let's go back to an example to guide us. Consider the hamiltonian

$$H = p^2 + g(1 - \cos \chi) \tag{82}$$

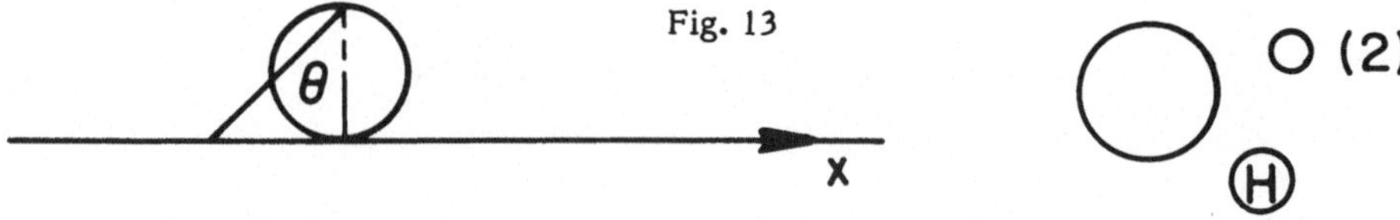

The classical theory has an infinite number of equilibrium states (ground states) at $\chi = \pm 2\pi q$, q = 0, 1, 2,... Which one shall we choose for the quantum ground state? In good old quantum mechanics we don't have to think twice about this since these classical ground states are connected by <u>tunneling</u> so a linear superposition

$$| \text{Vacuum} > \; = \sum_{q=-\infty}^{+\infty} a_q | q > \qquad , \qquad (83)$$

with $|q>$ a state centered about $\chi = 2\pi q$, is needed.

In field theory we must be a bit more careful. Consider the ϕ^4 theory

$$\mathcal{L} = \tfrac{1}{2}(\partial_\mu \phi)^2 + \frac{\mu^2}{2} \phi^2 - \lambda^2 \phi^4 \qquad (84)$$

so the potential is

$$V(\phi) = -\frac{\mu^2 \phi^2}{2} + \lambda^2 \phi^4 \qquad (85)$$

and has two equal energy minima at

$$\phi_{min} = \pm \mu/2\lambda \qquad . \qquad (86)$$

If the number of dimensions of space-time is greater than two, there is no tunneling between these classical ground states and either one forms the lowest state, in the sense,

$$\Psi_0[\phi] \; \alpha \, e^{-\int d^D x [\phi(x) - \phi_{min}]^2} \qquad , \qquad (87)$$

on which the hilbert space of states may be built. We are to choose one or the other of the classical vacuua, (86), and the physics should be much the same.

So we want to inquire if there is tunneling between the infinity of classical ground states in QCD. To look at this it is convenient to go over to imaginary time, { the lectures of S.J. Chang at this seminar indicate how to do all this directly in Minkowski space}, or Euclidian space. In quantum theory going under the barrier is an operation that can be described as a motion in imaginary time for a real co-ordinate since

$$\frac{dx}{dt} = \sqrt{2m(E - V(x))} = i\sqrt{2m(V(x) - E)} \quad , \tag{88}$$

or

$$\frac{dx}{d(it)} = \sqrt{2m(V(x) - E)} \quad . \tag{89}$$

We want to consider only solutions to the classical QCD equations, which are

$$D_\mu F^{\mu\nu} = 0 \tag{90}$$

where

$$D_\mu \psi = \partial_\mu \psi + [A_\mu, \psi] \quad , \tag{91}$$

which have zero $\theta_{\mu\nu}$. This solution can take us from one classical vacuum to another. Now

$$\theta_{\mu\nu} = \frac{1}{8} [(F_{\mu\lambda}{}^\alpha - \tilde{F}_{\mu\lambda}{}^\alpha)(F_{\lambda\nu}{}^\alpha + \tilde{F}_{\lambda\nu}{}^\alpha) + \mu \leftrightarrow \nu] \quad , \tag{92}$$

with

$$\tilde{F}_{\mu\nu}{}^\alpha = \tfrac{1}{2} \epsilon_{\mu\nu\sigma\tau} F_{\sigma\tau}{}^\alpha \quad . \tag{93}$$

So $\theta_{\mu\nu} = 0$ requires

$$F_{\mu\nu}{}^\alpha = \pm \tilde{F}_{\mu\nu}{}^\alpha \quad . \tag{94}$$

Take, say, $F_{\mu\nu} = \tilde{F}_{\mu\nu}$. This means

$$F_{12} = F_{34} \;, \quad F_{13} = -F_{24} \;, \quad \text{and} \quad F_{14} = F_{23} \qquad . \tag{95}$$

Suppose we introduce (Yang,[12] Belavin and Zakhorov[13])

$$y = (x_1 + ix_2)/\sqrt{2} \;, \quad z = (x_3 - ix_4)/\sqrt{2} \qquad , \tag{96}$$

then the condition $F_{\mu\nu} = \tilde{F}_{\mu\nu}$ means

$$F_{yz} = 0 \longrightarrow A_y = M^{-1}\partial yM \;, \quad \det M = 1 \tag{97}$$

and

$$F_{y\bar{y}} + F_{z\bar{z}} = 0 \qquad . \tag{98}$$

This last equation may be written in terms of the hermitean matrix

$$H = MM^+$$

as

$$-M^+ \left\{ \sum_{j=y,z} \partial_{\bar{j}}(H^{-1}\partial_j H) \right\} (M^{-1})^+ = 0 \tag{99}$$

Under a gauge transformation $M \to Mg$, $g^+ = g^{-1}$, so H is gauge invariant. The general self-duality condition reads

$$\sum_{j=y,z} \partial_{\bar{j}}(H^{-1}\partial_j H) = 0 \qquad . \tag{100}$$

The general solution of this is not known. Particular solutions are known for SU(2). The first is due to Belavin, et al.[11] and reads

- 36 -

$$A_\mu(x) = \frac{x^2}{x^2 + 1}\, g^{-1}(x)\partial_\mu g(x) \qquad , \qquad (101)$$

with

$$g(x) = (x_4 + i\vec{\sigma}\cdot\vec{x})/\sqrt{x_4^2 + \vec{x}^2} \qquad , \qquad (102)$$

an element of SU(2). This solution has $q = 1$ where we characterize solutions by the topological charge

$$q = \frac{1}{32\pi^2} \int d^4x\, F_{\mu\nu}^{\ \alpha}\, \tilde{F}_{\mu\nu}^{\ \alpha} \qquad , \qquad (103)$$

and can connect classical ground states with label q to states with $q \pm 1$. These classical field configurations with non-trivial topological charge are called instantons.

In SU(2) a set of solutions with $q = N$ have been given by 't Hooft[14]; Corrigan and Fairlie[15]; and Jackiw, Nohl, and Rebbi.[16] They are

$$A_\mu^{\ a}(x) = -\eta_{\mu\nu}^{\ a}\partial_\nu \log \phi(x) \qquad (104)$$

where $\eta_{\mu\nu}^{\ a}$ can be read off from (101) and (102) when $N = 1$. $\phi(x)$ satisfies

$$\partial^2 \phi(x) = 0 \qquad (105)$$

and for $q = N$ has the form

$$\phi(x) = \sum_{j=1}^{N+1} \lambda_i/(x - x_i)^2 \qquad . \qquad (106)$$

The existence of these tunneling solutions give a classical path with real co-ordinate $A_\mu^{\ a}$ and imaginary time which takes us from classical vacuum $|q\rangle$ to

$|q \pm N>$, $N = 1, 2,...$ The transition probability from $|q>$ to $|q + N>$ is finite and approximately of order

$$\exp -N(8 \pi^2/g^2) \qquad , \qquad (107)$$

with g the QCD coupling constant.

This all means that the quantum mechanical ground state in QCD is a superposition of classical vacuua

$$|\Omega> = \sum_{n=-\infty}^{\infty} c_n |n> \qquad , \qquad (108)$$

whereas in QED $|\Omega> = |0>$. We go from $|n>$ to $|n + 1>$ by a gauge transformation since in state $|n>$ we have $F_{\mu\nu} = 0$ and

$$A_\mu^{(n)}(x) = g_n^{-1} \partial_\mu g_n \qquad (109)$$

where the gauge transformation g_1 takes us from $|n>$ to $|n + 1>$. This gives rise to a periodicity similar to that in crystals and Bloch's theorem tells us

$$|\Omega> = \sum_{n=-\infty}^{\infty} e^{in\theta} |n> \qquad . \qquad (110)$$

Some immediate physics may be extracted from this vacuum structure:

1. Perturbation theory about $A_\mu^a = 0$ is not strictly correct since there are tunneling amplitudes of order $\exp -(g^{-2})$ to other field configurations. Presumably one wants to perturb about $|\Omega>$.

2. The so-called $U_A(1)$ problem is solved by this. We'll come to this.

3. It may be connected with quark confinement.

Let us look now at the $U_A(1)$ problem.[17] In massless QCD with F flavors the symmetry of the theory is the gauge group and

$$G_{flavor} = SU(F)_A \otimes SU(F)_V \otimes U_V(1) \otimes U_A(1) \quad . \tag{111}$$

The current associated with $U_A(1)$ is

$$\mathscr{A}_\lambda(x) = \sum_{f=1}^{F} \bar{q}_f \gamma_\lambda \gamma_5 q_f \quad . \tag{112}$$

The other currents are conserved to all orders in g, while the axial current has an anomaly so

$$\partial^\lambda \mathscr{A}_\lambda(x) = cg^2 \sum_\alpha F_{\mu\nu}{}^\alpha \tilde{F}_{\mu\nu}{}^\alpha \quad , \tag{113}$$

$$= cg^2 \partial^\lambda x_\lambda \quad , \tag{114}$$

where we note that $F_{\mu\nu}{}^\alpha \tilde{F}_{\mu\nu}{}^\alpha$ is a total divergence. Now it looks like

$$Q_5 = \int d^3x \, \mathscr{A}_0(\vec{x}, t) \tag{115}$$

is conserved because of the total divergence structure of $F_{\mu\nu}{}^\alpha \tilde{F}_{\mu\nu}{}^\alpha$, but because of the instanton fields with $q \neq 0$ there are matrix elements of $\partial \cdot \mathscr{A}$ of the form

$$\langle \Omega | (\bar{q}q \, \partial \cdot \mathscr{A}) | \Omega \rangle \ \alpha \ e^{-8\pi^2/g^2} \tag{116}$$

which vanish in all orders of perturbation theory. Chiral symmetry breaking which leads to massless pions would, we usually expect, lead to a ninth Goldstone boson with quantum numbers of $\partial \cdot \mathscr{A}$. But since $\mathscr{A}_\lambda$ is not conserved really, the problem is absent.

So we see that $|\Omega\rangle$ is not chirally invariant. Indeed because of the anomaly

$$[g_1, Q_5] = - 2F \, g_1 \tag{117}$$

and the real invariance of QCD is

$$U_{Vector}(1) \otimes SU_V(F) \otimes SU_A(F) \tag{118}$$

How $SU_A(F)$ is broken to produce massless pions is still not agreed upon in detail.

We began our excursion into instantonology because we wanted to learn about the QCD vacuum for confinement purposes. Polyakov[18] has given convincing arguments how instantons give rise to confinement in two space and one time dimension, but in 3 space, one time a similar hope has not yet been realized.

Indeed the various approaches to confinement have tried to populate the vacuum with various objects. Mandelstam[19] has considered filling the vacuum with magnetic monopoles which will crowd electric field lines into "strings" (much in the way Cooper pairs allow Abrikosov flux lines in type II superconductivity) which will then bind color. So far these attacks have been very illuminating, but not yet conclusive. No doubt at the next Topical Seminar held here in Tubingen we will hear conclusive progress in this matter. I know the generous hospitality and excellent organization of our hosts makes all of us plan on returning for that seminar before the university here passes its second 500 years.

References

[1] F.J. Gilman, Proceedings of the SLAC Summer Institute on Particle Physics, SLAC Report No. 191, Martha C. Zipf, ed.; Nov., 1975.

[2] O.W. Greenberg, Phys. Rev. Lett. 13, 122 (1964); W.A. Bardeen, et al., CERN Preprint TH-1538, 1972.

[3] G. 't Hooft, unpublished; D. Gross and F. Wilczek, Phys. Rev. Lett. 30, 1343 (1973); H.D. Politzer, Phys. Rev. Lett. 30, 1346 (1973).

[4] V.A. Novikov, L.B. Okun, M.A. Shifman, A.I. Vainshtein, M.B. Voloshin, and V.I. Zakharov, ITEP Preprint Nos. 79 and 83, 1977.

[5] K. Lane, Phys. Rev. D10, 2605 (1974).

[6] A.A. Anselm, JETP (Sov. Phys.) 9, 608 (1959).

[7] V.G. Vaks and A.I. Larkin, HETP (Sov. Phys.) 13, 979 (1961).

[8] D. Gross and A. Neveu, Phys. Rev. D10, 3235 (1974).

[9] H.D.I. Abarbanel, Nuc. Phys. B130, 29 (1977).

[10] K.G. Wilson, Phys. Rev. D10, 2445 (1974).

[11] A. Belavin, A.M. Polyakov, A. Schwartz, and Y. Tyupkin, Phys. Lett. B59, 85 (1975).

[12] C.N. Yang, Phys. Rev. Lett. 38, 1377 (1977).

[13] A. Belavin and V. Zakharov, Landau Institute for Theoretical Physics, (1977), to be published.

[14] G. 't Hooft, Coral Gables Conference, 1977.

[15] F. Corrigan and D. Fairlie, Phys. Lett. B67, 69 (1977).

[16] R. Jackiw, et al., Phys. Rev. D15, 1642 (1977).

[17] G. 't Hooft, Phys. Rev. Lett. 37, 8 (1976).

[18] A.M. Polyakov, Nuc. Phys. B120, 429 (1977).

[19] S. Mandelstam, Paper given at the American Physical Society meeting in Washington, D.C., April, 1977.

Composite Hadrons and Relativistic Nuclei

R. Blankenbecler

Stanford Linear Accelerator Center, Stanford University, Stanford,
California 94305

I. Introduction

In these lectures I would like to describe a model of hadronic scattering at large momentum transfer, either transverse or longitudinal. This model emphasizes in this regime the importance of forces involving the interchange of constituents of the hadrons, hence the name, CIM, or constituent interchange model. [1,2] As will be shown, this model should not be thought of as being different from quark-quark scattering models, or from QCD (quantum chromodynamics) but contains both of them. Omission of the CIM diagrams is not a consistent approximation. The CIM is, in fact, a rearrangement of standard perturbation theory to take into account the fact that the binding force is very strong in color singlet states. We could call this <u>Singlet</u> <u>Dominance</u>. For example, if one demands that an anti-quark, or $\bar{q}$, be found a large distance from the center of a baryon, the easiest way for it to propagate to such distances is via intermediate states involving light mesonic (singlet) states. But more of this later on when the CIM contributions and their absolute normalization will be discussed.

Before discussing the physically complicated case of hadrons, it is helpful to discuss constituent models in a regime where we know they must apply—i.e. nuclear scattering. [3] In addition to developing our intuition and methods of approximation, we can extend the usual description of nuclei into the relativistic domain—that is, a regime where particle production occurs but does not dominate, where the finite size of nucleons plays an important role, and where the motions of the nucleons must be described by relativistic kinematics. All of this is clearly possible to include in a theoretical model but the important point is its simplicity and usefulness, and the ease with which predictions can be made. We will find a remarkably simple model that works quite well for certain experimental cross sections. Further tests are required before its general validity can be assessed.

II. The Hard Scattering Expansion

For applications to both the hadronic and nuclear case, an expansion of
the full scattering amplitude must be made in order to compute anything. In the
hadronic case the relevant expansion is called the hard scattering expansion,
whereas in the nuclear case it is called the impulse approximation. Let us
review some of the basic assumptions used in this familiar expansion. This is
particularly important since by trying to add ad hoc features to the first term,
it is possible to define nonsense that doesn't appear to be nonsense at first glance.

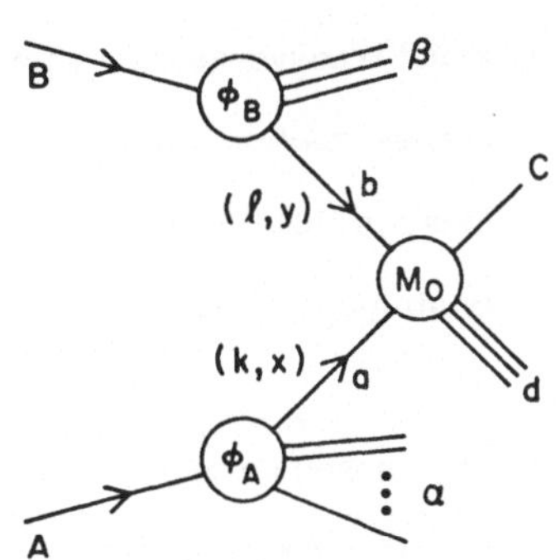

Fig. 1

The basic hard scattering model diagram
with the notation used in the text.

The hard scattering expansion for an inclusive reaction is illustrated in
Fig. 1 and written as [2]

$$E_C \frac{d\sigma}{d^3C} (AB \to CX) = \sum_{a,b,d} \int dx d^2k_T \, dy d^2\ell_T \, G_{a/A}(x, k_T) G_{b/B}(y, \ell_T)$$

$$r(s, s^1, x, y) \, E \frac{d\sigma}{d^3C} (ab \to Cd; s^1, t^1, u^1), \qquad (2.1)$$

where the a,b,d sum is over <u>incoherent</u> final states. If this is the case, then the
formula takes on a simple probabilistic meaning. The ratio $r = \lambda(s^1, k_\mu^2, \ell_\mu^2)/$
$xy \lambda(s, A^2, B^2)$ is the ratio of the internal (off shell) and external (on shell)
phase space factors and $\lambda(s, t, u)$ is the usual quatratic form,

$$\lambda(s, t, u)^2 \quad s^2 + t^2 + u^2 - 2(st + su + tu).$$

If one keeps k_T and ℓ_T small, one finds $r \cong 1$.

Our discussion will be based in the diagram shown in Fig. 1, which represents the inclusive process $A + B \rightarrow C + X$. Here the interaction takes place through the emission of virtual subsystems ($\underline{a}$ and $\underline{b}$), which are the ones that scatter in an internal basic process $a + b \rightarrow C + d$, where C is the detected particle. M_0 is the amplitude for this basic interaction, and the amplitudes for the emission of the $\underline{a}$ and $\underline{b}$ subsystems will be contained in distribution functions $G(x, \vec{k}_T)$, to be defined precisely in a moment. However, the interpretation of the various factors in Eq. (1) is clear. The factor $G_{a/A}(x, \vec{k}_T)$ is the probability of finding a consti- tuent of type a in nucleus A with fractional "momentum" x and transverse momenta $\vec{k}_T$. A similar interpretation holds for $G_{b/B}$. The basic cross-section factor that actually produces the detected particle C also has a clear probabilistic meaning. We have neglected any final state decay to C for simpliticy, but such can easily be added in.

The sum over the intermediate and final states must be chosen so that <u>Eq. (2.1) is a sum over incoherent final states</u>. This means that the simplest way to classify the terms is according to the final state configuration of particles, not the possible intermediate states that can contribute. The sum over a, b, and d must be chosen with the final state configurations in mind. For example, this means that one particular Feynman graph contributes to several terms in the sum depending upon the disposition of the final state particles, that is which particular particle (or particles) is recoiling against the large (transverse) momentum of particle C, for example.

The internal amplitude M_0 that describes the process $ab \rightarrow Cd$ must be predicted by the model or fit to experiment (as in the nuclear case).

The "momentum" fraction x(and y) is not exactly the momentum fraction of component a in A. It is defined rather as an "infinite" momentum fraction $x = (a_0 + a_z)/(A_0 + A_z)$. In detail, we define the awkward looking momenta

$$A = \left(P_1 + \frac{A^2}{4P_1} \ , \ \vec{O}_T, \ P_1 - \frac{A^2}{4P_1} \right) \tag{2.2}$$

$$B = \left(P_2 + \frac{B^2}{4P_2} \ , \ \vec{O}_T, \ -P_2 + \frac{B^2}{4P_2} \right)$$

where a particle's name and four-momentum are denoted by the same symbol except for the off-shell particles a and b. A and B have been defined in a general set of frames along the interaction axis. A specific frame in this set is selected by relating P_1 and P_2. For example, the center-of-mass frame is defined by the conditions

$$P_1 - \frac{A^2}{4P_1} = P_2 - \frac{B^2}{4P_2}$$

and

$$\sqrt{s} = P_1 + \frac{A^2}{4P_1} + P_2 + \frac{B^2}{4P_2} \ .$$

The other momenta that are on-shell are defined as

$$\alpha = \left((1-x)P_1 + \frac{\alpha^2 + k_T^2}{4(1-x)P_1} \ , \ -\vec{k}_T, \ (1-x)P_1 - \frac{\alpha^2 + k_T^2}{4(1-x)P_1} \right)$$

$$\tag{2.3}$$

$$\beta = \left((1-y)P_2 + \frac{\beta^2 + \ell_T^2}{4(1-y)P_2} \ , \ -\vec{\ell}_T, \ -1(-y)P_2 + \frac{\beta^2 + \ell_T^2}{4(1-y)P_2} \right)$$

This rather cumbersome set of variables will greatly simplify our later discussion. For example, note that with these parametrizations, the phase space integrals are of the form

$$d^4\alpha = d^2 k_T \ \frac{dx}{2|1-x|} \ d\alpha^2$$

and then the α^2 integral, due to the corresponding on-mass-shell δ-function, is trivial. The off-shell momenta are calculated by momentum conservation:

$$a = \left(xP_1 + \frac{k^2 + k_T^2}{4xP_1} \ , \ \vec{k}_T, \ xP_1 - \frac{k^2 + k_T^2}{4xP_1} \right)$$

$$b = \left(yP_2 + \frac{\ell^2 + \ell_T^2}{4yP_2} \ , \ \vec{\ell}_T, \ -yP_2 + \frac{\ell^2 + \ell_T^2}{4yP_2} \right) ,$$

$$(2.4)$$

where

$$k^2 = \left(x(1-x)A^2 - x\alpha^2 - k_T^2 \right)/(1-x)$$

$$\ell^2 = \left(y(1-y)B^2 - y\beta^2 - \ell_T^2 \right)/(1-y).$$

$$(2.5)$$

Note that with these parametrizations, as mentioned before,

$$x = \frac{a_0 + a_3}{A_0 + A_3}$$

which is the usual light-cone variable, and hence x can only have values between zero and one.

Since the G's are probability functions, they must be related to the square of a wave function; a careful analysis shows that

$$G_{a/A}(x, \vec{k}_T) = \frac{1}{2(2\pi)^3} \frac{x}{(1-x)} |\psi(x, \vec{k}_T)|^2 \qquad (2.6)$$

where ψ is the bound state Bethe-Salpeter wave function with one leg (α) on-shell. We will see in our analysis that the distribution functions are explicitly measured in the experiments we are considering. For this reason it is important to have a reasonably good prediction or description of their properties. We shall analyze these functions in detail and get information about them from limiting cases, like the non-relativistic limit and the short distance or ultrarelativistic behavior.

It is also straightforward to derive an equation for the electromagnetic form factor of the state A in terms of ψ and the result is

$$F_A\left(q_T^2\right) = \sum_a F_a(q^2) \int \frac{dx\, d^2 k_T}{2(2\pi)^3} \; \frac{x}{(1-x)} \; \psi^*(x, \vec{k}_T)\, \psi\left(x, \vec{k}_T - (1-x)\vec{q}_T\right) \qquad (2.7)$$

where the integral multiplying F_a is the body form factor of the nucleus. This can be used to predict the form factor when a sufficiently promising wave function $\psi(x, k_T)$ is developed. We are particularly interested in possible relations between the large q_T^2 behavior of the form factor and inclusive measurements in the forward direction.

III. Incoherence Problems

Let us illustrate some of the problems associated with keeping the various terms in Eq. (2.1) incoherent. For simplicity, we shall consider the production of a massive lepton pair (the Drell-Yan process) since the gauge invariance of the photon more or less forces us to a consistent treatment.[4] In QCD, the gauge invariance with respect to the gluons will certainly force one to a similar complete treatment with similar results.

Consider the two diagrams in Fig. 2 which correspond to $a=\bar{q}$, $b=q$, and $b=(qq)$, respectively. The first diagram is the natural one in the Drell-Yan theory[5] in order to produce large Q_T pairs. The large transverse momentum

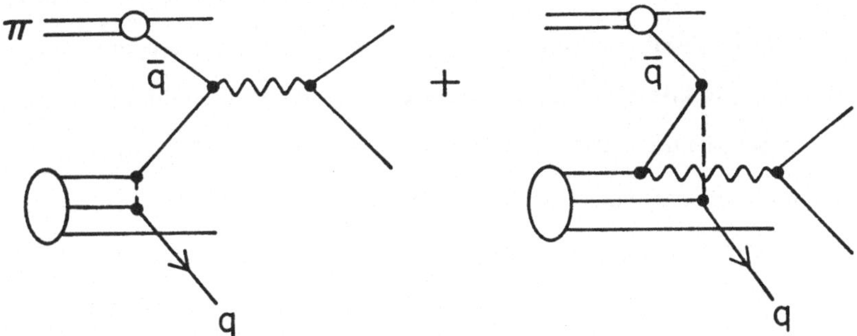

Fig. 2

Two coherent contributions to the Drell-Yan process.

of the pair arises from the wave function or structure function $G_{q/p}(y, \ell_T)$. However, this is an incorrect treatment since the second $b=(qq)$ term is coherent with the first and is even of the same order in all couplings. It is necessary to treat these two terms together and to include them both. They can be described as an initial and a final state interaction, respectively. From the relevant quark's point of view, they are the direct and crossed graph for the process glue + quark $\rightarrow$ photon + quark. At large Q_T, one finds that these two terms cancel to leading order, as will be discussed in more detail later.

That this is not an unexpected feature of initial and final state interactions can be seen by considering the theorem derived in potential scattering by Amado and Woloshyn.[6] They proved that the term with the leading behavior of the wave function at large relative momentum actually cancels in a general class of break-up reactions. This cancellation is due (essentially) to the orthogonality of the bound and (ingoing) scattering states and is therefore expected to be a very general phenomena.

We have seen therefore that simply adding a large k_T spectrum to the initial state quarks and then using the Drell-Yan formula is incorrect in principle. This should come as no surprise since D-Y stated in their original paper that their model was not gauge invariant if large transverse momenta were allowed. It is easy to see that if an intermediate quark carries a large k_T, its $(\text{mass})^2$ is of order $(-k_T^2/(1-x))$. Thus it is not possible to make such a contribution gauge invariant to this order unless the photon is attached to both ends of this far offshell propagator. This leads us naturally to the initial and final state interaction effects described earlier. One way to avoid troubles here is to demand that the intermediate particles a and b remain near their respective mass shells and hence that they not carry a large transverse momentum. This forces all large momentum transfers to occur in the central process. This in turn allows the photon to be attached in all necessary orderings to insure gauge invariance, as well as the proper incoherence properties of the hard scattering expansion.

We discussed final state photons above only because a proper treatment of gauge invariance forces us to a reasonable result. The same is true of reactions in which ordinary particles are produced —one must still satisfy the conditions used to define the hard scattering expansion. <u>This is especially true in QCD where the gluons must be treated in a gauge invariant manner.</u> Since we are now forwarned, let us return to a treatment of nuclear (or heavy-ion) scattering processes.

IV. Nuclear Wave Functions and Counting Rules

We shall demand that our theory join onto the familiar nonrelativistic treatments when the energies and momenta are small. In particular, the G function must be closely related in this limit with the square of the nonrelativistic wave function. This requirement will allow us to achieve a clearer understanding of these functions and their expected behavior, and also to explore the way masses should enter into our formalism.

First we want to see the meaning of the x-variable in a nonrelativistic limit. For momenta small with respect to the masses, and in the rest frame of the nucleus, x becomes

$$x \simeq \frac{a}{A} + \frac{k_z}{A} \; , \tag{4.1}$$

and hence is related to the longitudinal momentum, measuring deviations with respect to a central value $\frac{a}{A}$. Since on the average, and in the rest frame of the nucleus, we expect $k_Z = 0$, this means that on the average $x = \frac{a}{A}$. In other words, each nucleon carries the same fraction of the total momentum of the nucleus. A very reasonable result in the weak binding limit.

Recall that G is the probability of finding a constituent of A with longitudinal momentum x and transverse momentum $\vec{k}_T$. This means that G must have a maximum at $x \simeq \frac{a}{A}$ (the average nucleon longitudinal momentum) and at $\vec{k}_T=0$. Considering the definition of G, and using the Bethe-Salpeter equation we see that

$$G \sim \frac{\phi^2 x(1-x)}{\left[k_T^2 + M^2(x)\right]^2} \quad , \tag{4.2}$$

where ϕ is defined as the vertex function and

$$M^2(x) = (1-x)(a^2-k^2)-k_T^2 = (1-x)a^2 + x\alpha^2 -x(1-x)A^2 \quad .$$

This form implies that G has a maximum at $\vec{k}_T=0$ and at $x=x_0$, where $M^2(x)$ is a minimum. We find

$$x_0 = \frac{A^2+a^2-\alpha^2}{2A^2} \cong \frac{a}{A} \quad , \tag{4.3}$$

and as expected, the constituents prefer to share the momentum according to their mass.

In the limit of small momenta one then finds

$$k_T^2 + M^2(x) \cong 2a\epsilon\left(\frac{A-1}{A}\right) + \vec{k}^2 \tag{4.4}$$

where it has been assumed that the binding energy for nucleon ϵ is the same for both A and α. The G function becomes

$$G \sim |\psi_{NR}(\vec{k})|^2 \sim \frac{\phi^2_{NR}(\vec{k})}{\left[a\epsilon + \vec{k}^2\right]^2} \quad ,$$

where (A=2 for simplicity)

$$\phi^2_{NR} \sim x_0(1-x_0)\phi^2 \quad .$$

In order to have a better understanding of the function ϕ_{NR}, consider the Schrodinger equation in momentum space

$$\psi_{NR}(\vec{k}) = (a\epsilon + \vec{k}^2)^{-1} \int d^3p \, V(\vec{k} - \vec{p}) \, \psi_{NR}(\vec{p}) = (a\epsilon + \vec{k}^2)^{-1}\phi_{NR}(\vec{k}) \quad ,$$

so that the vertex function expresses more or less directly the behavior of the potential V. The falloff of ϕ is related to the softness (or hardness) of the potential. As a simple example consider a general Hulthen model of the nuclear wave function:

$$\psi_{NR} = (a\epsilon + \vec{k}^2)^{-1}(a\epsilon_1 + \vec{k}^2)^{\frac{1-g}{2}} \quad .$$

In the familiar Hulthen deuteron case, one usually chooses g=3, and $\epsilon_1 \sim 36\epsilon$. The second factor is then much flatter in $\vec{k}^2$ than the first.

A relativistic version of this wave function can be achieved by writing

$$\psi = \frac{N(x)}{(k^2-a^2)(k_1^2-a_1^2)^{\frac{g-1}{2}}} \quad , \tag{4.5}$$

where N(x) is slowly varying for x near 1, and where we choose

$$(1-x)(a_1^2 - k_1^2) = M^2(x) + \delta^2 + k_T^2 \tag{4.6}$$

since the second factor should have a minimum at the same place as the first, i.e. $x = x_0$.

The form factor for the type of wave function of (4.5) is easily seen to fall as

$$F^2\left(q_T^2\right) \sim \left(q_T^2\right)^{-g-1} \tag{4.7}$$

for large q_T^2. Thus the falloff of the form factor and the behavior of G for large k_T^2 are closely related and also we shall see that the behavior of G for x near 1 is also closely related to the form factor falloff. This latter relation is the Drell-Yan-West relation.

For general x, the relativistic G function can then be written as

$$G(x, \vec{k}_T) = \frac{N^2(x) \, x(1-x)^g}{2(2\pi)^3} \left[M^2(x) + k_T^2 \right]^{-2} \left[M_1^2(x) + k_T^2 \right]^{1-g} . \qquad (4.8)$$

For $x \simeq x_0$, the denominator factors are rapidly varying and as has been discussed, this reduces to a familiar nonrelativistic Hulthen form. For $x \gg x_0$, the numerator factors control the behavior of G, and

$$G(x, \vec{k}_T) \sim (1-x)^g$$

while its large k_T^2 behavior is $\left(k_T^2 \right)^{-g-1}$.

In our analysis, the behavior of G for $x \gg x_0$ will be especially important. Note that this is new information not directly contained in the nonrelativistic wave function. We can also discuss quasielastic scattering which explores the G function for $x \sim x_0$ as well. Let us now turn to a discussion of the calculation of the power g in selected theories of the nucleon-nucleon interaction.

Our main contact with experimental data is through the structure functions $G(x, k_T)$. A helpful tool for expressing the predictions of specific theories is in terms of "counting rules." [7,8] These allow one to characterize the asymptotic behavior of G in terms of the number of constituents <u>and</u> the short distance behavior of the basic interactions of the theory.

The procedure here is to extract the leading behavior from the lowest order diagram in perturbation theory. For "soft" theories, one can show that the higher orders either are small compared to the leading term or have the same behavior.

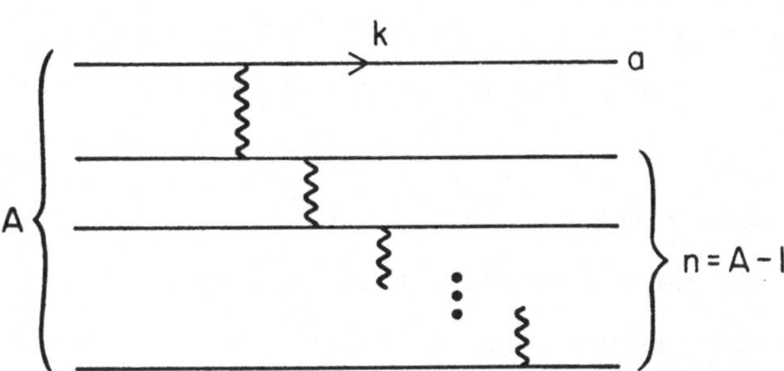

Fig. 3

The wave function diagram used to compute the probability functions.

Consider the wave function (or structure function) diagram given in Fig. 3, where k is the momentum of particle a and is defined in Eq. (2.4). We shall assume scalar particles for simplicity. Note that A now also means the atomic number of particle A.

For a renormalizable interaction between the constituents, such as vector exchange with point interactions, the falloff of the vertex function arises solely from the constituent propagators. One finds

$$\phi \sim \left(k_1^2 - a_1^2\right)^{1-n} \quad .$$

where the masses in k_1 (see Eq. (II-15)) depend on detailed properties of the force. The wave function is

$$\psi \sim \left(k^2 - a^2\right)^{-1}\left(k_1^2 - a_1^2\right)^{1-n}$$

Comparison with Eq. (II-14) immediately tells us that

$$g = 2A - 3$$

As a perhaps more relevant example, consider a nucleon-nucleon interaction mediated by the exchange of vector mesons, such as rhos or omegas, with a monopole form factor at each vertex (vector dominance would assume such a behavior to fit the dipole nucleon form factor). One finds

$$\phi \sim \left(k_1^2 - a_1^2\right)^{1-n}\left(k_2^2 - a_2^2\right)^{-2n}$$

where the masses in the form factors and/or gluon propagtors are chosen to be the same for simplicity. The final result in this case is

$$g = 6A - 7 \quad .$$

This is the same result as one would get by counting quarks. While one might expect that the quark degrees of freedom become relevant at ultrahigh energies where they can be excited, we see that one gets the same prediction for g in this theory when the nucleon form factor effects play a role. These, of course, may in turn be due to internal structure, but the internal degrees of freedom need not be fully excited.

For more general structure functions $G_{a/A}$, where the state a is a bound state of $\underline{a}$ nucleons, a similar analysis can be carried through. One finds in this case

$$g = 2T(A-a) - 1 \tag{4.9}$$

where T (= 1, 3, etc.) depends upon the basic nucleon–nucleon interaction as discussed earlier. Again, we have assumed full breakup of the nucleus after a is extracted. If the nucleus is not fully broken up, then the calculation is extremely complicated but one might conjecture that g should be replaced by

$$g = 2T(A-a)_{eff} - 1 \quad , \tag{4.10}$$

where $(A-a)_{eff}$ is the effective number of fragments that the remaining nucleons produce. One may also expect additional nonscaling behavior as well if some of the fragments remain bound.

Now we shall predict the behavior of inclusive yields in order to test the theory and to see if one value of one value of T fits all the inclusive scattering reactions $\underline{and}$ the elastic form factors.

V. Interaction Between Nuclei

As a first application we will consider inclusive scattering of nuclei at high energies and get simple predictions that can easily be compared with experiment without extensive numerical calculation.[3] First, consider the situation in which the energy per nucleon is large compared to the nucleon mass. The kinematics for this regime is quite simple:

$$s' = xys$$
$$t' = yt$$
$$u' = xu \tag{5.1}$$
$$d^2 = xys + yt + xu$$

and

$$C_T^2 = \frac{ut}{s} = C_T'^{\,2}$$

The condition $d^2 > 0$ restricts the range of x and y that contribute for fixed values of s, t, u, and all finite masses have been neglected. Note that the internal reaction can be inclusive ($d^2 > 0$) or exclusive ($d^2 = 0$). This last situation is also called quasielastic scattering.

All inclusive basic processes of interest to us here will be parametrized as

$$\left[E_C \frac{d\sigma}{d^3C}\right]' = E(s')\left[1 - \left(|x_F'|\right)^H\right] f(k_T') \tag{5.2}$$

and exclusive processes as

$$\left[E_C \frac{d\sigma}{d^3C}\right]' = E(s')\delta\left[(k+\ell-C)^2 - d^2\right] f(k_T') \quad . \tag{5.3}$$

where $k_T' = C_T - k_T - \ell_T$ and $E(s')$ is assumed slowly varying. H will be assumed to be constant. The function $f(k_T')$ is strongly peaked at $k_T' = 0$ (for small k_T' it could be written as an exponential $e^{-r^2 k_T'^2}$, for example).

Now we will go into a discussion of our model in some regions of phase space in which it is easy to get predictions.

First define in terms of the missing mass m

$$\epsilon = \frac{m^2}{s} = 1 - x_R \cong 1 - \frac{|\vec{C}_{cm}|}{|\vec{C}_{max}|}$$

$$x_T = \frac{C_T}{C_{max}}$$

$$x_L = \frac{C_L}{C_{max}} \cong \frac{t-u}{s} \quad , \qquad x_R = \sqrt{x_T^2 + x_L^2} \qquad (5.4)$$

and for the most part we will concentrate in the region ϵ not near one.

When t is fixed (and s, u large), one finds ($x_R = x_F$, Feynman's variable) in this the projectile fragmentation region,

$$1 - x_F' = 1 + \frac{yt + xu}{xys} \quad ,$$

and hence $x_F' \cong x_F/y$. The condition $d^2 > 0$ becomes $y > x_F$. In this regime, the inclusive cross section becomes

$$\sum_{ab} \int_0^1 dx d^2 k_T \, G_{a/A}(x, \vec{k}_T) \int_{x_F}^1 dy d^2 \ell_T \, G_{b/B}(y, \vec{\ell}_T) \left[E_C \frac{d\sigma}{d^3 C} \right]' \quad . \qquad (5.5)$$

First consider an <u>inclusive basic process</u>. Since $f(\vec{k}_T')$ in formula (III-2) is strongly peaked in $k_T'^2$, we can approximate the k_T and ℓ_T integrals by replacing k_T^2 and ℓ_T^2 in the G's by the mean value K^2 which should be of the order of C_T^2. The inclusive cross section is then proportional to

$$\propto \int_{x_F}^1 dy \, \frac{N^2(y) \, y(1-y)^{g_B}}{\left[K^2 + M^2(y)\right]^2 \left[K^2 + M_1^2(y)\right]^{g-1}} \, (1 - x_F/y)^H \quad .$$

Note that the distribution for the target $G_{a/A}$ has integrated out in this limit, and it depends on A through its normalization only.

If $C_T^2 \gtrsim M_1^2(x_0)$, and if x_F is not small compared to x_0, then the main variation in the integrand is from the factors of $(1-y)$ and $(1-x_F/y)$. The first factor cuts off the integrand near $y = 1$ and the other near $y = x_F$. If only their variation is retained, and the denominators taken constant, we have

$$E \frac{d}{d^3C} \sim (1-x_F)^{g_B+H+1} \; . \qquad (5.6)$$

A more accurate treatment is possible but the above will suffice for our purposes. In the target fragmentation region, where u is fixed and s, t large, the above arguments can be repeated with the result that

$$\sim (1+x_F)^{g_A+H+1} \; , \qquad (5.7)$$

where g_A is the power behavior of the target distribution function $G_{a/A}$. This result could also have been achieved by simply interchanging the target and beam particles in the previous result. These predictions will be compared to data in a later section.

One can estimate the range of validity in x_F of the above formulas by a simple argument. The momentum fraction x_F must be large enough so that the particle is out of the "quasielastic" peak where the denominator factors in Eq. (III-6) are rapidly varying. The average momentum fraction x_B of particle B is $\langle x_B \rangle = 1/B$. The average x retained by the detected particle is roughly $\simeq 1/(H+2)$. Therefore, the behavior given by Eq. (5.6) should hold reasonably well for $x_F \gtrsim 1/B(H+2)$.

For an exclusive basic process, which yields the familiar quasielastic formula, the calculation is also simple. Using Eq. (III-3) and expanding the arguments of the delta function for the case $b+n \rightarrow C+n$, where b and C are nucleons, one finds that a reasonable approximation is:

$$\left[E \, \frac{d\sigma}{d^3 C} \right]' = E(s') \, \delta\left[x(s - A^2 - B^2)(x_F - \Delta - y) \right] \, f(k_T')$$

where the shift Δ has to be calculated using more exact kinematics. At high energies $\Delta \to 0$; for the experimental data to be discussed shortly, $\Delta \sim 0.06$.

Again the x integral is not restricted and the full inclusive cross section is

$$E \, \frac{d\sigma}{d^3 C} \sim G_{C/B}\left(x_F - \Delta, K^2 \right) \left(\frac{d\sigma}{dt} \right)' \quad , \tag{5.8}$$

where the prime means that the elastic cross section is evaluated at reduced kinematics. Thus, the quasielastic peak should occur at $x_F = C/B + \Delta$. This is slightly larger than the naive expectation C/B, the most likely momentum in the state B. This shift will be included in all our numerical calculations. Eq. (5.8) can be interpreted as a relativistic generalization of the Glauber approximation but with a more precise definition of the covariant wave function.

For our particular model, we require, of course, that the same value of T fit both the inclusive (say pion production) and exclusive case.

Another situation in which we can get simple predictions from the model is when the produced particle (C) has <u>large transverse momentum</u> (s, t, u large, and masses negligible). In this regime ($\theta_{cm} \cong 90^o$), we have that

$$x_L \cong 0$$

$$x_R \cong x_T$$

and

$$1 - x_R' \cong \frac{d^2}{s'} = \frac{x_T}{2} \left(\frac{1}{x} + \frac{1}{y} \right) \quad . \tag{5.9}$$

<u>Inclusive basic process</u>. The condition $d^2 > 0$ means that in the limit of primary interest to us x and y in formula (II-7) are going to be both close to one. By a similar analysis as in Section A, the $\vec{k}_T$ and $\vec{l}_T$ can then be integrated out and we can write approximately ($G(x) \sim (1-x)^g$)

$$\sim \int_0^1 dx \int_0^1 dy (1-x)^{g_a} (1-y)^{g_b} \left[R'_C (a+b \to C+X) \right] \theta(d^2) \quad,$$

which can be applied, for example, to pion and proton (not including the quasi-elastic peak) production at large transverse momentum.

We will consider two different parametrizations for the internal cross section:

$$\sim (1-x'_R)^H f(C'^2_T) s'^P$$

and then one finds, using Eq. (III-12):

$$E_C \frac{d\sigma}{d^3C} \sim \epsilon^{F+H+1} (C^2_T)^P f(C^2_T) \quad, \tag{5.10}$$

where $F = 1 + g_a + g_b$ and where ϵ is close to zero. For ϵ not near zero, there is a function which is slowly varying in ϵ that multiplies this result.

We see that this regime gives us information about the distribution functions of both target and projectile. It also follows that the C^2_T dependence is the same for the complete process as it is for the internal interaction, except for the $(C^2_T)^P$ factor, but P is normally small and this factor is essentially constant.

In certain cases (quasielectric scattering), the internal basic process is going to be dominated by elastic scattering. This means that the cross section can be written as $(G(x) \sim (1-x)^g)$

$$\int_0^1 dx \int_0^1 dy\, (1-x)^{g_a} (1-x)^{g_b} \left[R'_C (C + b \to C + d) \right] \quad,$$

where $R'_C = \delta(d^2) \frac{s'}{\pi} \frac{d\sigma}{dt} (C + b \to C + d)$ and R'_C now contains a factor $\delta(d^2)$.
Writing the elastic cross section as

$$\frac{d\sigma}{dt} \sim s'^P f(C^2_T)$$

gives the prediction

$$E_C \frac{d\sigma}{d^3C} = I(\epsilon)\, \epsilon^F (C^2_T)^P f(C^2_T) \quad. \tag{5.11}$$

Note that in this case the multiplying function $I(\epsilon)$ must be rapidly varying in order to produce the quasielastic peak. Right at the quasielastic peak we have $\langle y \rangle \simeq 1/B, \langle x \rangle \simeq 1/A$, where A and B are the atomic number of the states A and B respectively. Using $d^2 = 0$, we get for the peak value (neglecting kinematic mass effects)

$$\epsilon_Q \simeq 1 - \frac{2}{A+B} \quad .$$

One expects that $(I(\epsilon)\, \epsilon^F)$ will be rapidly varying and will lead to a quasielastic peak for $\epsilon \sim \epsilon_Q$. For $\epsilon > \epsilon_Q$, $I(\epsilon)$ should vary much less, and the main ϵ dependence should come from the ϵ^F factor.

In the previous section we have presented a detailed analysis of our relativistic model and the predictions it gives when applied in specific regions of phase space (projectile fragmentation, target fragmentation and large p_T). While because of their simplicity we have considered only these special cases, it is possible to obtain a generalization of these results that is valid for all angles. In fact, it can be shown that

$$E_C \, \frac{d\sigma}{d^3C} = I(\epsilon)\, \epsilon^{F+1+H}(1-x_R z)^{-F-}\, (1+x_R z)^{-F+}\, J(C_T^2) \quad , \qquad (5.12)$$

where $z = \cos\theta_{cm}$, $F- = 1 + g_a$, $F+ = 1 + g_b$. As before $F = 1 + g_a + g_b$, and $I(\epsilon)$ is a slowly varying function of ϵ above any quasielastic peak. This result is valid for an inclusive internal process parametrized in the form

$$\sim (1-x_R')^H \, J(C_T'^2) \quad .$$

If need be, this form can be easily generalized. Although Eq. (III-20) was derived assuming $|z|$ not near one (outside the forward and backward cones), we see that it also has the correct limit inside those regions. This expression can then be used to characterize <u>inclusive nuclear reactions at all angles</u>. Furthermore since we expect a smooth transition from the regions of validity of Eq. (5.12) to other regions of the Peyrou plot; for example, the central region, this equation can be used to fit the data everywhere with effective powers $F, F-, F_+$.

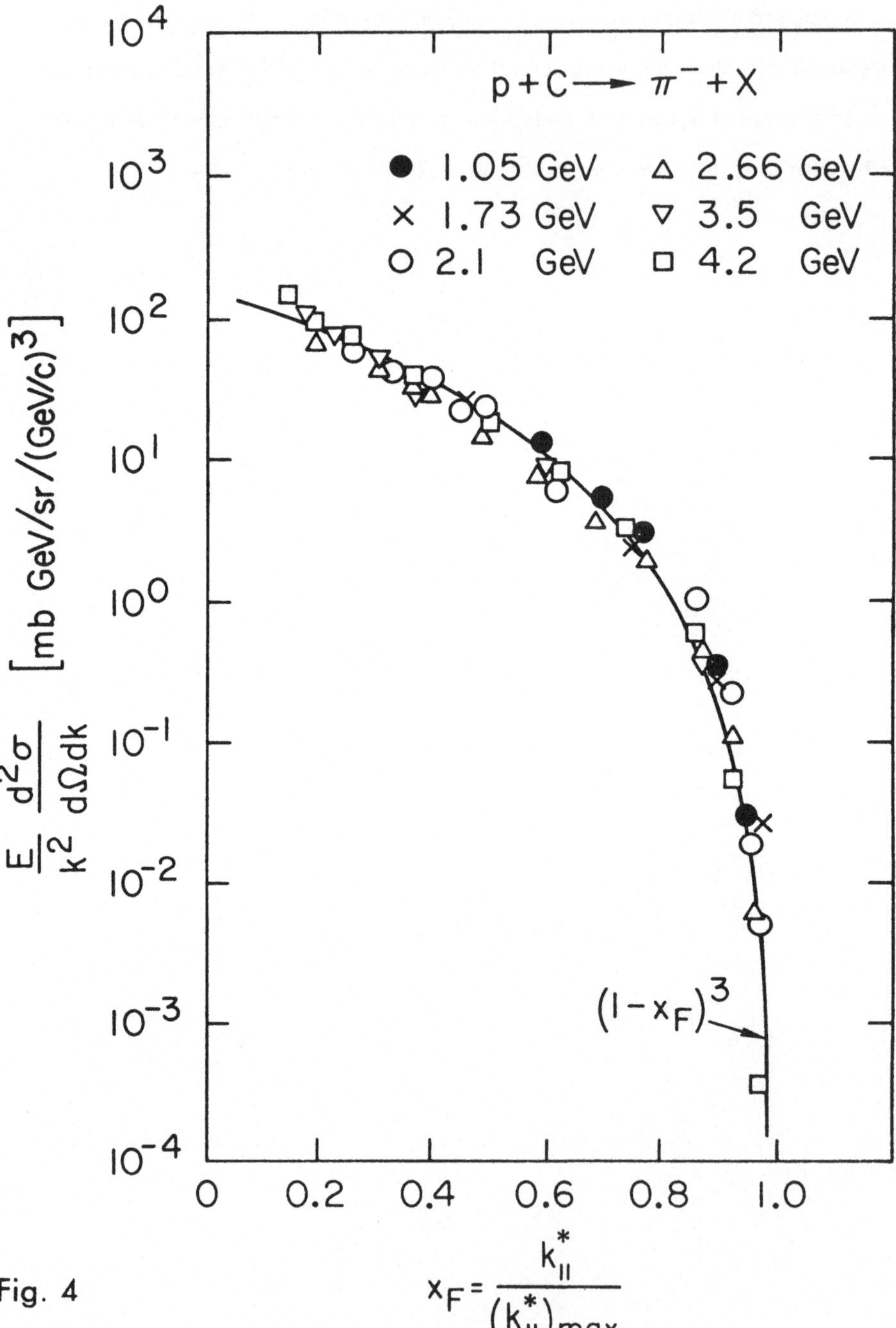

Fig. 4

The x_F spectrum compared to the carbon data illustrating scaling and the value of H.

VI. Pion and Proton Yields and Form Factors

As the first application of the model, consider π^- production in the projectile fragmentation region. The small angle data in Fig. 4 taken from J. Papp
et al.,[9] clearly supports the value H=3. The value of T now can be determined
by looking at deuteron beam data. The fit for T=3 is compared with the data in
Fig. 5. Now all parameters are fixed and the prediction for alpha beams is
compared with the data[2] in Fig. 6.

The proton yield in the forward and backward direction is fully determined
since it depends only on T (H=-1). For these reactions the data is not as extensive
in the variable x_F as in the pion data so the tests are not as severe. The reaction
$d+C \rightarrow p+X$ is predicted to go as ϵ^5. The data is in reasonable agreement with
this behavior but it does not fall quite this fast for the largest x_F values. The
prediction for the proton yield from carbon-carbon collisions is compared with
the data in Fig. 7.

It will be very interesting to extend the above discussions to other energies
and reactions and to compare theory with experiment. Of particular importance
here is measurements over the entire angular range to check the validity of
Eq. (5.12). These will be forthcoming, I understand, from the Berkeley group.
NOTE: In particular reactions, and especially at lower energies, the nuclei may
not completely fragment so that a value of $A_{eff} < A$ may have to be used to fit the
data here. Thus even at low energies it may be useful to fit the form (5.12) to the
data and to extract values of F, F_+, and F_-. If they can be interpreted in terms
of an A_{eff}, this interpretation can be checked by comparing with other yields
(the pion to the proton for example).

Let us now discuss the nuclear form factor arising from the model wave
functions that fit the inclusive data (T=3). We are concentrating on the short

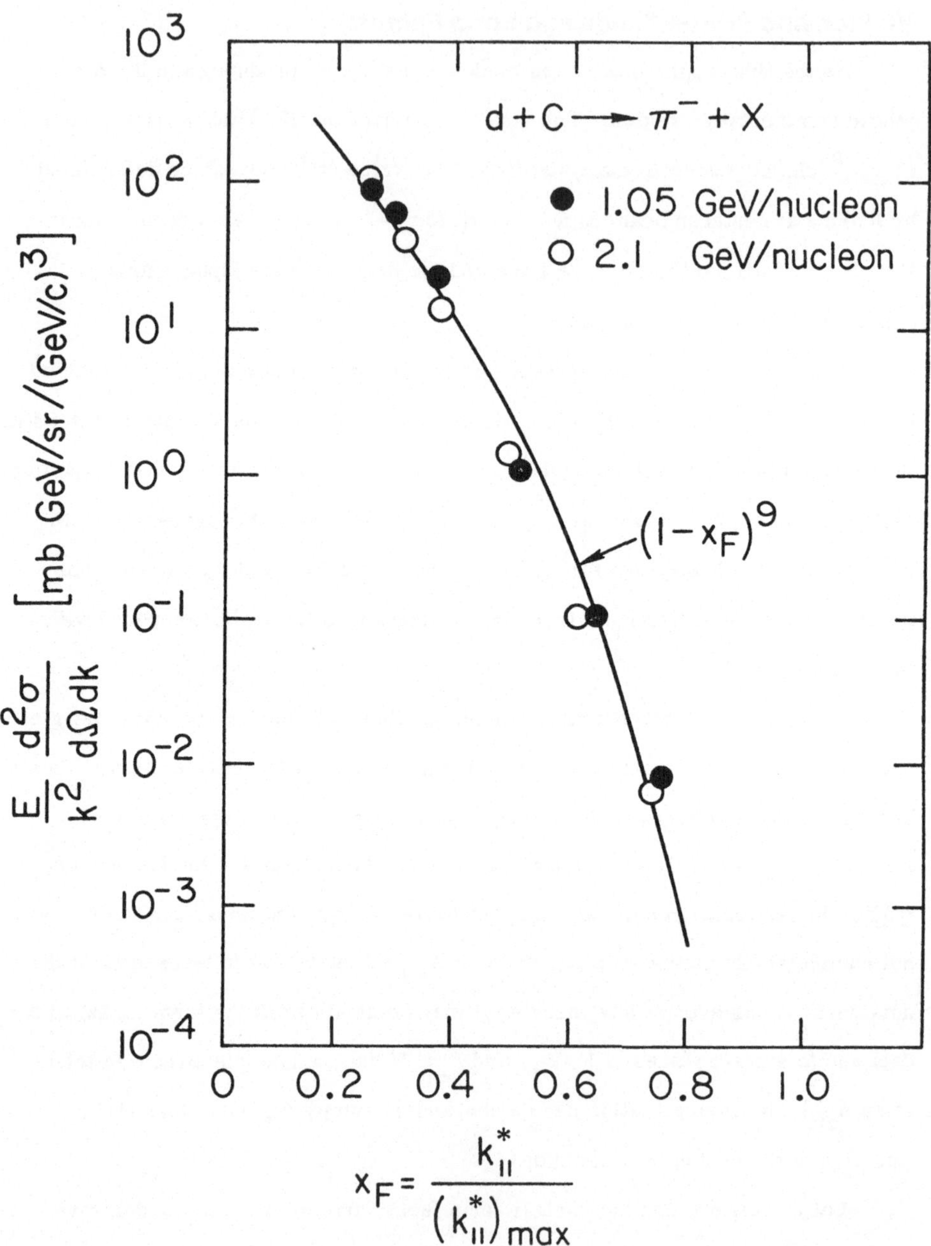

Fig. 5

The prediction for T=3 compared to the data of Ref. 9 for a deuteron beam.

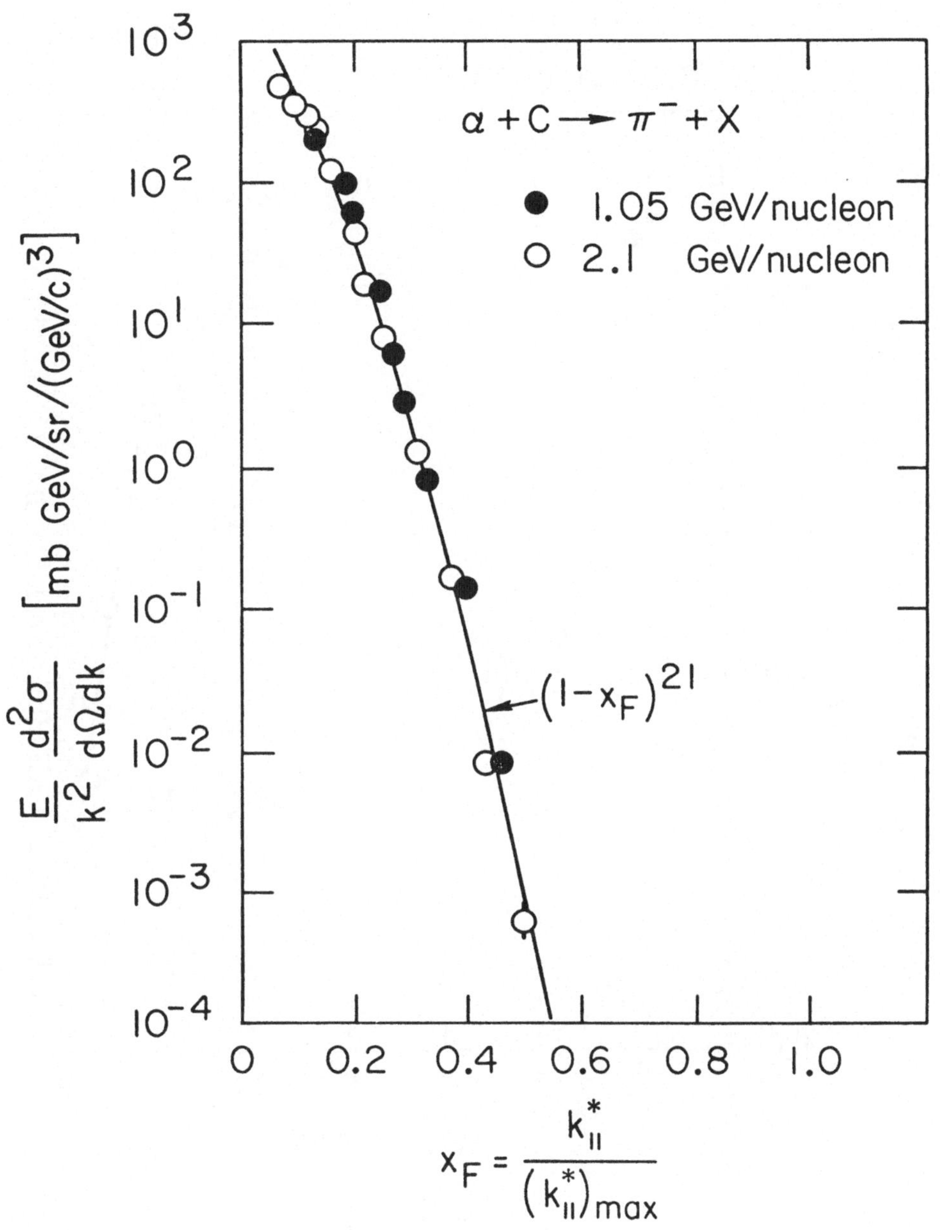

Fig. 6

The prediction for T=3 compared to the data of Ref. 9 for an alpha particle beam.

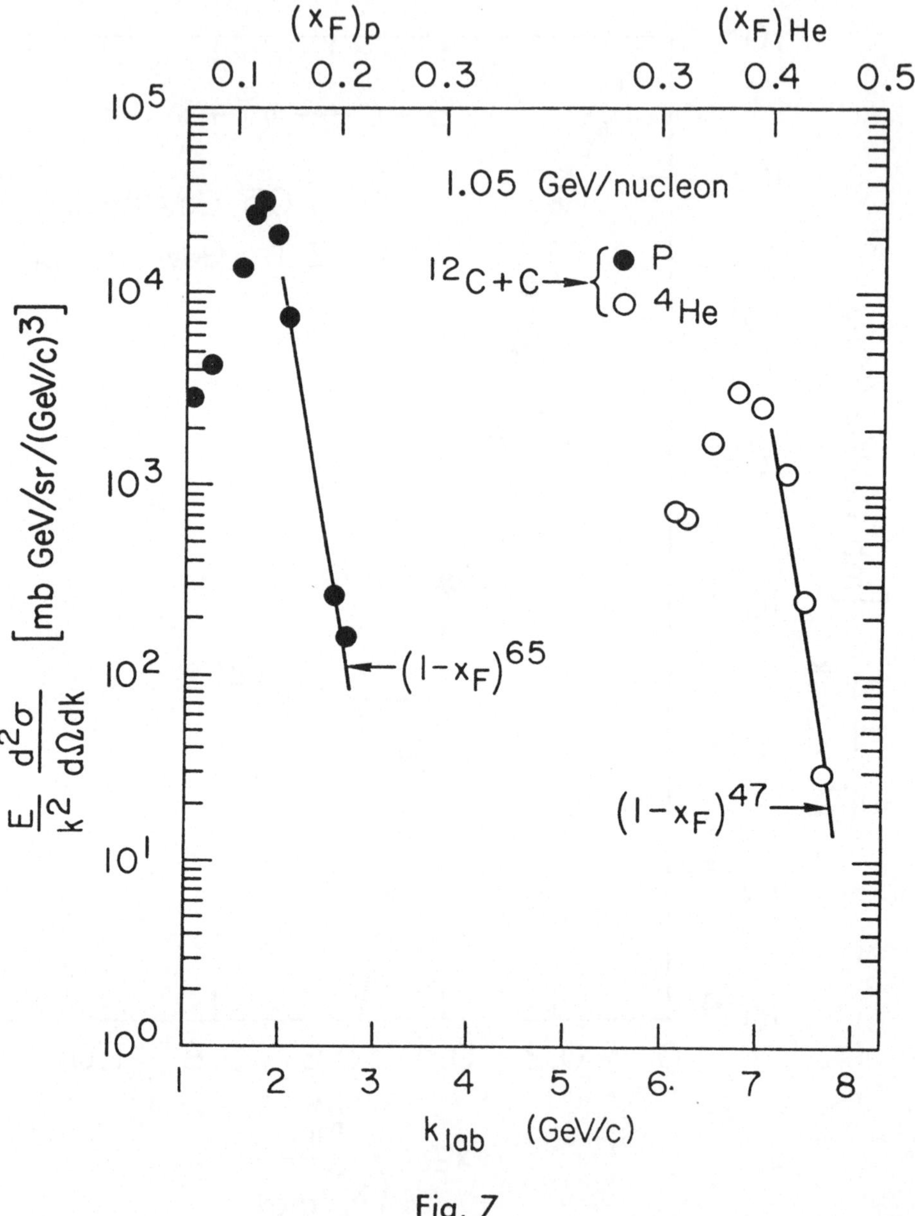

Fig. 7

Two inclusive processes for a carbon beam illustrating the counting rules
and the positions of the quasielastic peaks.

distance or large q^2 behavior, and hence will make no attempt to fit diffraction

etc., that arise from effects of the edge of the nucleus. The form

factor is written as

$$F_A(q^2) = \sum_a F_a(q^2) \int \frac{dx\, d^2 k_T}{2(2\pi)^3} \frac{x}{(1-x)} \psi_a^*(x, \vec{k}_T + (1-x)\vec{q}_T)\, \psi_a(x, \vec{k}_T) \quad ,$$

where the sum runs over the nucleons (protons and neutrons) in the nucleus A.
$F_a(q^2)$ has been replaced by its on-shell value, and the integral multiplying it

is then the intrinsic body form factor of the nucleus.

A very plausible wave function ψ which we saw before gives a G function

with many correct properties, is

$$\psi(x, \vec{k}_T) = \frac{N(x)(1-x)^{\frac{g+1}{2}}}{\left[k_T^2 + M^2(x) \right] \left[k_T^2 + M^2(x) + \delta^2 \right]^{\frac{g-1}{2}}}$$

where for the case of the deuteron g=5 for T=3 (the exchange of vector mesons

with monopole form factors at each vertex).

First, since ψ describes one off-shell and one on-shell particle, neither
ψ nor G are necessarily symmetric around $x = 1/2$. Isospin symmetry implies

that $G_{p/D}(x) = G_{n/D}(x)$, not that $G_{p/D}(x) = G_{n/D}(1-x)$. However, this is a good

approximation at not too high energies, when we consider a deuteron as composed

of only one proton and one neutron, which means

$$\int dx\, d^2 k_T\, G_{a/D}(x, \vec{k}_T) = 1 \quad .$$

Note that then this is equivalent to a momentum normalization condition. The

symmetry of G around $x = 1/2$ fixes the function $N(x) = N_0 x^2$. The deuteron form

factor can now be computed. A fit that can be achieved for our spinless model

is given in Fig. 8 for the value

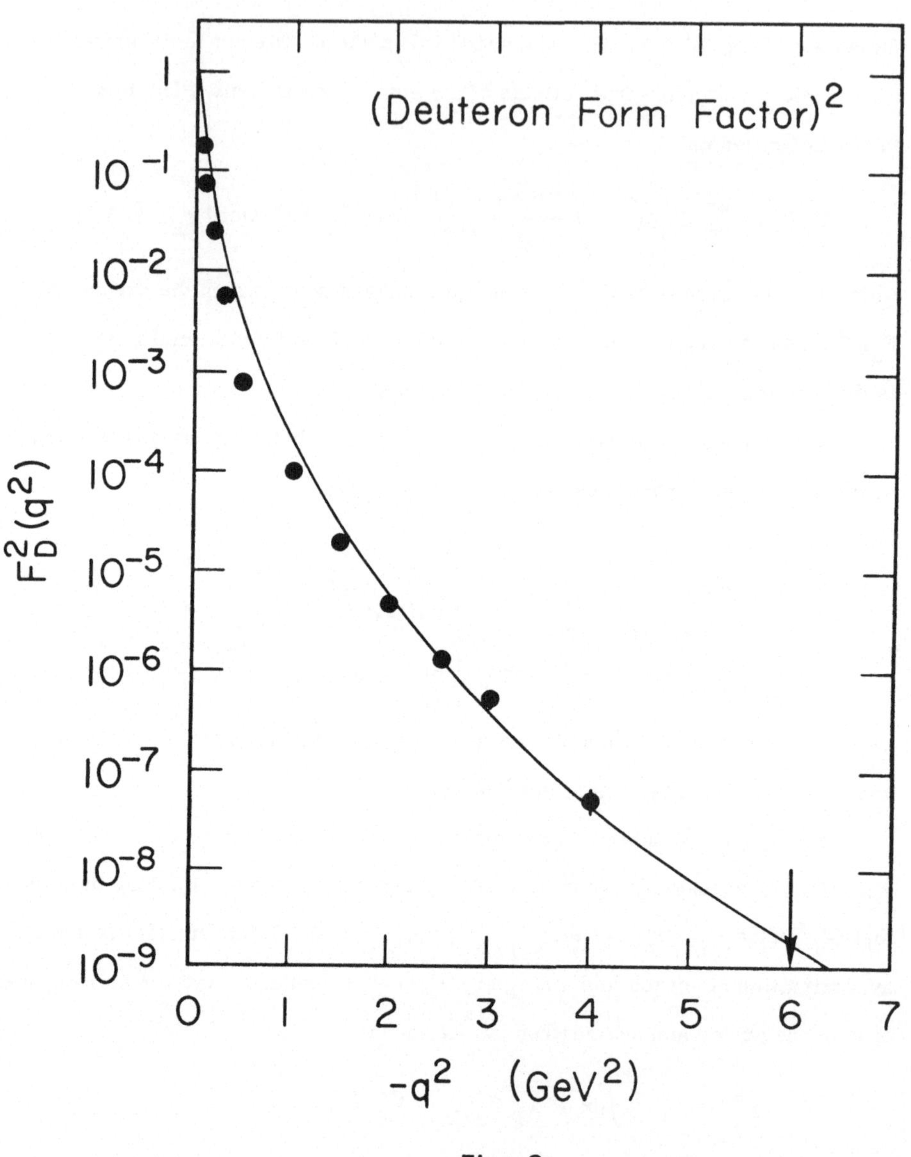

Fig. 8

Fit to the (deuteron form factor)^2 data of Ref. 10.

$$\delta_D^2 \simeq 200 \, M \, \epsilon \quad ,$$

where M is the nucleon mass and ϵ is the binding energy of the deuteron. Here the isoscalar form factor was taken to be equal to the proton form factor. This is the same δ_D^2 value that was used to fit quasielastic scattering discussed earlier. The data is from Ref. 10. If spin were put intc the model, and especially if D-state effects were then included, the fit could be made much better since the quadrupole contribution naturally gives a shape that is similar to that of the data points.

Predictions have been made by I. Schmidt[3] for the He^3 and He^4 form factors out to a $(-q^2)$ of 6 $(GeV/c)^2$. These will soon be measured at SLAC, and will provide new tests for the model. Finally, note that T=3 is the same behavior as expected by quark counting,[11] as was pointed out before.

VII. Structure Functions and Nonscaling

Since we now have a relativistic wave function for the deuteron that fits both inelastic inclusive scattering and the elastic deuteron form factor, it can be used to describe deep inelastic electron scattering from the deuteron. A separation of the neutron and proton structure functions can then be performed with some confidence in regions where one would not trust a nonrelativistic treatment.[12] To discuss inelastic scattering we simply return to Fig. 1 and Eq. (2.1) and set B=b=C=electron, and consider the various choices for a and d. A classification of the terms contributing to the deuteron structure function is given in Fig. 9. In Fig. 9c, the two nucleons recoil coherently, sharing the q^2 of the virtual photon. This is small for large q^2. In the second term,

Fig. 9b, one has d=nucleon (or baryon resonance), and it recoils with momentum q. This term may be important for moderate q values, and is given by terms of the form (a = proton or neutron)

$$\sim x_D\, G_{a/D}(x_D)\, F_a^2(q^2) \quad , \tag{7.1}$$

where $F_a(q^2)$ is the dipole form factor of the nucleon.

The first term Fig. 9a is inelastic scattering from the nucleon and is given in terms of their structure function $F_{2a}(x, q^2)$ (which may not scale):

$$\sim \int_{x_D}^{1} dy\, F_{2a}(x_D/y, q^2)\, G_{a/D}(y) \quad , \tag{7.2}$$

which clearly vanishes faster than (7.1) as $x_D \to 1$. Note that if $G_{a/D}(y)$ is strongly peaked at y=1/2, which is certainly true in the limit of zero binding, this contribution becomes

$$\sim F_{2a}(2x_D, q^2)\, \theta(1-2x_D) \quad ,$$

which has a simple physical interpretation.

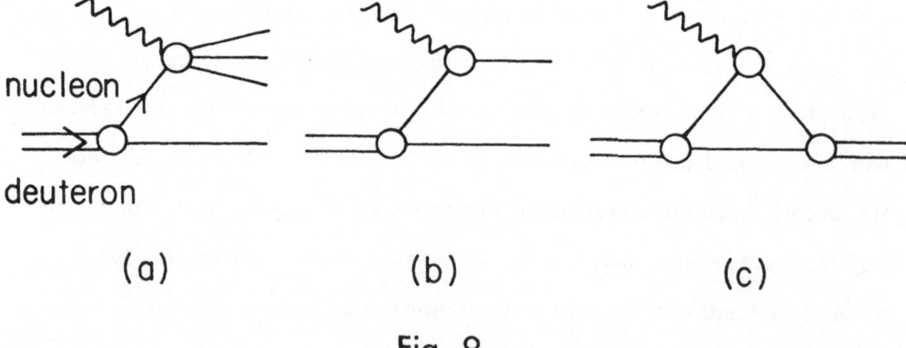

Fig. 9

Contributions to the deuteron structure function, with partons of one of the nucleons (a), one nucleon (b), and both nucleons (c) recoiling coherently.

These contributions have been used to describe inelastic scattering from the deuteron from low to high q^2 values and to extract the neutron structure function.[13] We will not go into this in detail but will use the physical picture described above in which coherent recoils in the final state lead to nonscaling terms that fall in q^2 but vanish less rapidly as $x \to 1$. Examples are shown in Fig. 10 for the nucleon. Figure 10a has a single recoil quark, 10b has a recoil diquark and meson system, and 10c depicts a recoil baryon (or baryon resonance) system. The last three terms are not important in the range of large to moderate q^2 and large x and will be neglected.

Using our previous rules, (T=1 of course), and dimensional counting for the form factors, the diquark term is q of the form

$$F_{2a}^{d}(x,q^2) = A_d \, F_d^2(q^2) \, x^2(1-x) \qquad (7.3)$$

where

$$F_d(q^2) = d^2/(d^2-q^2)$$

is the diquark form factor. The valence, or large x, part of Fig. 10a will be written in the form

$$\sim A(x)(1-x)^3 \quad ,$$

where $A(x)$ is finite at x=1 and slowly varying.

The total nucleon structure function will be written in the form

$$F_{2a}(xq^2) = A^a(x)(1-x)^3 + A_d^a \, F_d^2(q^2) x^2(1-x) \quad , \qquad (7.4)$$

and the main question is whether or not this will fit the data for large x, say x>0.2, with an A (x) that does not depend on q^2. If this is possible, then we have possibly identified important contributions to nonscaling that have a very simple and expected physical origin. Note that the sea is in $A(x)$—that is, there is a term that vanishes as $(1-x)^4$ that measures the amount of $(q\bar{q})$ sea present. It is not important for x>0.2, our region of primary interest.

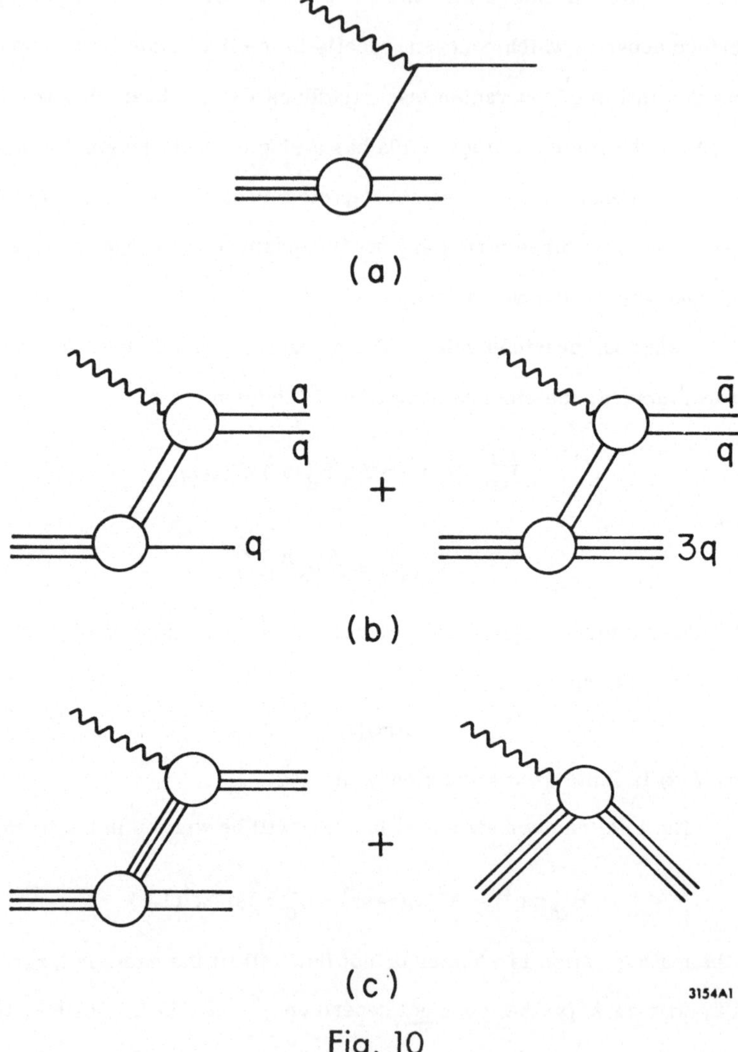

Fig. 10

Contributions to the nucleon structure function, with one (a), two (b), and three (c) quarks recoiling coherently.

Suffice it to say, the fits are very good for all x and for $x > 0.2$, $A(x)$ is independent of q^2. The final results are shown in Fig. 11 for the proton structure function. Since the deuteron was treated earlier, one may extract the neutron structure function which is shown also in Fig. 11. The diquark term is quite small, but fits all the scale breaking for $q^2 > 2$, $x > 0.2$.

Note that the ratio of $A(x)$ for the proton to that for the neutron is $\approx 3/2$, which is the ratio of the squares of the valence quark charges. Also, the ratio of the diquark term for proton/neutron is ≈ 3, which is the ratio of the squares of the diquark charges! This agreement is an amusing feature of our model and show its consistency if not correctness. There are, however, other ways to fit this data, using QCD, for example, and its asymptotic freedom behavior but the particular method as applied seems questionable to me (i.e., believing in a leading log expansion and even a leading log log expansion), and the resultant mass scale (the Λ^2 in the coupling constant) seem highly artificial. However, taste aside, this may be correct. What seems more likely to me is that each is $\approx 1/2$ the truth.

Finally these functions can be applied[12] to neutrino data and fit quite well the main features of the data without any change in the parameters (for example, the lack of nonscaling seen in the new data for $x < 0.2$).

Recalling the comments in Section III, we see that here too we disagree with the conventional QCD calculations for the structure functions. For example, if a virtual gluon makes a $q\bar{q}$ pair, the $\bar{q}$ structure function is dominated by antiquarks near their mass shell (if one does not take too seriously the leading log approximation). Thus the $\bar{q}$ is around a long time (by the uncertainty principle), interacts with the other quarks, "thermalizes", and produces a $(1-x)^7$ behavior in $G_{\bar{q}/p}^8$.

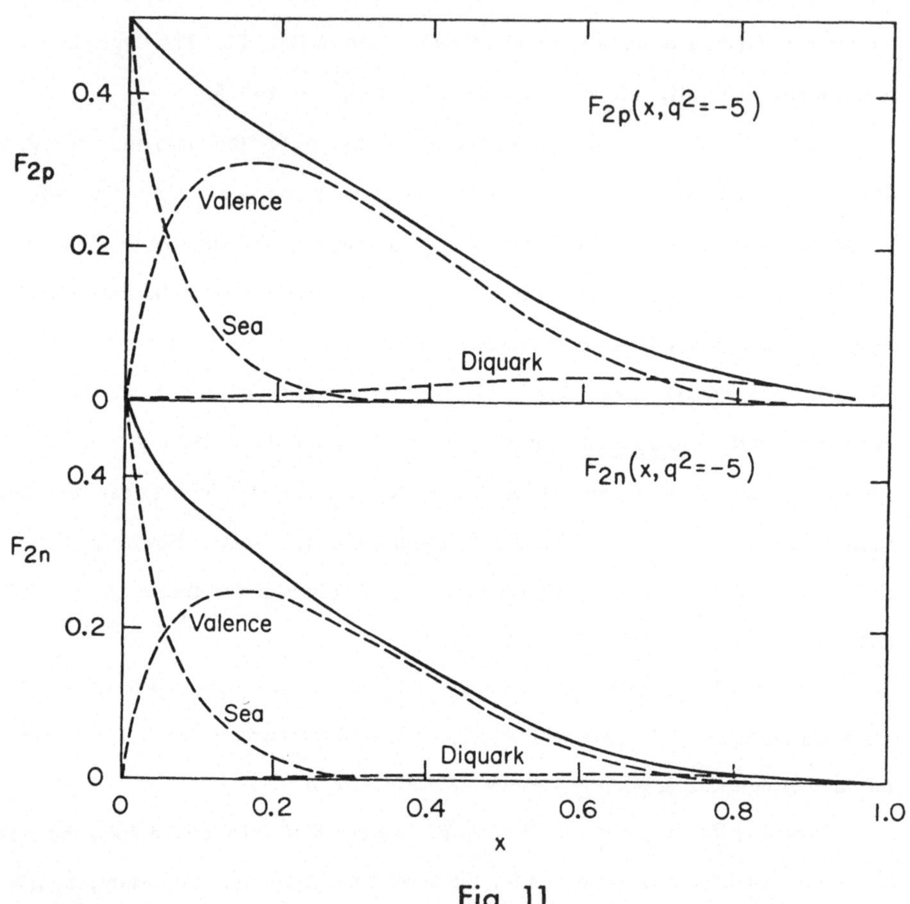

Fig. 11

The contributions (valence, diquark and sea) and the total structure functions of proton and neutron (for $q^2 = -5$ $(GeV)^2$).

VIII. Massive Lepton Pairs

Before considering purely hadronic processes, let us consider an intermediate one—the production of a massive lepton pair by hadron beams (inclusive inverse "photo"-production, where the virtual photon produces the lepton pair). Let us reiterate the points made in Section III about coherence and the hard scattering expansion. The total yield is a sum over intermediate states a and b (and the final state d). These terms must be incoherent—and this requires, for example, that a beam fragment in one term (in the sum over a and b) not be allowed to end up in the same part of phase space as an identical fragment from the central process of another term in the sum (all other particles being the same). This is a difficult requirement to enforce with mathematical precision. Clearly one can easily make a mistake in this regard if large momentum transfer scattering is allowed in both the beam or target vertex (or structure) function and the central scattering process. Simply adding a broad transverse momentum to the beam fragmentation function can lead to double counting (as well as trouble with gauge invariance in the present reaction). To avoid this problem we shall insist that all large momentum transfer scatterings occur in the central process only. In this way we can avoid double counting and coherence problems but yet can include all possible diagrams.

The full cross section of a beam particle A on target particle B is found to be

$$Q^4 \frac{d\sigma}{d^4Q} (AB \to \ell^+\ell^-X) = \sum_{a,b,d} \int dx\, dy\, G_{a/A}(x) G_{b/B}(y) Q^4 \frac{d\sigma}{d^4Q} (ab \to \ell^+\ell^- d; s't'u'; Q^2)$$

$$(8.1)$$

The easiest way to enforce gauge invariance (<u>not</u> the only way) is to assume that the basic process is meson-quark $\to (\ell^+\ell^-)$-quark.[4] This model allows the rate to be normalized in two different ways and yet in the correct physical limit, is identical to the Drell-Yan model of $q\bar{q}$ annihilation. In Ref. 4, the cross section for meson + quark $\to \ell^+\ell^-$ + quark, was shown to be (see Fig. 12b)

$$Q^4 \frac{d\sigma}{d^4Q} \, (Mq \to \ell^+\ell^- q) = \frac{1}{6\pi^2} \alpha^2 h^2 \, \delta(s+t+u-Q^2 -2M^2)\Sigma(s,t,u;Q^2) \qquad (8.2)$$

where Σ is a simple function given there.

We are not interested in a quantitative evaluation of the cross section but rather in qualitative behavior in different regions of phase space.

The large Q_T^2 and Q^2 distributions can be extracted from the above formulas by writing $xG_{a/A}(x) \propto (1-x)^{g_a}$, and similarly for b/B. By manipulations similar to those used to extract the large transverse momentum behavior in hard scattering models, but which are more involved because both Q^2 and Q_T^2 are large, it is possible to derive the form

$$Q^4 \frac{d\sigma}{d^4Q} \, (AB \to \ell^+\ell^- X) \cong K\left(Q_T^2, Q^2\right) \epsilon^F \, J(\epsilon, x_F) \qquad (8.3)$$

and

$$K\left(Q_T^2, Q^2\right) = \left(1 + \frac{Q_T^2}{\mu^2}\right)^{-2} \left(1 + \frac{Q_T^2}{\mu^2 + dQ^2}\right)^{-1} , \qquad (8.4)$$

where μ is a mass parameter related to internal masses in the model, d is a constant ($d \approx 1$), J is slowly varying for small ϵ, and $F=1+g_a+g_b$. This is a universal characterization of the Q_T distribution for all beam particles (since μ and d are the same). Finally

$$\epsilon = \frac{\mathcal{M}^2}{s} = 1 - \frac{t+u}{s} - \frac{Q^2}{s} , \qquad (8.5)$$

where $\mathcal{M}$ is the total missing mass with respect to the photon (hadron masses were neglected in the above kinematics). This form can be used to parametrize detailed numerical calculation and may prove useful in fitting data. After integrating over Q_T^2, the factors $\epsilon^F J(\epsilon, x_F)$ are simply related to the folding of structure functions in the Drell-Yan formula. The explicit ϵ^F factor characterizes the threshold behavior. After integrating over x_F and Q_T^2, which adds an extra factor of ϵ^2, the threshold behavior for the mass distribution $d\sigma/dQ^2$ is ϵ^{F+2}. For D-Y, this final power is 11 for pp scattering and 5 for πp scattering if one uses the dimensional counting predictions for the structure functions.[8] For the meson-quark scattering case, F=9 and 3, respectively (again using dimensional counting), hence the final ϵ power is the same in the two cases. This is not unexpected since the first diagram in Fig. 12b is clearly the same as the Drell-Yan mechanism. It is equally clear that the second diagram is required for gauge invariance. In fact, at large Q_T, it cancels the leading Q_T^{-4} term arising from the first diagram!

At large energies, where the ϵ dependence can be neglected, it is easy to see that $<Q_T>$ has the limiting values

$$<Q_T> \cong \frac{\pi}{4}\mu \qquad\qquad Q \ll \mu$$
$$\cong \frac{\pi}{2}\mu\,(1-2\mu/\sqrt{d}Q) \quad,\quad Q \gg \mu \tag{8.6}$$

a prediction that can be experimentally checked but the present data is not over a large enough range in Q. The ϵ dependence produces an energy dependence in $<Q_T>$ in addition to the dependence on Q. A <u>very</u> rough estimate yields

$$\frac{<Q_T>_s}{<Q_T>_{s'}} \cong \left[1 + \pi\mu\left(\frac{1}{\sqrt{s'}} - \frac{1}{\sqrt{s}}\right)\right]^F \quad, \tag{8.7}$$

which can be a growth of $\approx 30\%$ over the ISR energy range.

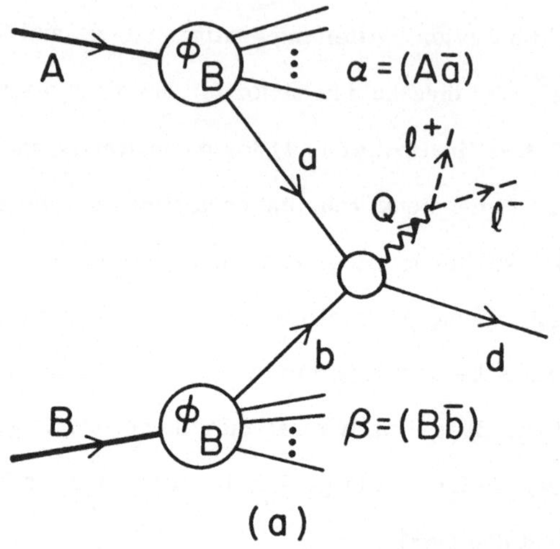

Fig. 12

Massive lepton pair production.

IX. Hadrons at Large Transverse Momentum

In order to treat hadrons, we simply consider quarks as the constituents with a normal gluon interaction (thus T=1) and apply our previous formulas. The main difference is that is is customary to interchange A and B in our previous discussion and to consider A as the beam particle. This changes only the notation, not the physics! First a review of some relevant experimental results and numbers.

The fixed angle (90° center of mass) exclusive two-body cross sections at large s will be parametrized in the form

$$\frac{d\sigma}{dt}\bigg|_{90^o} = Es^{-n} \quad .$$ (9.1)

Throughout the paper pure GeV units will be used. Inclusive large p_T cross sections at 90° center of mass from FNAL[1,2] and the ISR[3,4] can be fit to the form ($\epsilon = 1 - 2 p_T/\sqrt{s} = 1 - x_T$)

$$E\frac{d\sigma}{d^3p}\bigg|_{90^o} = I\epsilon^F \left(p_T^2\right) \qquad \begin{cases} x_T > 0.2 \\ 2 < p_T < 8 \text{ GeV/c} \end{cases}$$ (9.2)

Table I summarizes the values for I, N, F, and E, n for the various well-known cross sections of interest.[14] For N and F we have chosen the nearest integer values, and then fit the normalization constant I.

The reactions $pp \to K^- X$ or $\bar{p} X$ near 90° can be fit by the following behavior ($\epsilon < 1$):

$$\frac{E\frac{d\sigma}{d^3p} (pp \to K^- X)}{E\frac{d\sigma}{d^3p} (pp \to K^+ X)} \approx 1.0\,\epsilon^{3.5}$$ (9.3)

$$\frac{E\frac{d\sigma}{d^3p} (pp \to \bar{p}X)}{E\frac{d\sigma}{d^3p} (pp \to pX)} \approx 0.3\,\epsilon^{6.9\pm 2.4} \left(p_T^2\right)^{1.4\pm 1.3}$$ (9.4)

Table I

(GeV Units)

Reaction	E	n	
$pp \rightarrow pp$	1.2×10^9	10	
$\pi^\pm p \rightarrow \pi^\pm p$	2.5×10^4	8	
$\pi^- p \rightarrow \pi^o n$	2.5×10^4	8	
$\gamma p \rightarrow \pi^+ n$	1.2×10^1	7	
__Reaction__	$\underline{I}$	$\underline{N}$	$\underline{F}$
$pp \rightarrow \pi^{+,0,-} X$	$(9, 8, 7)$	4	9
$pp \rightarrow K^+ X$	5	4	9
$pp \rightarrow pX$	500	6	7
$\pi^\pm p \rightarrow \pi^o X$	3.5	4	7

By analyzing data on the momentum distribution of particles balancing a large p_T trigger, one can estimate that between 1/2 and 1/4 of the trigger particles are "prompt" as opposed to those that arise from the decay of produced resonances. We shall attempt to compute the yield of the prompt component only, which we take to be roughly 1/3 of the values of I given in Table I.

An example of the quality of the fit is shown in Fig. 13 taken from Cronin et al., Ref. 14.

We first need to choose simple forms for the G functions in order to compute the yields[15]. The G's are constrained to satisfy the spectator counting rules[6] as $x \rightarrow 1$ (with T-1) and to have a reasonable shape for small x (i.e., some flattening off in xG). A form (Fig. 14) with these properties which yields simple integrals in later calculations is

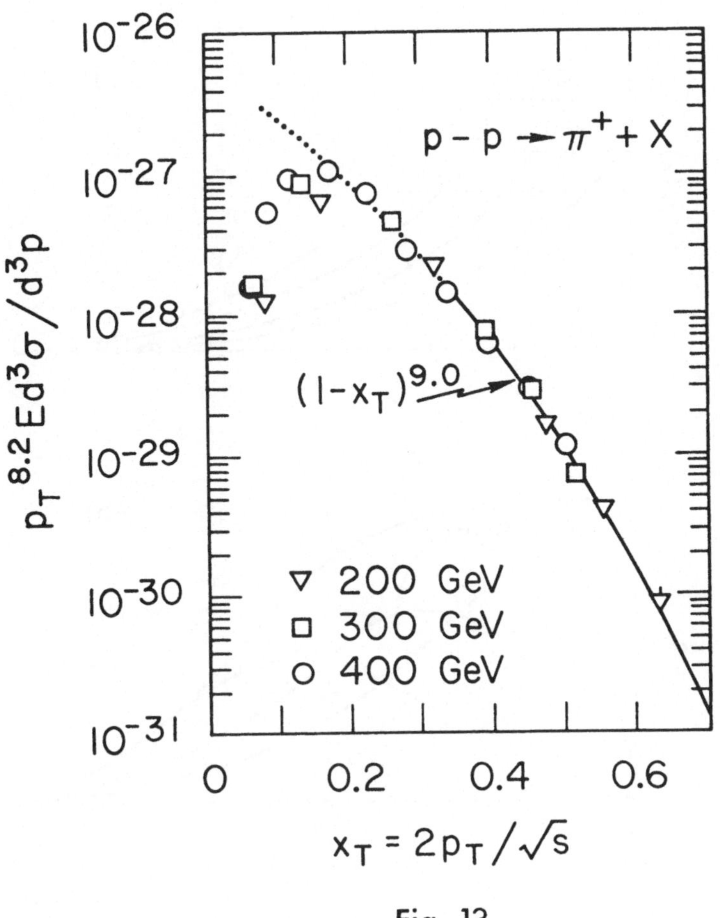

Fig. 13

Fit to data given by Cronin, et al., Ref. 14.

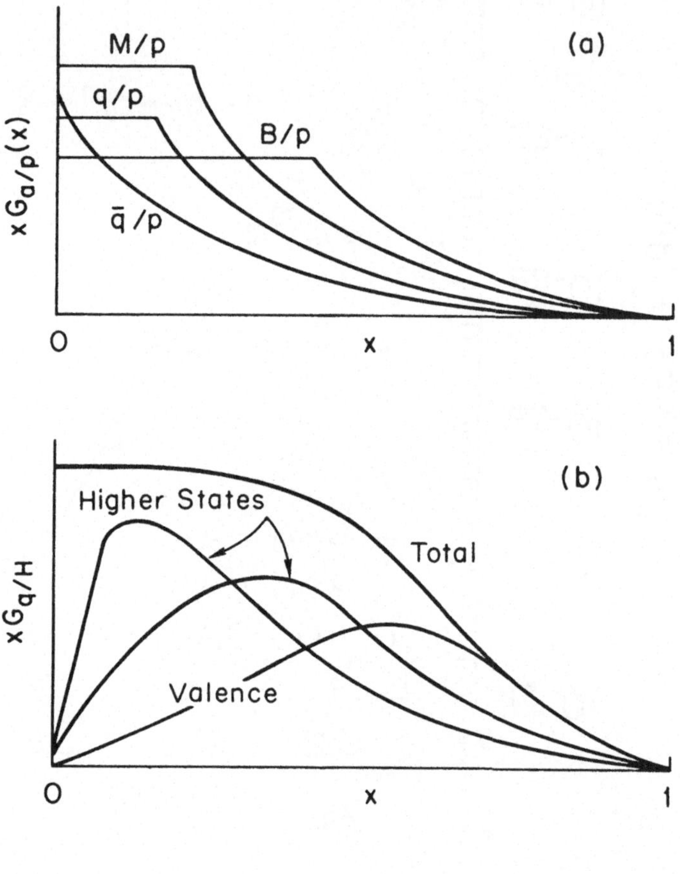

Fig. 14

Simple model for G's and their more realistic G functions.

$$xG_{a/A}(x) = (1 + g_a)f_{a/A} N(a/A)(1-x)^{g_a} \qquad x > \hat{x}_a$$

$$\tag{9.5}$$

$$= (1 + g_a)f_{a/A} N(a/A)(1-\hat{x}_a)^{g_a} \qquad x < \hat{x}_a \quad ,$$

where

$$g_a = 2n(\bar{a}A) - 1$$

and $n(\bar{a}A)$ is the minimum number of quarks in the spectator system. Here $f_{a/A}$ is the fraction of total momentum carried by a in A

$$f_{a/A} \equiv \int_0^1 dx\, xG_{a/A}(x) \tag{9.6}$$

and

$$N(a/A) = \left[(1-\hat{x}_a)^{g_a} (1 + g_a \hat{x}_a)\right]^{-1} \tag{9.7}$$

As an example, reasonable values for u or d quarks in a proton are $g_a=3$ and $\hat{x}_a = 0.25$. $N(a/A)$ adjusts for the shape dependence of the structure function relative to a pure $(1-x)$ power and approaches 1 as $\hat{x}_a \to 0$. Throughout these lectures if "a" refers to a quark, it will be a quark of a given color. This means that there will be many odd looking factors of 3 in our formulas that are due to color sums and normalizations.

In order to assign values to the momentum fractions f (of mesons in a proton, for example), it is necessary to use the convolution formula

$$G_{a/A}(x) = \sum_n{}' \int_x^1 \frac{dz}{z} G_{a/n}^V \left(\frac{x}{z}\right) G_{n/A}(z) \quad . \tag{9.8}$$

Note that to avoid double counting one of the G's must be "irreducible," i.e., a valence distribution function, and the sum, $\sum'$, is restricted to non-overlapping intermediate particle states (n) having no quarks in common. A simple integration yields

$$f_{a/A} = {\sum_n}' f^V_{a/n} f_{n/a} \quad . \qquad (9.9)$$

Using the above formulas and experimental data wherever possible, such as for the quark and antiquark distributions, one arrives at the values given in Table II. These will be used in our later calculations of note.

In a scale invariant theory all exclusive differential cross sections at large momentum transfers in the fixed angle regime can be written as a sum of terms of the form

$$\frac{d\sigma}{dt} = \pi \mathscr{D} \, s^{T+U-N} (-t)^{-T} (-u)^{-U} \quad , \qquad (9.10)$$

where $\mathscr{D}$ contains the relevant coupling constants. This parametrization is appropriate for general processes involving quarks, gluons and hadrons.

There are two critical coupling constants for quark-hadron scattering which we now define. For the coupling of a meson to its simplest valence two-quark component, we define a standard coupling, $g/\sqrt{3}$, of the π^+ to a u and $\bar{d}$ quark of one color. We also define $h/\sqrt{3}$ to be the coupling of a quark of one color to a baryon. Defining standard values

$$\alpha_M = \left(\frac{1}{3} \frac{g^2}{4\pi} \right) \qquad \alpha_B = \left(\frac{1}{3} \frac{h^2}{4\pi} \right) \quad , \qquad (9.11)$$

we give in Table III the cross section forms for all elementary processes of interest for quarks interacting with $J^P = 0^-$ mesons and $J^P = \frac{1}{2}^-$ baryons. One should take special note that the $\gamma q \to Mq$ cross form given in the table incorporates three additional diagrams, other than the one drawn, as required by gauge invariance. All cross sections are those for quarks or diquark systems of one given color. The spin 1/2 quark spin 1 vector gluon structure of QCD is reflected in the tabulated results. For instance the $qM \to qM$ cross section in a scalar quark ϕ^4 model is proportional to $1/s^2 u^2$ instead of $1/su^3$ as found for spin 1/2 quarks. The latter value is in the table.

Table II

Distribution Function Parameters (per color)

$$xG_{a/A}(x) = (1 + g_a)f_{a/A}N(a/A)(1-x)^{g_a} \quad (x > \hat{x}_a)$$

a/A	g_a	$\hat{x}_a$	$f_{a/A}$	N(a/A)
u/p	3	.2	.1	1.22
d/p	3	.2	.067	1.22
$\bar{q}/p$	7	0	.01	1
(2q)/p	1	.6	.1	1.6
M/P	5	.3	.1	2.4
K^-/P	9	0	.024	1
B/P	3	.5	.12	3.2
q/π	1	.3	.083	1.1
$\bar{q}/\pi$	1	.3	.083	1.1
M/π	3	.4	$.\bar{1}$	2.1

$$f^V_{u/p} = 2 f^V_{d/p} = .04 \qquad\qquad f^V_{q/M} = f^V_{\bar{q}/M} = .033$$

$$\sideset{}{'}\sum_M f_{M/p} = .1 \qquad\qquad \sum_B f_{B/p} = .18$$

$$\sum_M f_{M/p} \cong .4 \qquad \sum_B f_{B/p} \cong .72 \qquad \sideset{}{'}\sum_M f_{M/\pi} \cong .8$$

$$\sum_{q=u,V} f_{q/p} = .17 \qquad \sum_{\bar{q}} f_{\bar{q}/p} = .03 \qquad \sum_q f_{q/\pi} = \sum f_{\bar{q}/\pi} = .083$$

$$\sum_{(2q)} f_{(2q)/p} = .3$$

TABLE III

ELEMENTARY CIM SUB PROCESSES (Spin and Color Averaged)

$$\frac{d\sigma}{dt} = \frac{\pi\mathcal{D}}{s^2} \times \overline{\sum}\mathcal{M}^2$$

Subprocess		$\mathcal{D}$	$\sum\mathcal{M}^2$	Subprocess		$\mathcal{D}$	$\sum\mathcal{M}^2$
$qM \to qM$ (ut)		α_M^2	$\dfrac{s}{u^3}$	$d\bar{d} \to p\bar{p}$	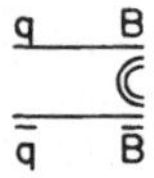	α_B^2	$\dfrac{s^2+u^2}{s^2 t^4}$
$qM \to qM$ (st)		α_M^2	$\dfrac{u}{s^3}$	$\bar{d}d \to p\bar{p}$		α_B^2	$\dfrac{s^2+t^2}{s^2 u^4}$
$q\bar{q} \to M\bar{M}$	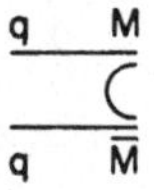	$\frac{1}{2}\alpha_M^2$	$\dfrac{u}{t^3}$	$\bar{d}p \to p\bar{d}$		α_B^2	$\dfrac{t^2+u^2}{u^2 s^4}$
$Mq \to \gamma q'$ (ut)	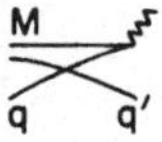	$2\alpha\alpha_M$	$\dfrac{1}{s^2 t}(s^2+u^2)\left(\dfrac{\lambda_{\bar{q}}}{u}-\dfrac{\lambda_{q'}}{s}\right)^2$	$dp \to pd$	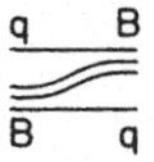	α_B^2	$\dfrac{s^2+u^2}{u^2 t^4}$
$\gamma q \to Mq'$ (ut)		$\alpha\alpha_M$	$\dfrac{1}{s^2 t}(s^2+u^2)\left(\dfrac{\lambda_{\bar{q}'}}{u}-\dfrac{\lambda_{q}}{s}\right)^2$	$q(2q) \to MB$	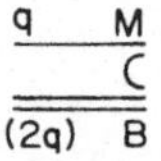	$\frac{1}{2}\alpha_B\alpha_M$	$\dfrac{u}{t^5}$
$dp \to dp$ (ut)	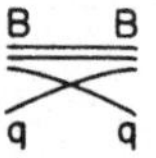	α_B^2	$\dfrac{s^2+t^2}{t^2 u^4}$	$(2q)M \to Bq$	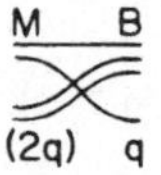	$\alpha_B\alpha_M$	$\dfrac{s}{u^5}$

11-77

3303C19

The exclusive scattering process $AB \to CD$ (see Fig. 3a) can be considered as the scattering of $A \to C$ off of a quark constitutent of the target B. Since the constituent in general has some fraction, x, of the target momentum, the basic subprocess occurs at a reduced energy and one readily shows from this quark interchange diagram that

$$\frac{d\sigma}{dt}(AB \to CD) = F_{BD}^2(t)N_{coh}^2 \frac{d\hat{\sigma}}{dt}(Aq \to Cq; \hat{s} = <x>s, \hat{t} = t, \hat{u} = <x>u)$$

$$+ \text{ permutations } (A, C \to B, D) \quad , \tag{9.12}$$

where the form factor reflects the sticking probability in the final state and the mean value theorem has been used to replace x under the integral by $<x>$, N_{coh} is the number of coherently interfering diagrams which contribute. Using the standard form for $d\hat{\sigma}/dt$, Eq. (9.10), defining $n = 4 + N$, and taking $F_B(t) = \left(1 - t/M_V^2\right)^{-2}$ for a proton target, we obtain at large s

$$s^n \frac{d\sigma}{dt}\bigg|_{90^\circ} = E_{e\ell}(AB \to CD) = \pi \ <x>^{T-N} N_{coh}^2 \ 2^{4+T+U}(M_V)^8 \quad , \tag{9.13}$$

which is consistent with dimensional counting. One expects that $<x> \overset{\sim}{>} 1/3$ should characterize scattering from a valence component of the proton target.

Using the above formula, and the values of T and U from Table III, one finds by fitting the data for E that

$$\alpha_M = 1.2 \text{ GeV}^2, \quad \alpha_B = 10 \text{ GeV}^4 \tag{9.14}$$

which we adopt as standard values. These values also give reasonable numbers when one computes momentum fractions such as $f_{\bar{q}/M}$ and $f_{q/p}$. This consistency further supports the model.[15]

It is apparent that the _direct_ inclusive process, $AB \to CX$, in which the beam does _not_ radiate prior to interacting with a constituent of the target, is obtained from the previously quoted exclusive scattering formula by replacing the form factor by the relevant target structure function. A simple calculation then leads to the result

$$E_c \frac{d\sigma}{d^3 p_c} (AB \to CX) = \frac{1}{\pi} \frac{s}{s+u} \, 3 \sum_{b,d} x \, G_{b/B}(x) \cdot$$

$$\frac{d\hat{\sigma}}{dt} (Ab \to Cd', s' = xs, \; t'=t, \; u'=xu) \quad . \tag{9.15}$$

The sum of b, d is over quark flavors and the explicit 3 results from the sum over colors. The structure function G and $d\sigma/dt$ are for quarks of one given color. The variables used to describe inclusive scattering subprocess cross section $A + B \to C + X$ are

$$s + t + u = \mathcal{M}^2 = \epsilon s,$$

where $\mathcal{M}$ is the total missing mass and

$$x_1 = -u/s = \frac{1}{2} x_R(1 + z)$$

$$x_2 = -t/s = \frac{1}{2} x_R (1 - z) \tag{9.16}$$

where

$$\epsilon = 1 - x_R, \; x_R \cong E^{cm}/E^{cm}_{max} \quad ,$$

and $z \, (= \cos \theta)$ is the cosine of the center-of-mass scattering angle. The on mass-shell condition for particle C determines x in Eq. (9.15) to be

$$x = x_2/(1-x_1) \quad .$$

For $x > \hat{x}_b$ one may substitute the simple forms for $G_{b/B}$ and $d\sigma/dt$ from (3.1) and (3.6) and obtain

$$E_c \frac{d\sigma}{d^3 p_c} = 3 \mathcal{D} \sum_{b,d} (1+g_b) f_{b/B} N(b/B) \, x_1^{N-U} (1-x_1)^{N-T-1-g_b} K(g_b, N) \tag{9.17}$$

where the dominant dynamical variation in ϵ and p_T is contained in

$$K(F, N) = \epsilon^F (p_T^2 + M^2)^{-N} \quad , \tag{9.18}$$

and the effective mass scale M is less than ~ 1 GeV.

The double bremsstrahlung process depicted in Fig. 1 is easily evaluated using the G functions and $d\sigma/dt$ forms already discussed. The result can be written in the form

$$E_C \frac{d\sigma}{d^3 p_C} = 3 \sum_{a,b} I(a,b) K(F, N; F, {}^+F^-) J(\epsilon, z) \quad . \tag{9.19}$$

Here we have employed (as appropriate for all our CIM applications) the presence of one quark-loop color sum. The sum is over the flavors of the interacting constituents, and

$$I(a,b) = \mathscr{D} f_{a/A} f_{b/B} N(a/A) N(b/B) \, 2^{F^+ + F^-} \frac{\Gamma(2 + g_a)\Gamma(2 + g_b)}{\Gamma(2 + g_a + g_b)} \quad . \tag{9.20}$$

The main dynamical behavior is contained in the K function

$$K(F, N; F^+, F^-) \equiv \epsilon^F (p_T^2 + M^2)^{-N} (1 + x_R z)^{-F^+} (1 - x_R z)^{-F^-} \tag{9.21}$$

Note that the effective power of $\epsilon = (1-x_R)$ changes as one approaches $z = \pm 1$. Also note the similarity to formula (5.12) which we derived in the nuclear case. The parameters in the above are given by

$$F = 1 + g_a + g_b$$

$$F^+ = 1 + U + g_a - N \tag{9.22}$$

and

$$F^- = 1 + T + g_b - N \quad ,$$

all of which can be easily calculated by quark counting and reference to Table III for T and U values.

For generality let us write the structure functions in a slightly more general form than Eq. (9.5), namely

$$xG_a(x) = f_a N_a (1+g_a)(1-x)^{g_a} R_a(x) \tag{9.23}$$

where, as in our (9.5) example, $R_a(x) = 1$ for $x > \hat{x}_a$, and $R_a(x) = \left((1-\hat{x})/(1-x)\right)^{g_a}$ for $x < x$. Using this form, the slowly varying factor J becomes

$$J(\epsilon, z) \equiv \frac{\Gamma(2+g_a+g_b)2^{-F}}{\Gamma(1+g_a)\Gamma(1+g_b)} \int_{-1}^{1} d\eta (1+\eta)^{g_a}(1-\eta)^{g_b} H(\eta) \quad , \tag{9.24}$$

where

$$H(\eta) = \left[\frac{1+zx_R}{1+zx_R+\epsilon\eta}\right]^{F^+} \left[\frac{1-zx_R}{1-zx_R-\epsilon\eta}\right]^{F^-} \quad .$$

$$\cdot\, R_a\left(\frac{x_R(1+z)}{1+zx_R+\epsilon\eta}\right) R_b\left(\frac{x_R(1-z)}{1-zx_R-\epsilon\eta}\right) \quad .$$

One can see that for $x_R > \frac{1}{2}(\hat{x}_a + \hat{x}_b)$, and for z not too near ± 1, $J(\epsilon, z) \approx 1$. We have found that for the relevant values of g_a, etc., J can differ from one by typically 20 or 30% in the interesting experimental range, say $\epsilon < 0.7$.

Setting $J=1$ allows us to make simple predictions for the prompt rate of production of mesons, baryons, antibaryons, etc. All constants are now known.

For meson production from a proton beam, we will write down the complete answer (the only time I dare!). Now the leading subprocesses which contribute to large p_T inclusive reactions are those which have the minimum p_T and $\epsilon \to 0$ fall-off and the largest overall normalization. In this case the dominant contributions based on quark-hadron interactions arise from quark-meson scattering $(qM^* \to qM)$ and the fusion process $(q\bar{q} \to M\bar{M}^*)$. The contributions from the ut diagram Fig. 15a, and fusion diagram Fig. 15b yield

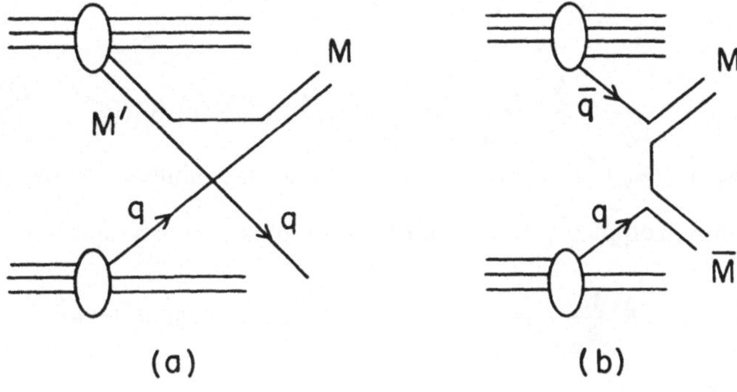

Fig. 15

Diagrams contributing to inclusive meson production.

$$E \frac{d\sigma}{d^3p} (pp \to MX) = \alpha_M^2 \, 3 \sum_{M^*q} f_{M^*/p} f_{q/p} N(M^*/p) N(q/p) 2^6 \frac{\Gamma(7)\Gamma(5)}{\Gamma(10)} K_s(9,4;5,0)$$

$$+ \frac{3}{2} \alpha_M^2 \sum_{q,\bar{q}\epsilon M, \bar{M}^*} f_{\bar{q}/p} f_{q/p} N(\bar{q}/p) N(q/p) 2^7 \frac{\Gamma(9)\Gamma(5)}{\Gamma(12)} K_s(11,4;7,-1)$$

$$+ \frac{3}{2} \alpha_M^2 \sum_{\bar{q}, q\epsilon M, \bar{M}^*} f_{\bar{q}/p} f_{q/p} N(\bar{q}/p) N(q/p) 2^7 \frac{\Gamma(9)\Gamma(5)}{\Gamma(12)} K_s(11,4;3,3),$$

where $K_s = \frac{1}{2}(K(;F^+F^-) + K(;F^-,F^+))$. Using the numbers in the tables, the yield at 90°, for prompt π^+, from the $qM \to qM$ plus fusion graph is

$$E \frac{d\sigma}{d^3p} (pp \to \pi^+ x) = K(9,4) \left[2.2 + 0.06n(\bar{M}^*) \epsilon^2 \right] \qquad (9.26)$$

where $n(\bar{M}^*)$ is the number of states in the $\bar{M}^*$ sum (~ 3 to 4). This equation includes the contribution from the st topology diagram of Fig. 15c which has the angular distribution $K_s(9,4;1,0)$; at 90° it contributes only 1/4 of that of the dominant ut contribution. Using a factor of 3 to account for the total/prompt effect, the experimental rate is roughly 9K(9,4), compared to the prediction of 6.6 K(9.4).

The rate for π^- production is somewhat smaller than for π^+ since $G_{u/p}/G_{d/p}$ increases as x increases. The π^+ and π^- rates must be equal, however, at $x_T = 0$, in the Feynman scaling limit, so the π^+/π^- ratio must decrease as x_T decreases to zero. As x_T increases, however, it should rise and saturate to a constant value in the symmetric quark model used here.

The dominant K^- cross section for $\epsilon \to 0$ arises from the fusion term. However, for moderate ϵ it is vital to retain various contributions with higher ϵ powers, for example ϵ^{13}, arising from the ut and st topology $qK^- \to qK^-$ graphs. Combining all these terms, see Fig. 15, our estimate for the K^-/K^* ratio is

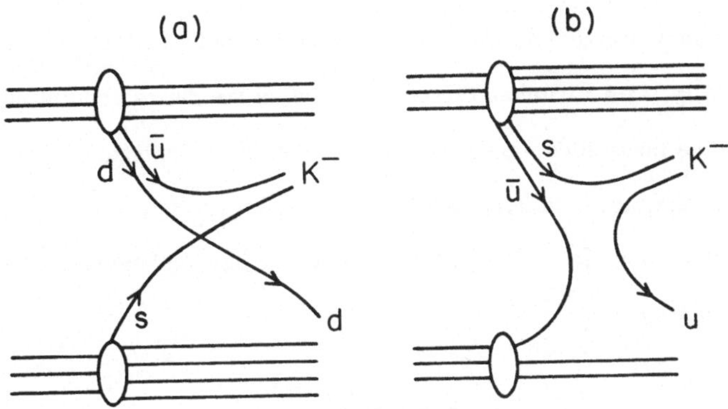

Fig. 16

Dominant contributions to K^- production.

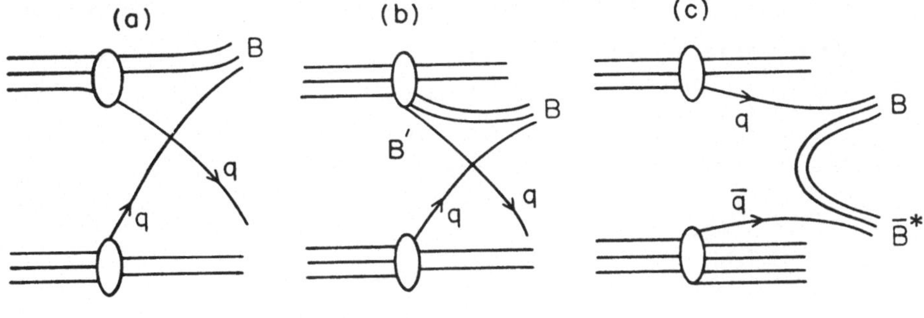

Fig. 17

Important graphs for B and $\bar{B}$ production.

$$\frac{d\sigma(pp \to K^-)}{d\sigma(pp \to K^+)} = 0.1 \ \epsilon^2 \ \frac{1+4 \ \epsilon^2}{1+0.07 \ \epsilon^2} \simeq 0.1 \ \epsilon^2 (1+4 \ \epsilon^2) \qquad (9.27)$$

Recall that the numerical approximations used are not valid for $\epsilon \to 1$ ($x_T \to 0$). Experimentally, this ratio has the same shape as the above prediction but with about twice the magnitude. The fusion term dominates only for $x_T > 0.6$. The agreement of the model is quite good.

It is also possible to do estimates for other baryon beams. One very interesting example is

$$\frac{d\sigma(\bar{p}p \to \pi x)}{d\sigma(pp \to \pi x)} = \frac{1+0.03 \ n(\bar{M}^*) \ \epsilon^{-2}}{1+0.03 \ n(\bar{M}^*) \ \epsilon^2}$$

$$\cong 1+0.12 \ (\epsilon^{-2} - \epsilon^2) \ , \qquad (9.28)$$

using $n(\bar{M}^*) = 4$. For example, at $x_T = 0.3$, the ratio is predicted to be 1.20. In quark scattering models, it should be unity, thus there is not much difference until x_T is large. In a model with fusion only, the ratio would be ϵ^{-4}, which is 4 at $x_T = 0.3$.

We havenot time to go through all possible estimates of yields, however a sampling of possible tests of our model include (these are prompt yields—multiply by 3 for total rates—see also Fig. 17):

$$E \frac{d\sigma}{d^3p} (pp \to px) = 120 \, x_T^2 \, K(3.6)$$

$$+ 130 \, K(7,6) + 1.4 \, n(\bar{B}^*) \, K(11,6) \quad ,$$

$$E \frac{d\sigma}{d^3p} (pp \to \bar{p}x) = 5.6 \, K(11,6) \quad ,$$

$$\frac{d\sigma(pp \to \gamma)}{d\sigma(pp \to \pi^o)} \cong \frac{2}{3} \frac{5\alpha}{3\alpha_M} (p_T^2 + M^2) \sim 0.007 \, (p_T^2 + M^2)$$

$$\frac{d\sigma\pi^+p \to \pi^o)}{d\sigma(pp \to \pi^o)} = \epsilon^{-2} \frac{3}{4} \frac{\sum_q \left[(f_{q/\pi} + f_{\bar{q}/\pi}) N(q/\pi) + \frac{2}{5} f_{q/p} N(q/p) \right]}{\sum_q f_{q/p} N(q/p)}$$

$$\cong 1.0 \, \epsilon^{-2} \quad .$$

$$E \frac{d\sigma}{d^3p} (\pi^+p \to \pi^o X) = K(7,4) \left[1.6 + 0.016 \, n(\bar{M}^*) \, \epsilon^{-2} + 0.35 \, x_T \, \epsilon^{-4} \right]$$

and

$$\frac{d\sigma(\pi p \to \pi)}{d\sigma(\pi p \to p)} = \frac{\alpha_M^2}{\alpha_B^2} \epsilon^2 \left(p_T^2 + M^2 \right) \frac{2^6 \cdot 5}{7} \frac{\sum_M f_{M/p} N(M/p)}{\sum_B f_{B/p} N(B/p)}$$

$$\sim 0.27 \, \epsilon^2 \left(p_T^2 + M^2 \right).$$

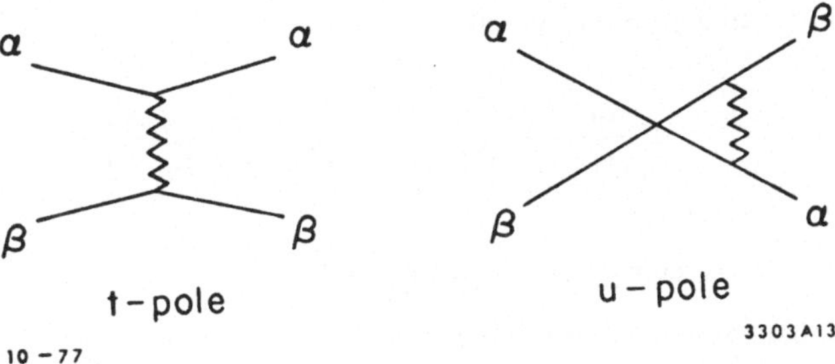

Fig. 18

Quark-quark scattering diagrams.

X. Quark-Quark Scattering

What about <u>quark-quark</u> (as well as quark-gluon and gluon-gluon) scattering?[16, 17, 18] The relevant diagrams are shown in Fig. 18 and are easily calculated. However, in order to compute the cross section for the production of a specific hadron h one must incorporate the final state fragmentation function $D_{h/q}$ of the quark

$$E \frac{d\sigma}{d^3p} (p_1 p_2 \quad hx) = \sum_q \int_{x_R}^1 \frac{dz}{z^2} D_{h/q}(z) \, E \frac{d\hat\sigma}{d^3p} (p_1 p_2 \rightarrow q; s, t/z, u/z).$$

(10.1)

Using

$$D = d \, \frac{(1-z)^f}{z} \quad \text{and} \quad E \frac{d\hat\sigma}{d^3p} = I K(F, N) \quad ,$$

one finds

$$E \frac{d\sigma}{d^3p} (pp \rightarrow h) = \sum_q d \, I K(F+f+1, N) \, \hat J(x_T) \quad ,$$

(10.2)

where $\hat J$ is essentially constant,

$$\hat J \simeq \Gamma(1+f)\Gamma(1+F) \Big/ \Gamma(2+f+F) \quad .$$

For the π^+ yields, a reasonable fit to the quark fragmentation functions gives $D_{\pi^+/u}$ (or $\bar d$)$(z) = 1.0 \, (1-z)/z$. Eq. (9.8) then gives at 90°

$$E \frac{d\sigma}{d^3p} (pp \rightarrow q \rightarrow \pi^+) = K(9, 2) \, \alpha_s^2(0.035) \quad .$$

where we evaluated $J(x_T)$ at $x_T = 0.3$. The total yield can be succinctly written in the form (adding in Eq. (9.26)

$$E \frac{d\sigma}{d^3p} = A \left[\left(\frac{10}{p_T}\right)^8 + \left(\frac{\alpha_s}{0.15}\right)^2 \left(\frac{10}{p_T}\right)^4 \right] \epsilon^9 \quad ,$$

(10.3)

which may prove convenient in fitting the large p_T data. Thus for $\alpha_s = 0.15$, the CIM diagrams dominate for $p_T < 10$, and one has a p_T^{-8} behavior, whereas for $p_T > 10$, the cross section should show a p_T^{-4} behavior. Actually, quark-gluon scattering is also important and should be added to the above estimate. Roughly speaking it doubles the second term (at least !).

XI. Final Remarks

In conclusions, I think we have shown that there is a simple model for the scattering of composite systems that works well for nucleus-nucleus scattering and for hadronic reactions. In the former case, we have been able to extend the descriptions of nuclei away from the usual and familiar nonrelativistic limit in a remarkably simple way. This application allows us to check as well the physical interpretations that we make in the considerably more complex nuclear case.

In the hadron case, I would argue that the CIM diagrams must be included in any complete treatment and to omit them would be internally inconsistent. They are important because up till now, experiments are not at sufficiently large values of p_T, and their couplings (α_M and α_B) are large (compared to α_s). In this regard, it is useful to distinguish three regions in transverse momentum for hadronic inclusive reactions:

(1) The asymptotically scale-free, large p_T region (above $p_T \sim 7$ GeV for single particles, and $p_T \sim 5$ GeV for jets), where the simple perturbation theory contributions for QCD are expected to dominate if $\alpha_s \cong 0.3$. In this region, strong interactions take their most elementary form,

(2) The intermediate p_T zone, where the CIM diagrams are predicted to dominate giving scaling law contributions of the form p_T^{-8}, p_T^{-12} ... at fixed x_T, depending on the detected particle. In this region (roughly $2 < p_T < 7$ GeV for single particle reactions), one can trace the quantum number flow characteristic of duality diagrams. In the case of exclusive reactions, Regge behavior takes its most basic form, with trajectories $\alpha(t)$ receding to negative integers (or half-integers), or in the case of Compton scattering to a J=0 fixed pole.

(3) The most complicated region is at low p_T where the cross sections Feynman-scale and many different coherent, diffractive, Regge, and resonance/ cluster phenomena operate.

In the general CIM approach several different areas of hadron phenomenology become interconnected: (a) form factors, (b) large t and u exclusive reactions (c) Regge behavior at large t, (d) particle yields for x_L near $\pm$ 1 at low t, and (e) large p_T inclusive reactions. We have shown here that the normalization of the various CIM contributions to inclusive scattering are fixed by external con- straints and are not arbitrary. They are of a reasonable size to explain the moderate transverse momentum single particle yields.

A theory of short distance hadronic processes patterned after asymptotically free QCD is in reasonable agreement with data, however CIM processes based on quark-hadron scattering are required for theoretical completeness and to describe. the experimental data at intermediate p_T.

ACKNOWLEDGEMENTS

I wish to acknowledge my collaborators, I. A. Schmidt, S. J. Brodsky, M. Duong-van, and J. F. Gunion, without whom this work would not have been done. I wish to thank Dr. W. Dittrich for the kind hospitality and stimulating intellectrual environment of this meeting.

References

1. R. Blankenbecler, S. J. Brodsky, and J. F. Gunion, Phys. Lett. $\underline{B39}$, 649 (1972); $\underline{42}$, 461 (1973); Phys. Rev. $\underline{D12}$, 3469 (1975). Spectator counting rules for structure functions are derived in R. Blankenbecler and S. J. Brodsky, Phys. Rev. $\underline{D10}$, 2973 (1974); J. Gunion, ibid. $\underline{10}$, 242 (1974). The fusion reactions $q\bar{q} \rightarrow M\bar{M}$ have been particularly emphasized by P. V. Landshoff and J. C. Polkinghorne, Phys. Rev. $\underline{D10}$, 891 (1974). See also M. K. Chase and W. J. Stirling, Cambridge U. preprint, DAMTP-77/15 and references therein.

2. For a general review, see D. Sivers, S. J. Brodsky, R. Blankenbecler, Physics Reports $\underline{23C}$, No. 1 (1976). See also Refs. 13, 15, and S. J. Brodsky, R. Blankenbecler, J. F. Gunion, XII Recontre de Moriond, Flaine, France, March 1977.

3. I. A. Schmidt and R. Blankenbecler, Phys. Rev. $\underline{D15}$, 3321 (1977). I. A. Schmidt, SLAC Report No. 203, 1977.

4. M. Duong-Van, K. V. Vasavada, and R. Blankenbecler SLAC-PUB-1882, Phys. Rev. $\underline{D16}$, 1389 (1977), and M. Duong-Van and R. Blankenbecler, SLAC-PUB-2017 (Phys. Rev., in print).

5. S. D. Drell and T. M. Yan, Phys. Rev. Lett. $\underline{25}$, 316 (1970); Ann. Phys. $\underline{6b}$, 578 (1971).

6. R. D. Amado and R. M. Woloshyn, IAS preprint 77-0374 (1977). To be published in Phys. Rev.

7. There is some confusion in the literature on this point. Counting rules giving the energy dependence at fixed angle for exclusive scattering were derived by S. J. Brodsky and G. Farrar, Phys. Rev. Lett. $\underline{31}$, 1153 (1973); Phys. Rev. $\underline{D11}$, 1309 (1975). V. Matveev, R. Muradyan, and A. Tavhelidze, Nuovo Cimento Lett. $\underline{7}$, 719 (1973).

8. Spectator counting rules giving the behavior of general G functions wer first derived by R. Blankenbecler and S. J. Brodsky, Phys. Rev. $\underline{D16}$, 2973 (1974).

9. J. Papp et al., Phys. Rev. Lett. $\underline{34}$, 601 (1975). J. Papp, Ph.D. Thesis, University of California, Berkeley, Report LBL-3633 (1975).

10. R. G. Arnold et al., Phys. Rev. Lett. $\underline{35}$, 776 (1975). Lower energy data is taken from: J. Elias et al., Phys. Rev. $\underline{177}$, 2075 (1969); and S. Galster et al., Nucl. Phys. $\underline{B32}$, 221 (1971).

11. S. J. Brodsky and B. T. Chertok, Phys. Rev. $\underline{D14}$, 3003 (1976).

12. I. A. Schmidt and R. Blankenbecler, Phys. Rev. $\underline{D16}$, 1318 (1977). See also A. Fernandez-Pacheco, J. A. Grifols and I. A. Schmidt, SLAC-PUB-1993, (1977).

13. Compare, for example, with W. B. Atwood and G. B. West, Phys. Rev. $\underline{D7}$, 773 (1973).

14. The data are from, for example, J. W. Cronin et al., Phys. Rev. $\underline{D11}$, 3105 (1975). D. Antveasyan et al., Phys. Rev. Lett. $\underline{38}$, 112 (1977); G. Donaldson et al., Phys. Rev. Lett. $\underline{36}$, 1110 (1976); F. W. Busser et al., Nucl. Phys. $\underline{B106}$, 1 (1976); and B. Alper et al., Nucl. Phys. $\underline{B100}$, 237 (1975). See ref. 2 for a review of elastic data, and A. Eide et al., Nucl. Phys. $\underline{B60}$, 173 (1973). See also H. I. Miettinen, SLAC-PUB-1813, presented at the Third European Symposium on Anti-nucleon-Nucleon Reactions, Stockholm (1976); J. L. Stone et al., Phys. Rev. Lett. $\underline{38}$, 1315, 1317 (1977); V. Chabaud et al., Phys. Lett. $\underline{41B}$, 209 (1972); A. Donnachie and P. R. Thomas, Nuovo Cimento $\underline{19A}$, 279 (1974).

15. Our discussion here follows work reported in ref. 2 and in R. Blankenbecler S. J. Brodsky and J. F. Gunion, SLAC-PUB-(to be published). For comparison see B. Pire, Phys. Rev. $\underline{D15}$, 3475 (1977).

16. S. M. Berman and M. Jacob, Phys. Rev. Lett. $\underline{25}$, 1683 (1970); S. M. Berman, J. D. Bjorken, and J. B. Kogut, Phys. Rev. $\underline{D4}$, 3388 (1971); J. Bjorken, Phys. Rev. $\underline{D8}$, 3098 (1973). See also R. F. Cahalan, K. A. Geer, J. Kogut, and L. Susskind, Phys. Rev. $\underline{D11}$, 1190 (1975); M. Bander, M. Barnett, D. Silverman, Phys. Lett. $\underline{48B}$, 243 (1974).

17. In particular see D. Sivers and R. Cutler, Phys. Rev. $\underline{D16}$, 679 (1977) and B. Combridge, J. Kripfganz, and J. Ranft, Phys. Lett. $\underline{70B}$, 234 (1977).

18. In contrast see R. D. Field and R. P. Feynman, Phys. Rev. $\underline{D15}$, 2590 (1977); R. D. Field, R. P. Feynman, and G. C. Fox, CALT-68-595 (1977); G. C. Fox, CALT-68-573 (1977).

Vacuum Tunneling in Minkowski Space

Shau-Jin Chang*

Department of Physics, University of Illinois at Urbana-Champaign
Urbana, Illinois 61801

I. Introduction

The recent discovery of classical Euclidean solutions to non-abelian gauge theories[1-5] and their possible interpretation as tunnelings among vacua with different winding numbers[6-10] stimulate much excitement and speculation. However, the interpretation of a classical Euclidean solution as a realization of vacuum tunneling is not completely transparent. Since the tunneling process is well understood in a quantum mechanical potential scattering theory, it is conceptually satisfactory to reduce a vacuum tunneling process in a quantum field theory to an effective one particle quantum mechanical potential scattering problem. In these lectures, I shall present the vacuum tunneling in Minkowski space[10] as a change of field configurations from one ground state to another. By properly choosing the intermediate field configurations, we obtain an effective one parameter quantum mechanical system in which the tunneling between two different vacua take place. In the first lecture (Sec. 2 below), I shall work out a simple example describing the vacuum tunneling in a two-dimensional ϕ^4 field theory. In the next lecture (Secs. 3 and 4), I shall work out the vacuum tunneling associated with the simplest one-instanton solution in a Yang-Mills theory.

Most of the material given here is based on a paper written in collaboration with Professor K. Bitar.[11]

II. Tunneling in a Φ^4 Field Theory

The example presented here is related to the vacuum tunneling problem studied earlier by M. B. Voloshin et al[12, 13]. Consider a two-dimensional ϕ^4 theory described by the Lagrange function,

$$L = \frac{1}{2}\dot{\phi}^2 - \frac{1}{2}(\frac{\partial\phi}{\partial x})^2 - \frac{1}{4}g(\phi^2-c^2)^2$$

$$+ B(c^2\phi - \frac{1}{3}\phi^3) + \frac{2}{3}Bc^3 \qquad (2.1)$$

with a small and positive B. Field $\phi(x)$ obeys the field equation

$$- \ddot{\phi} + \frac{\partial^2\phi}{\partial x^2} - g(\phi^2-c^2)\phi + B(c^2-\phi^2) = 0 \quad . \qquad (2.2)$$

Ignoring the $(\frac{\partial\phi}{\partial x})^2$ term, we can identify the potential energy density term as

$$\mathcal{V}(\phi) = \frac{1}{4} g(\phi^2-c^2)^2 - B(c^2\phi- \frac{1}{3}\phi^3) - \frac{2}{3}Bc^3 \quad . \qquad (2.3)$$

Graphically, $\mathcal{V}(\phi)$ is shown as a function of ϕ in Fig. 1a, and as a function of both $\phi(x)$ and x in Fig. 1b. From (2.3), we find that

$$\mathcal{V}(-c) = 0, \text{ and } \mathcal{V}(c) = -\varepsilon \qquad (2.4)$$

with

$$\varepsilon = \frac{4}{3}Bc^3 \quad . \qquad (2.5)$$

Quantum mechanically, the $\phi = -c$ state represents an unstable vacuum, and the tunneling to the stable vacuum state $\phi = c$ will occur. We can understand the tunneling process qualitatively as follows. Due to quantum fluctuations, a typical field configuration associated with the state at $\phi = -c$ actually looks like that in Fig. 2a. The field $\phi(x)$ fluctuates

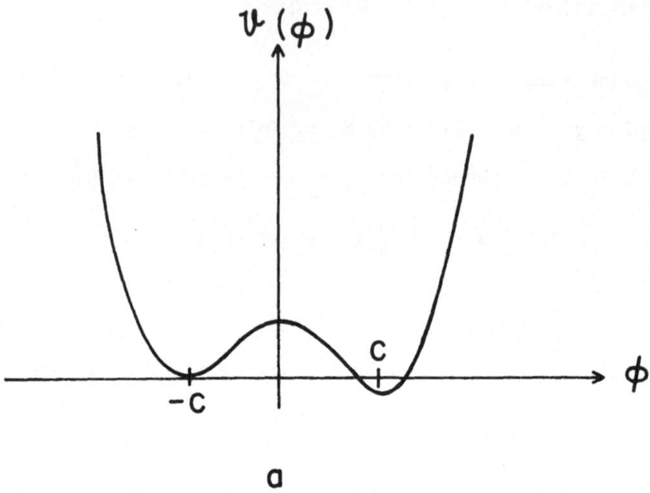

Potential energy density $\mathcal{V}(\phi)$ as a function of ϕ,

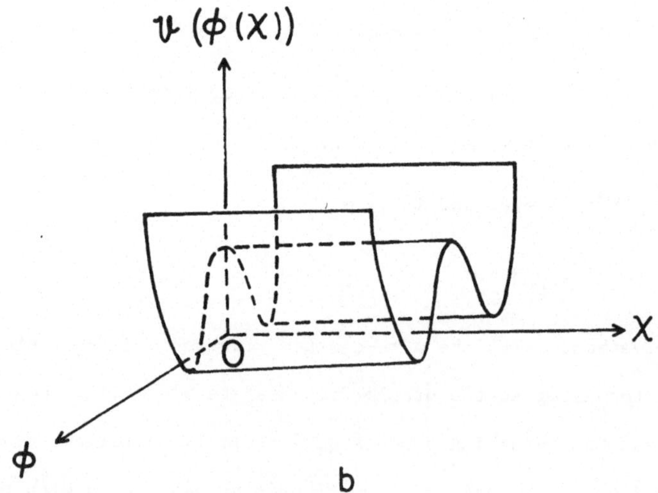

$\mathcal{V}(\phi(x))$ as a function of ϕ and x.

Fig. 1

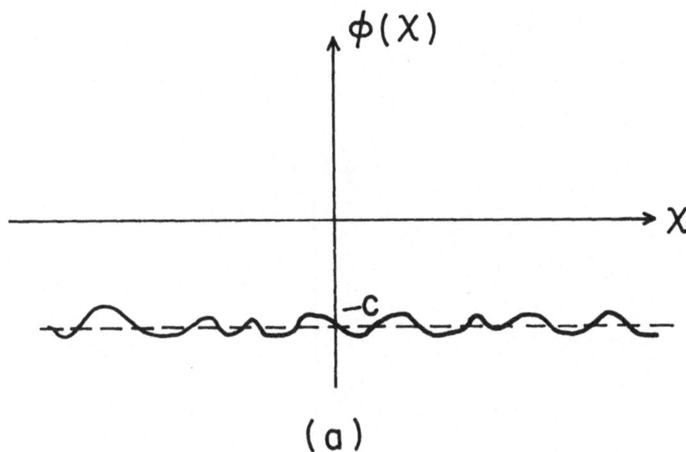

(a)

Field configuration associated with the unstable vacuum
$\phi = -c$,

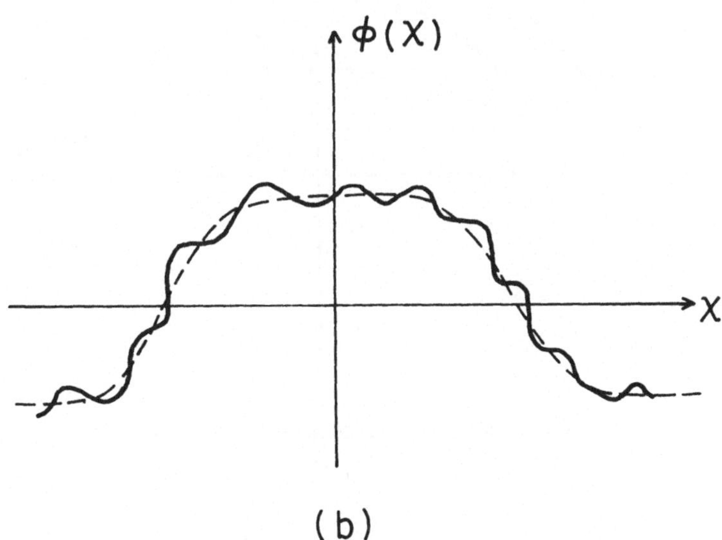

(b)

Field configuration associated with a kink and an antikink.
This field configuration can be energetically more favorable
than the previous configuration (a) if the $\phi = c$ region becomes large.

Fig. 2

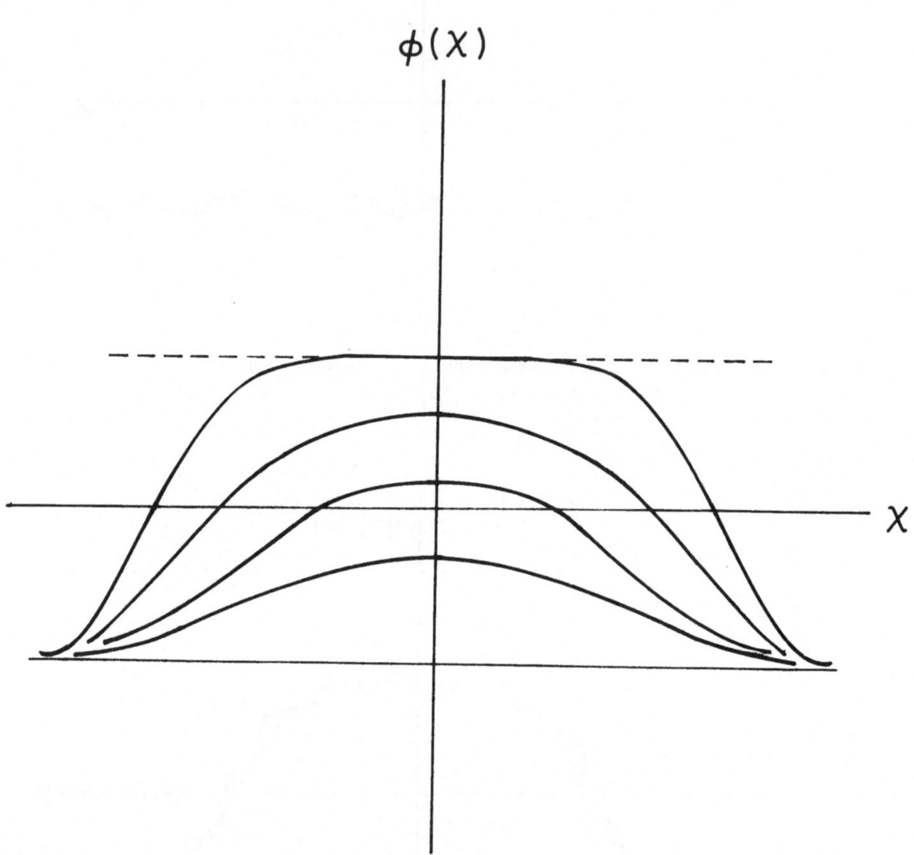

A typical set of field configurations connecting the $\phi = -c$ state to the kink-antikink state.

Fig. 3

around $\phi = -c$. Occasionally, the fluctuation may become rather large.
If the $\phi = -c$ state were a true ground state; these large fluctuations
would disappear shortly in time, and ϕ would return invariably to those
configuration associated with small fluctuations around $\phi = -c$ as
described in Fig. 2a. However, in the present case, the state $\phi = -c$
is <u>not</u> a true minimum. A fluctuation which creates a sizable section
of true ground state (i.e. the $\phi = c$ state) such as that described in
Fig. 2b may be energetically more favorable than the original field configuration 2a. These configurations can live indefinitely, and the $\phi = c$
region in these configurations will increase in size until it takes over
the whole space.

For simplicity, we ignore the small fluctuations of $\phi(x)$, and
consider a single mode describing the overall shape of $\phi(x)$. We can
describe the change of the overall shape of $\phi(x)$ in Fig. 3. To understand the tunneling process, we note that there are kink–antikink states
which are degenerate with the $\phi = -c$ state. The kink-antikink state
centered at the origin is described approximately at small B by

$$\phi = c \left[\tanh \tfrac{\mu}{2} (x + \lambda_c) - \tanh \tfrac{\mu}{2} (x - \lambda_c) - 1 \right] \tag{2.6}$$

with

$$\mu = \sqrt{2gc^2} , \tag{2.7}$$

$$\lambda_c = M/\varepsilon , \tag{2.8}$$

$$M = \frac{2\sqrt{2}}{3} g^{\frac{1}{2}} c^3 . \tag{2.9}$$

In (2.6)-(2.9), μ denotes the mass associated with small oscillations
around $\phi = \pm c$, $2\lambda_c$ is the separation of the kink and the antikink, and M
is the mass of the kink. The total energy associated with the field $\phi(x)$

given in (2.6) is

$$E = 2M - 2\varepsilon\lambda c = 0 \qquad\qquad (2.10)$$

as desired. Note that the kink-antikink state described above is not
the only state which is degenerate with the excited ground state $\phi = -c$.
However, it is the final state which allows a maximal tunneling to
take place.

We shall work out the tunneling amplitude between the $\phi = -c$ state
and the kink-antikink state given by (2.6). Once the field $\phi(x)$ reaches
the configuration (2.6), a real kink-and -antikink pair is produced.
The kink and the antikink will move away from each other, and leave be-
hind them an increasing region of true ground state $\phi = c$. Thus, the
tunneling between the $\phi = -c$ state and the kink-antikink state represents the vacuum tunneling that we are looking for.

The intermediate field configuration associated with the vacuum
tunneling is

$$\phi = c \left[\tanh \frac{\mu}{2} \frac{x + \lambda}{\sqrt{1-\dot\lambda^2}} - \tanh \frac{\mu}{2} \frac{x - \lambda}{\sqrt{1-\dot\lambda^2}} -1 \right] . \qquad (2.11)$$

It describes a kink-antikink configuration (see Fig. 4) located at
$\pm \lambda$, moving with velocity $\pm \dot\lambda$. The $\dfrac{1}{\sqrt{1-\dot\lambda^2}}$ factor denotes the "Lorentz
contraction" of the moving kinks. In the tunneling process, $\dot\lambda$ is imaginary,
and $\dfrac{1}{\sqrt{1-\dot\lambda^2}} < 1$. Thus, it actually describes a Lorentz expansion.
Ignoring terms $0(B^2)$ or higher, we obtain the effective Lagrangian,

Fig. 4

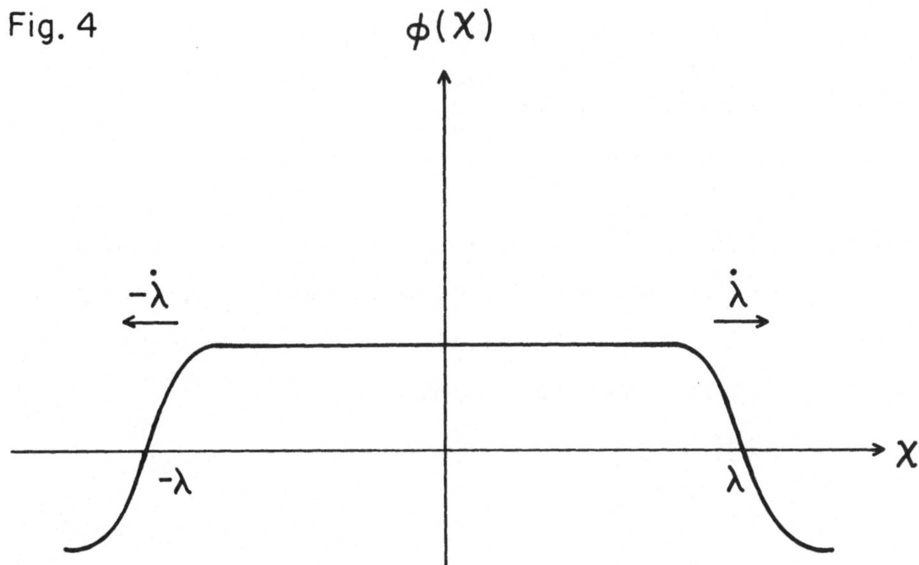

Field configuration associated with a kink and an antikink located at $\pm\lambda$, and moving with velocities $\pm\,\dot\lambda$.

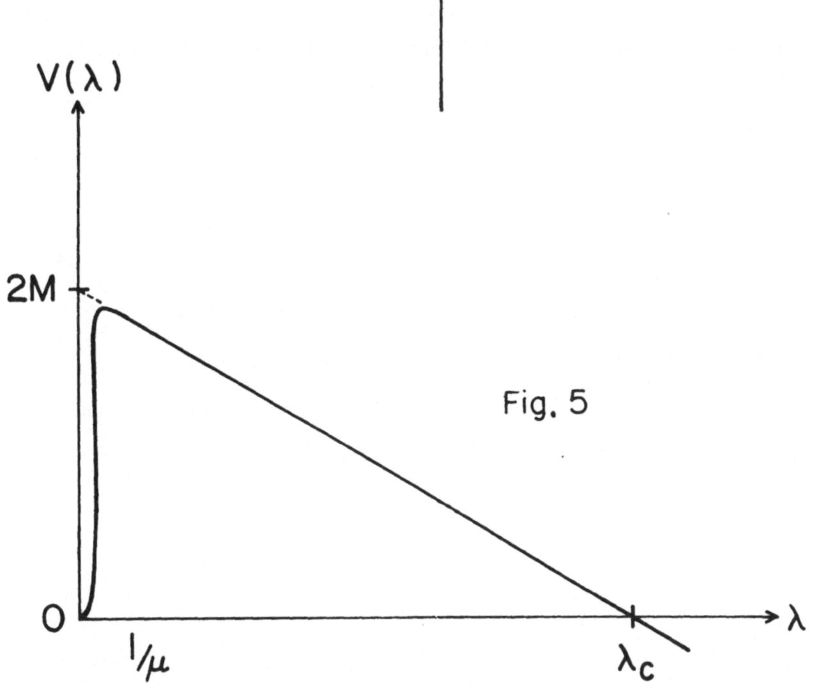

The potential barrier as a function of the kink-antikink separation in a two dimensional ϕ^4 theory. The variable λ denotes one half the kink-antikink separation.

$$L = \int dx \left[\frac{1}{2} \dot{\phi}^2 - \frac{1}{2} (\frac{\partial\phi}{\partial x})^2 - \frac{1}{4} g (\phi^2-c^2)^2 \right.$$
$$\left. + B (c^2\phi - \frac{1}{3} \phi^3 + \frac{2}{3} c^3) \right]$$
$$= -2 M \sqrt{1-\dot{\lambda}^2} + 2\varepsilon\lambda . \tag{2.12}$$

The first term in (2.12) denotes the free Lagrangian of a kink and an antikink moving with velocity $\pm\dot{\lambda}$. The second term $2\varepsilon\lambda$ describes the volume energy of the $\phi = c$ region between the kink and the antikink. The Hamiltonian associated with the tunneling is

$$H = \frac{\partial L}{\partial\dot{\lambda}} \dot{\lambda} - L$$
$$= \sqrt{p^2 + 4M^2} - 2\varepsilon\lambda \tag{2.13}$$

where

$$p \equiv \frac{\partial L}{\partial\dot{\lambda}} = \frac{2M}{\sqrt{1-\dot{\lambda}^2}} \dot{\lambda} \tag{2.14}$$

is the momentum conjugate to λ. The kinetic energy K and the potential energy V are

$$K = \sqrt{p^2 + 4M^2} - 2M \tag{2.15}$$

and

$$V = 2M - 2\varepsilon\lambda \tag{2.16}$$

respectively. In Fig. 5, we plot V as a function of λ. Then, V (λ) decreases linearly to zero due to the $-2\varepsilon\lambda$ term.

We now treat our system as a one-dimensional quantum system, and obtain the WKB tunneling amplitude as $P=e^{-R}$ with

$$R = \int_0^{\lambda_c} \sqrt{4M^2 - (2\varepsilon\lambda)^2} \, d\lambda$$
$$= \frac{1}{2} \pi M \lambda_c = \pi M^2/2\varepsilon . \tag{2.17}$$

As a quantum mechanical system, λ and p obey the Heisenberg's equations:

$$\dot{\lambda} = i \left[H, \lambda \right] = \frac{p}{\sqrt{p^2 + 4M^2}} \qquad (2.18)$$

and

$$\dot{p} = i \left[H, p \right] = 2\varepsilon . \qquad (2.19)$$

Eq. (2.18) is equivalent to (2.14). Eq. (2.19) describes the equation of motion for a relativistic particle moving under the influence of a constant force ε,

$$\frac{d}{dt} \left[\frac{\dot{\lambda}}{\sqrt{1-\dot{\lambda}^2}} \right] = \frac{\varepsilon}{M} = \frac{2Bc}{\mu} . \qquad (2.20)$$

The region $\lambda < \lambda_c$ is forbidden classically by the energy conservation. However, vacuum tunneling can occur quantum-mechanically, and it takes place precisely in the forbidden region $\lambda < \lambda_c$. Heisenberg equations of motion (2.18) (2.19) apply equally well to the tunneling process as well as to the real motions of the kink and antikink after tunneling.

I wish to mention several results without going into details:

(1) If we identify operators p and λ by the WKB values

$$p = i \sqrt{4M^2 - (2\varepsilon\lambda)^2},$$

$$\dot{\lambda} = i \sqrt{\frac{4M^2}{(2\varepsilon\lambda)^2} - 1} \ ;$$

we obtain a classical Euclidean solution in two-dimensional ϕ^4 theory which is the analog of Coleman and Frampton's solution in four-dimensional ϕ^4 theory[12]; (2) If p and λ obey the Heisenberg's equation of motion as given in (2.18), (2.19); we can construct a slightly modified ϕ which obeys the full field equation (2.2)[14]; and (3) Since independent vacuum tunnelings can occur at spacelikely separated regions, the total decay rate is proportional to the total length L of the system

$$\Gamma = e^{-2R} L/a$$

where $\underline{a}$ stands for some length scale. The size $\underline{a}$ was estimated in Ref. 13.

III. Vacuum Tunneling in a Yang-Mills Theory

We consider a non-abelian (Yang-Mills) gauge theory[15] described by

$$L = -\tfrac{1}{4} f_{\mu\nu}^{(a)} f^{(a)\mu\nu} , \tag{3.1}$$

$$f_{\mu\nu}^{(a)} = \partial_\mu a_\nu^{(a)} - \partial_\nu a_\mu^{(a)} + g\, \varepsilon_{abc}\, a_\mu^{(b)} a_\nu^{(c)} \tag{3.2}$$

with a,b,c being isospin indices, and μ,ν Lorentz indices. In terms of the matrix notations,

$$F_{\mu\nu} = g\, f_{\mu\nu}^{(a)} \frac{\tau^a}{2} \tag{3.3}$$

$$A_\mu = g\, a_\mu^{(a)} \frac{\tau^a}{2} , \tag{3.4}$$

we have

$$L = - \frac{1}{2g^2}\, \mathrm{Tr}\, (F_{\mu\nu})^2 , \tag{3.5}$$

$$F_{\mu\nu} = \partial_\mu A_\nu - \partial_\nu A_\mu + \frac{1}{i}\, [A_\mu, A_\nu] . \tag{3.6}$$

In (3.3), $\tau^{(a)}$ (a=1,2,3) are Pauli matrices. It is often more convenient to introduce

$$E_i = F_{oi}, \quad B_i = \tfrac{1}{2}\, \varepsilon_{ijk}\, F_{jk} \tag{3.7}$$

where $\vec{E}$ and $\vec{B}$ are generalizations of the electric and magnetic fields. In terms of $\vec{E}$ and $\vec{B}$, we have the Lagrange density

$$L = \frac{1}{g^2}\, \mathrm{Tr}\, (\vec{E}^2 - \vec{B}^2), \tag{3.8}$$

and the Hamiltonian density

$$H = \frac{1}{g^2}\, \mathrm{Tr}\, (\vec{E}^2 + \vec{B}^2) \geq 0. \tag{3.9}$$

One can define the winding number difference between two space-like surfaces σ_1, σ_2 as[1]

$$Q = \frac{1}{4\pi^2} \ \text{Tr} \int_{\sigma_1}^{\sigma_2} d^4x \ \vec{E}\cdot\vec{B} \qquad\qquad (3.10)$$

If $F_{\mu\nu} = 0$ on σ_1, σ_2; then, Q is an integer. Eq (3.10) also tells us that if the vacua at σ_1 and σ_2 give rise to a nonvanishing Q, there must exist some regions between σ_1 and σ_2 such that $F_{\mu\nu} \neq 0$. Since the energy density H (see Eq. (3.9)) is positive-definite and vanishes only if $F_{\mu\nu} = 0$, the existence of such a region (region with $F_{\mu\nu} \neq 0$) can only occur during a tunneling process. For a sourceless classical theory, $F_{\mu\nu} = 0$ on σ_1 (or σ_2) implies $F_{\mu\nu} = 0$ everywhere. Thus, it is impossible to change the winding number of a classical ground state. However, as a quantum system, the winding number of the ground state may change due to quantum fluctuations. This phenomena can be understood as a tunneling in some collective mode in the field configuration. Recently, Gervais and Sakita studied the vacuum tunneling and its quantum corrections by means of the collective coordinate method.[10] The readers are referred to this paper for details.

For simplicity, we choose σ_1 and σ_2 as surfaces at equal-time (t_1, and t_2). We work in the Coulomb gauge, $A^o = 0$, We denote the vector potentials on σ_i and σ_2 as $\vec{A}^{(1)}$ and $\vec{A}^{(2)}$. We require that $A_\mu^{(1)}$ and $A_\mu^{(2)}$ are associated with pure gauge transformations such that $F_{\mu\nu} = 0$ on both σ_1 and σ_2. The winding number difference Q depends only on the boundary values $\vec{A}^{(1)}$ and $\vec{A}^{(2)}$, and is independent of A_μ inside. We choose $\vec{A}^{(1)}$ and $\vec{A}^{(2)}$ such that $Q \neq 0$.

The procedure for obtaining the maximal single mode tunneling amplitudes are: (1) We introduce a family of intermediate field configurations as

$$A_o = 0 \qquad (3.11)$$

$$\vec{A}(x,t) = \vec{f}(x, \lambda(t)) \qquad (3.12)$$

where λ is a parameter describing the field configuration within the family. We require that

$$\vec{f} = \vec{A}^{(1)} \qquad \text{at } \lambda = \lambda_1 \qquad (3.13)$$

$$\vec{f} = \vec{A}^{(2)} \qquad \lambda = \lambda_2 \qquad (3.14)$$

and that $\vec{f}$ varies continuously from $\vec{A}^{(1)}$ to $\vec{A}^{(2)}$ as λ varies from λ_1 to λ_2. Given (3.11) and (3.12), we obtain

$$E_i = \frac{\partial f}{\partial \lambda} \dot{\lambda} \qquad (3.15)$$

$$B_i = \tfrac{1}{2} \varepsilon_{ijk} \left(\partial_j f_k - \partial_k f_j - \frac{1}{i} [f_j, f_k] \right) . \qquad (3.16)$$

The field variables $\vec{E}$ and $\vec{B}$ for intermediate values of λ usually do not vanish.

(2) From $\vec{E}$ and $\vec{B}$, we construct a single-parameter Lagrangian and Hamiltonian as

$$L = \tfrac{1}{2} m(\lambda) \dot{\lambda}^2 - V(\lambda) \qquad (3.17)$$

and

$$H = \frac{p^2}{2m(\lambda)} + V(\lambda) \qquad (3.18)$$

where

$$m(\lambda) = \frac{2}{g^2} \int d^3x \; \text{Tr} \left(\frac{\partial \vec{f}}{\partial \lambda} \right)^2 , \qquad (3.19)$$

$$V(\lambda) \;=\; \frac{1}{g^2} \int d^3x \; \mathrm{Tr} \; \vec{B}^{\,2} \;, \qquad\qquad (3.20)$$

$$p \;\equiv\; \frac{\partial L}{\partial \dot\lambda} \;=\; m(\lambda)\dot\lambda \quad . \qquad\qquad (3.21)$$

A typical potential $V(\lambda)$ is shown in Fig. 6.

To obtain the tunneling amplitude, we treat (3.18) as a quantum-mechanical Hamiltonian. With appropriate orderings of the operators, we can make H Hermitian. We can now obtain the tunneling amplitude by solving an ordinary Schrodinger equation. At the weak coupling limit in which the tunneling amplitude is small, we can compute the tunneling amplitude by the WKB method as e^{-R} with,

$$R = \int_{\lambda_1}^{\lambda_2} d\lambda \; \sqrt{2m(\lambda)\; V(\lambda)}$$

$$= \frac{2}{g^2} \int_{\lambda_1}^{\lambda_2} d\lambda \; \sqrt{\mathrm{Tr}\int d^3x \left(\frac{\partial \vec{f}}{\partial \lambda}\right)^2 \cdot \; \mathrm{Tr}\int d^3x \; \vec{B}^{\,2}}$$

$$= \frac{2}{g^2} \int_{t_1}^{t_2} dt \; \sqrt{\mathrm{Tr}\int d^3x \; \vec{E}^{\,2} \cdot \mathrm{Tr}\int d^3x \; \vec{B}^{\,2}} \; . \qquad (3.22)$$

(3) By varying the field configurations $f(x,\lambda(t))$, we can find the conditions for obtaining the maximal tunneling amplitude.[16]

We can use Schwartz inequality to derive a necessary and sufficient condition for the WKB tunneling amplitude to be maximal. We interprete

$$\mathrm{Tr}\int d^3x \; \vec{a}(x) \cdot \vec{b}(x) \;\equiv\; a.b \qquad\qquad (3.23)$$

as a scalar product with a positive norm. i.e.

$$|a^2| \;\equiv\; a.a > 0$$

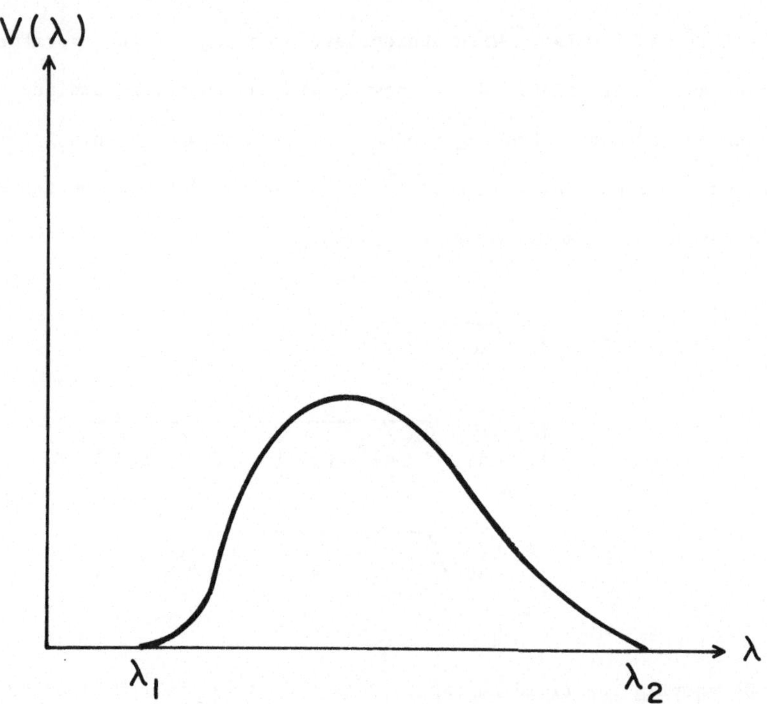

Fig. 6

A typical potential barrier associated with the vacuum tunneling.
The parameter λ denotes the change of field configurations.

for every $a \neq 0$. This implies that Schwartz inequality holds for every a and b,

$$\sqrt{|a^2| \; |b^2|} \; \geq \; |a.b| \; .$$

The Schwartz inequality implies in the present case

$$R \geq \frac{2}{g^2} \; |Tr \int_{\sigma_1}^{\sigma_2} d^4x \; \vec{E}.\vec{B}| \; = \; \frac{8\pi^2}{g^2} \; |Q| \; . \tag{3.24}$$

Thus, for a given winding number difference Q, R is bounded from below by $8\pi^2|Q|/g^2$; and hence, the WKB tunneling amplitude is bounded from above by

$$P \equiv e^{-R} \; \leq \; \exp \left[-\frac{8\pi^2}{g^2} \; |Q| \right] \; . \tag{3.25}$$

In addition, R reaches its minimal value $R = 8\pi^2|Q|/g^2$ (or P reaches its maximal value) if and only if $\vec{E}$ is proportional to $\vec{B}$ for all $\vec{x}$ and t with an x-independent proportionality constant. The problem of finding the fastest WKB tunneling rate is equivalent to finding a set of field configurations which gives rise to parallel $\vec{E}$ and $\vec{B}$. Note that even though $\vec{E}$ and $\vec{B}$ are gauge dependent quantities, the relevant quantities such as L, Q, and the condition $\vec{E} \mathbin{/\mkern-5mu/} \vec{B}$ are all gauge independent. Hence, the WKB tunneling amplitude is gauge invariant. In the following section, we shall present field configurations which give rise to the maximal tunneling rate for $Q = 1$.

An explicit family of field configurations which gives rise to the maximal vacuum tunneling amplitude for $Q = 1$ is

$$A_o = \frac{\vec{x}.\vec{\tau}}{\vec{x}^2 + \lambda^2 + a^2} \; \dot{\lambda}, \tag{3.26}$$

$$\vec{A} = \frac{\lambda\vec{\tau} + \vec{x} \times \vec{\tau}}{\vec{x}^2 + \lambda^2 + a^2} \; . \tag{3.27}$$

The vector potential A_μ given in (3.26)-(3.27) is not in the Coulomb gauge. We can transform it into the Coulomb gauge $A'_\mu = (0, \vec{A}')$ by a gauge transformation U,

$$A'_\mu = U^{-1} \left[(\partial_\mu - iA_\mu) U \right] \tag{3.28}$$

with

$$U = \exp \left[\frac{i\vec{x} \cdot \vec{\tau}}{\sqrt{\vec{x}^2 + a^2}} \; \tan^{-1} \frac{\lambda}{\sqrt{\vec{x}^2 + a^2}} \right] . \tag{3.29}$$

The Coulomb gauge vector potential is quite complicated. Since both the Lagrangian L and the winding number Q are gauge invariant, we can compute them in the original gauge specified by (3.26)-(3.27). We can evaluate the vacuum tunneling amplitude once L and Q are known. Using (3.26)-(3.27) we find

$$\vec{E} = - \frac{2a^2 \dot{\lambda}}{(\vec{x}^2 + \lambda^2 + a^2)^2} \; \vec{\tau} , \tag{3.30}$$

$$\vec{B} = - \frac{2a^2}{(\vec{x}^2 + \lambda^2 + a^2)^2} \; \vec{\tau} \tag{3.31}$$

which are indeed parallel to each other. The effective one-parameter Lagrangian L and the Hamiltonian H are given by (3.17)-(3.21) with

$$m(\lambda) = \frac{6\pi^2 a^4}{g^2 (\lambda^2 + a^2)^{5/2}} , \tag{3.32}$$

$$V(\lambda) = \frac{3\pi^2 a^4}{g^2 (\lambda^2 + a^2)^{5/2}} . \tag{3.33}$$

One can verify that the WKB tunneling amplitude is $P = e^{-R}$ with

$$R = \frac{2}{g^2} \int_{-\infty}^{\infty} d\lambda \; \frac{3\pi^2 a^4}{(\lambda^2 + a^2)^{5/2}} = \frac{8\pi^2}{g^2}$$

as anticipated.

IV. Winding Number Space

For intermediate values of $\lambda(t)$, we define the winding number variable $q(\lambda)$ as

$$q(\lambda) = \frac{1}{4\pi^2} \; \mathrm{Tr} \int_{-\infty}^{t} dt \int d^3x \;\; \vec{E}\cdot\vec{B}$$

$$= \frac{1}{2} + \frac{1}{4} \; \frac{\lambda}{\sqrt{\lambda^2+a^2}} \left(\frac{a^2}{\lambda^2+a^2} + 2 \right).$$

(4.1)

According to this definition, we have

$$q(-\infty) = 0, \quad q(\infty) = 1$$

(4.2)

which implies $Q = 1$. We can express the kinetic energy (K.E.) and the potential energy (P.E.) terms as functions of the winding number variable q

$$\text{K.E.} = \tfrac{1}{2}\, m'(q)\, \dot{q}^2, \quad \text{P.E.} = V(q)$$

with m' and V described in parametrical forms as

$$m' = \frac{32\pi^2}{3g^2 a^4} \; (\lambda^2+a^2)^{5/2}$$

(4.3)

and (3.33) respectively. We have plotted $V(q)$ and $m'(q)$ in Fig. 7.

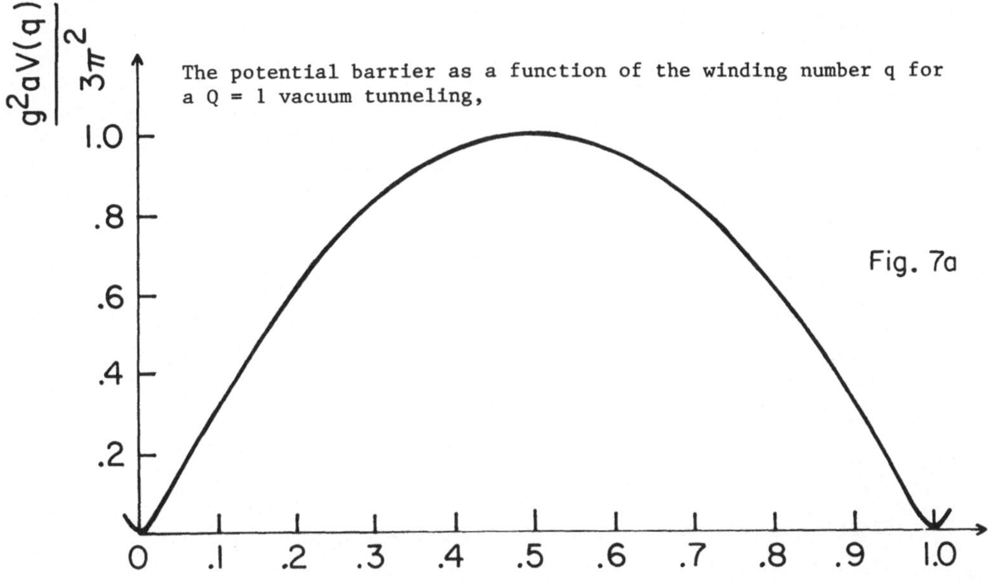

The potential barrier as a function of the winding number q for a Q = 1 vacuum tunneling,

Fig. 7a

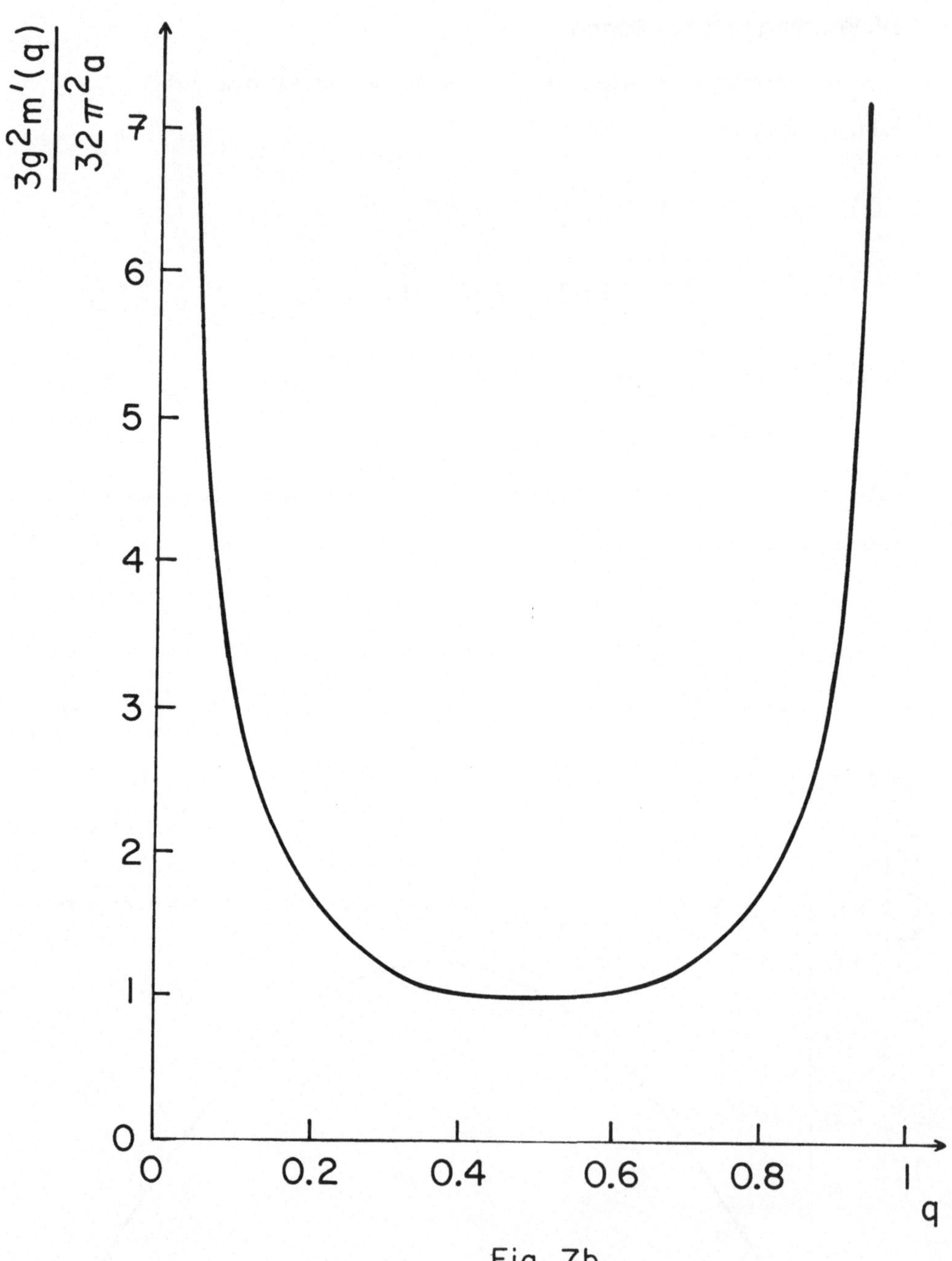

Fig. 7b

The effective mass as a function of q for a Q = 1 vacuum tunneling.

V. Concluding Remarks

(a) Relation to the instanton solution

Just as in the ϕ^4 theory, we can recover the Euclidean instanton solution by making the replacement

$$p \xrightarrow{\text{WKB}} i\sqrt{2mV} \quad . \tag{5.1}$$

Under this replacement, we have

$$\dot{\lambda} = i, \text{ or } \lambda = i\,(t-t_o) \, , \tag{5.2}$$

$$\vec{E} = i\vec{B} = -\frac{i2a^2}{(\vec{x}^2+\lambda^2+a^2)}\,\vec{\tau} \, , \tag{5.3}$$

which is precisely the analytic continuation of the instanton solution to the Minkowski space.

(b) Non-maximal tunneling solution

It is easy to construct solutions which do not give rise to the maximal tunneling amplitude. One simple example is

$$A_o = 0 \, , \tag{5.4}$$

$$\vec{A} = \frac{\lambda\vec{\tau} + \vec{x}\times\vec{\tau}}{\vec{x}^2+\lambda^2+a^2} \, . \tag{5.5}$$

$\vec{A}$ in (5.5) is the same as that given in (3.26), but A^o is different. This vector potential gives rise to the same $\vec{B}$, but a different $\vec{E}$. In particular, $\vec{E} \cancel{\parallel} \vec{B}$. We find that the WKB amplitude in the present case is $P = e^{-R}$ with

$$R = 15.8831 \, \pi^2/g^2 > 8\pi^2/g^2 \, .$$

(c)   Relation among classical, tunneling, and quantum-mechanical solutions[14]

Just as in the ϕ^4 theory, we believe that, with proper parametrization, the tunneling solution can obey the full field equation

$$F_{\mu\nu;\nu} = 0.$$

The difference among a classical, a tunneling, and a fully quantum-mechanical solution are their commutator relations. The classical solutions are c-number solutions. Their commutators are always zero. The fully quantum-mechanical solutions obey the equal-time canonical quantization relations. A tunneling solution is the minimal extension of a classical solution to a quantum solution. It has only one quantum mechanical degree of freedom. The equal-time commutation relation for the tunneling solution is

$$[\vec{E}, \vec{E}] = [\vec{A}, \vec{A}] = 0, \tag{5.6}$$

$$[E_i, A_j] = \frac{\partial f_i(x,\lambda)}{\partial\lambda} \, [\dot{\lambda}, f_j(x,\lambda)]$$

$$= i \, \frac{\partial f_i(x,\lambda)}{\partial\lambda} \, \frac{\partial f_j(x,\lambda)}{\partial\lambda} \, m(\lambda)^{-1} . \tag{5.7}$$

We can generalize this physical picture to multi-instanton solutions. In a N-instanton solution, the field equation $F_{\mu\nu;\nu} = 0$ are still satisfied, but there are N-pairs of dynamical variables being treated as quantum mechanical operators.

Acknowledgement

I wish to thank Professor and Mrs. Walter Dittrich for their hospitality.
Their unfailing kindness and considerations of others made this symposium
as well as my visit to Tübingen a most memorable one. The written version
of this lecture note was prepared during author's visit at the Aspen Center
of Theoretical Physics.

References

1. A. A. Belavin, A. M. Polyakov, A. S. Schwartz, and Yu S. Tyupkin, Phys. Lett $\underline{59}$B, 85 (1975).

2. E. Witten, Phys. Rev. Letters. $\underline{38}$, 121 (1977).

3. G. 't Hooft (unpublished).

4. R. Jackiw, C. Nohl, and C. Rebbi, Phys. Rev. D $\underline{15}$, 1642 (1977).

5. K. Bitar, Fermilab Preprint 77/15-THY.

6. A. M. Polyakov, Phys. Lett $\underline{59}$B, 82 (1975).

7. G. 't Hooft Phys. Rev. Letters $\underline{37}$, 8 (1976), and Harvard Univ. Preprint.

8. R. Jackiw, and C. Rebbi, Phys. Rev. Letters $\underline{37}$, 172 (1976).

9. C. G. Callan, R. F. Dashen, and D. J. Gross, Phys. Letters $\underline{63}$B, 334 (1976); and Princeton Preprint, "Structure of the Gauge Theory Vacuum".

10. J. L. Gervais and B. Sakita, "Collective Coordinates in SU(2) Yang-Mills Theory" CCNY-HEP-76/11. These authors studied the vacuum tunneling in the Minkowski space by means of the collective coordinate method.

11. K. Bitar and S. -J. Chang, Fermilab Preprint 77/67-THY (July 1977).

12. Many authors have studied the vacuum tunneling problem in scalar field theory. See, e.g. M. B. Voloshin, I. Yu. Kobzarev, and L. B. Okun, Sov. J. Nucl. Phys. $\underline{20}$, 644 (1975); S. Coleman, Harvard Preprint HUTP--77/A004 (1977), P. H. Frampton, Phys. Rev. Letters $\underline{37}$, 1378 (1976); M. Stone, Phys. Rev. D14, 3568 (1976).

13. H. J. Katz, "Lifetime of Metastable Vacuum States", Univ. of Illinois Preprint Ill-(TH)-77-15 (1977).

14. This interpretation is suggested to us by C. N. Yang.

15. C. N. Yang and R. Mills, Phys. Rev. $\underline{96}$, 191 (1954); E. S. Albers and B. W. Lee, Phys. Report $\underline{9}$ C, 1 (1973).

16. This is related to the idea of computing the most probable escape path as introduced by T. Banks, C. M. Bender, and T. T. Wu (Phys. Rev. D $\underline{8}$, 3346 (1973)).

Hartree Approximation in Field Theory

Shau-Jing Chang

Department of Physics, University of Illinois at Urbana-Chamgaign
Urbana, Illinois 61801, U.S.A.

I. Introduction

The Hartree approximation is well-known in non-relativistic many-body physics.[1] I shall review to you briefly the Hartree method in the above context. I shall then generalize the method to relativistic field theories. Most of the material presented here can be found in Refs. 2, 3, and 6.

II. Hartree Approximation in a Nonrelativistic Many-Body System

Consider a non-relativistic system described by

$$L = \int dx \left[i \phi^\dagger \frac{\partial \phi}{\partial t} - \frac{1}{2m} \vec{\nabla} \phi^\dagger \cdot \nabla \phi \right] - \frac{1}{2} \int dx_1 \, dx_2 \, \rho(x_1) \, U(x_1 - x_2) \, \rho(x_2) \qquad (2.1)$$

with

$$\hat{\rho}(x) = \phi^\dagger(x)\phi(x) \qquad (2.2)$$

being the density operator. By the variational method, we obtain the field equation,

$$i\frac{\partial \phi}{\partial t} = - \frac{1}{2m} \nabla^2 \phi + \int dx' \hat{\rho}(x') \, U(x' - x) \, \phi(x) \, , \qquad (2.3)$$

and a similar equation for $\phi^\dagger$. Field eq. (2.3), and its hermitian conjugate equation, together with the canonical quantization relations, specify the system completely. In the Hartree approximation, we replace the density operator $\hat{\rho}(x)$ by a c-number function $\rho(x)$. This c-number function is usually chosen to be the expectation value of $\hat{\rho}$ in an unperturbed theory. An improved approximation, known as the self-consistent Hartree approximation,[1] may be described as follows:

(1) Replace the density operator $\hat{\rho}$ by a c-number function $\rho(x)$ which will be determined self-consistently later.

(2) Solve (2.3) by assuming that ρ is a fixed external c-number function. Under this approximation, Eq. (2.3) reduces to an ordinary Schrödinger equation with a fixed non-local potential, and can be solved by standard methods. In particular, we can determine the c-number solutions ψ_n to (2.3). The field operator ϕ is related to these c-number wave functions (solutions) through

$$\phi(x) = \sum_n A_n \, \psi_n(x) \tag{2.4}$$

where A_n is the annihilation operator of state n.

(3) Determine ρ self-consistently by requiring

$$\rho(x) = \langle \hat{\rho} \rangle = \sum_{\substack{\text{all occupied} \\ \text{state}}} \psi_n^* \psi_n . \tag{2.5}$$

The Hartree approximation gives rise to very good results if the number of particles in the state is large. Then, we can replace $\hat{\rho}$ by its mean value $\rho(x)$.

III. Hartree Approximation Applied to Quantum Field Theory[2]

We consider a simple ϕ^4 scalar field theory described by

$$L = \frac{1}{2}(\partial_\mu \phi)^2 - \frac{1}{2} m^2 \phi^2 - \frac{g}{4} \phi^4 . \tag{3.1}$$

We separate ϕ into

$$\phi = \phi_c + \phi_q \tag{3.2}$$

where ϕ_c denotes the slow-varying part of ϕ and ϕ_q denotes the rapid fluctuations of ϕ around ϕ_c. We shall treat ϕ_c as a c-number function of x and t, and ϕ_q as a quantum operator whose expectation value vanishes. The exact form of ϕ_c depends on the physical system which we are interested in. For the ground state, for instance, ϕ_c is a constant. For a one-or multi-particle state ϕ_c may acquire explicit x and t dependences.

Using the action principle, we obtain the field equation

$$\partial^2 \phi + m^2 \phi + g\phi^3 = 0 \tag{3.3}$$

which can be written as

$$(\partial^2 + m^2)(\phi_c + \phi_q) + g(\phi_c^3 + 3\phi_c^2\phi_q + 3\phi_c\phi_q^2 + \phi_q^3) = 0. \tag{3.4}$$

Since ϕ_q fluctuates rapidly around $\phi_q = 0$, we expect that Eq. (3.4) is approximately symmetric under the reflection $\phi_q \to -\phi_q$. This implies that we can separate Eq. (3.4) into two independent equations according to the even and odd number of ϕ_q's:

$$(\partial^2 + m^2)\phi_c + g(\phi_c^3 + 3\phi_c\phi_q^2) = 0, \tag{3.5}$$

$$(\partial^2 + m^2)\phi_q + g(3\phi_c^2\phi_q + \phi_q^3) = 0. \tag{3.6}$$

It is straightforward to apply the Hartree approximation[2] to Eq. (3.5). We simply replace ϕ_q^2 in (3.5) by a c-number function ρ_q

$$\rho_q \equiv <\phi_q> \; . \tag{3.7}$$

and obtain

$$(\partial^2 + m^2 + 3g\rho_q)\phi_c + g\phi_c^3 = 0. \tag{3.8}$$

To implement the Hartree approximation to (3.6), we need to know how to approximate ϕ_q^3. A simple way is to consider ϕ_q^3 as a product of three different

- 133 -

operators, $\phi_q^{(1)} \phi_q^{(2)} \phi_q^{(3)}$. Each of the ϕ's is acting under the influences of the other two. The Hartree approximation is therefore given by

$$\phi_q^{(1)} \phi_q^{(2)} \phi_q^{(3)} \rightarrow <\phi_q^{(1)} \phi_q^{(2)}>\phi_q^{(3)} + <\phi_q^{(2)} \phi_q^{(3)}>\phi_q^{(1)}$$

$$+ <\phi_q^{(1)} \phi_q^{(3)}>\phi_q^{(2)}. \tag{3.9}$$

Setting $\phi_q^{(1)} = \phi_q^{(2)} = \phi_q^{(3)} = \phi_q$, we have the Hartree result

$$\phi_q^3 \rightarrow 3<\phi_q^2>\phi_q = 3\rho_q \phi_q. \tag{3.10}$$

Substituting (3.10) into (3.6), we obtain

$$(\partial^2 + m^2 + 3g\phi_c^2 + 3g\rho_q)\phi_q = 0. \tag{3.11}$$

Eqs. (3.7), (3.8) and (3.11) describe the Hartree approximation.

To solve these equations, we can express ϕ_q in terms of creation and annihilation operators

$$\phi_q = \sum_n (\psi_n^* A_n^\dagger + \psi_n A_n) \tag{3.12}$$

with ψ_n obeying the c-number equation,

$$\left(\frac{\partial^2}{\partial t^2} - \nabla^2 + 3g\phi_c^2 + 3g\rho_q\right) \psi_n = 0. \tag{3.13}$$

For the ground state, we have

$$\rho_q = \sum_n \psi_n^*(x)\psi_n(x). \tag{3.14a}$$

For an excited state, we have

$$\rho_q = 2 \sum_n (N_n + \tfrac{1}{2})\psi_n^* \psi_n \tag{3.14b}$$

where $\{N\}$ denotes the occupation numbers of the state. Given $\{N\}$, we can obtain ϕ_c, ψ_n, and ρ_q by solving the coupled c-number equations (3.8), (3.13), (3.14).

IV. Variational Principle

Even though we can in principle solve Eqs. (3.8), (3.13), (3.14). However, this is very impractical. These are coupled nonlinear differential equations, and they are infinite in number. For a time-independent ϕ_c, we can reformulate our problem as a variational calculation.[2] By a proper choice of trial functions, we can obtain a very good approximate solution to the above equations.

The variational calculation works in the following way. We first guess a $\langle \phi^2 \rangle$ with

$$\langle \phi^2 \rangle \equiv \phi_c^2 + \rho_q. \tag{4.1}$$

Once $\langle \phi^2 \rangle$ is known, we can obtain ψ_n in (3.13) by solving a potential scattering problem. From (3.14), we obtain both ρ_q and, via (4.1), ϕ_c.

The energy of our system is (See Ref. 2)

$$E = \int dx \left[\frac{1}{2}\dot{\phi}_c^2 + \frac{1}{2}\left(\frac{\partial \phi_c}{\partial X}\right)^2 + \frac{1}{2}m^2\phi_c^2 + \frac{g}{4}\phi_c^4 \right] + \sum_n (N_n + \frac{1}{2})\omega_n$$
$$- \frac{3g}{4}\int dx (\rho_q(x))^2 \tag{4.2}$$

where ω_n is the energy eigenvalue associated with ψ_n. In (4.2), the first term denotes the energy associated with the classical field ϕ_c; the second term denotes the energy associated with various oscillator modes in ϕ_q; and the last term in (4.2) is due to the nonlinear effect of the ϕ_q^4 term.

The correct ϕ_c and ρ_q are those which minimalize the energy E. i.e.

$$\frac{\delta E}{\delta \phi_c} = 0 \tag{4.3}$$

V. Renormalization

It is known that $(\phi^4)_2$ and $(\phi^4)_3$ theories give rise to divergent results due to vacuum and self-energy diagrams. To obtain a finite theory, we need to include renormalization counter terms. For instance, in a $(\phi^4)_2$ theory, we have

$$L = \frac{1}{2}(\partial_\mu\phi)^2 - \frac{1}{2}m^2\phi^2 - \frac{g}{4}\phi^4 - \frac{1}{2}B\phi^2 - \text{constant} \tag{5.1}$$

where $B\phi^2$ is the mass counter term.

The only mass counter term that we need in a $(\phi^4)_2$ theory is shown in Fig. 1. With this counter term, the Hartree field equations become

$$\partial^2\phi_c + (g\phi_c^2 + m^2 + 3g\Delta\langle\phi_q^2\rangle)\phi_c = 0, \tag{5.2}$$

$$\partial^2\phi_q + (3g\phi_c^2 + m^2 + 3g\Delta\langle\phi_q^2\rangle)\phi_q = 0 \tag{5.3}$$

with

$$B = -3g\langle\phi_q^2\rangle_{\phi_c=0} = -3g\int\frac{d^2k}{(2\pi)^2}\frac{i}{k^2 - m^2 + i\epsilon}, \tag{5.4}$$

$$\Delta\langle\phi_q^2\rangle = \langle\phi_q^2\rangle_{\phi_c} - \langle\phi_q^2\rangle_{\phi_c=0}. \tag{5.5}$$

It is easy to see that $\Delta\langle\phi_q^2\rangle$ is a finite quantity. The renormalized energy is

$$E_{ren} = \int dx\left[\frac{1}{2}\dot{\phi}_c^2 + \frac{1}{2}\left(\frac{\partial\phi_c}{\partial x}\right)^2 + \frac{m^2}{2}\phi_c^2 + \frac{g}{4}\phi_c^4\right]$$

$$+ \frac{1}{2}\sum_n [(2N_n + 1)\omega_n(\phi_c) - \omega_n(0)]$$

$$+ \frac{1}{2}B\int dx[\phi_c(x)^2 + \Delta\langle\phi_q^2\rangle] - \frac{3g}{4}\int dx[\Delta\langle\phi_q^2\rangle]^2, \tag{5.6}$$

which is also finite. The variational principle can now apply to E_{ren}.

Fig. 1

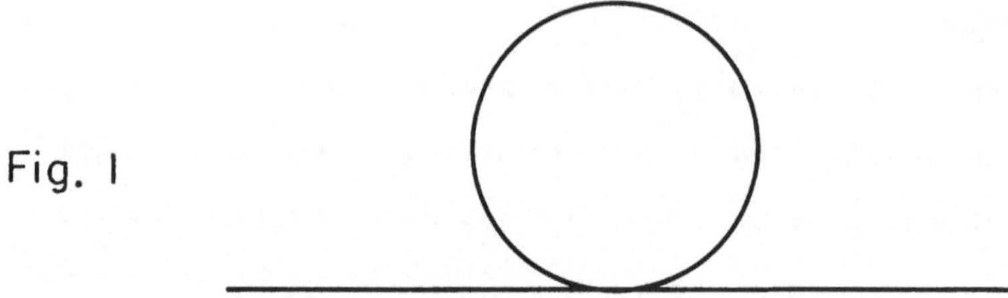

The only divergent self-energy diagram in a $(\phi^4)_2$ theory.

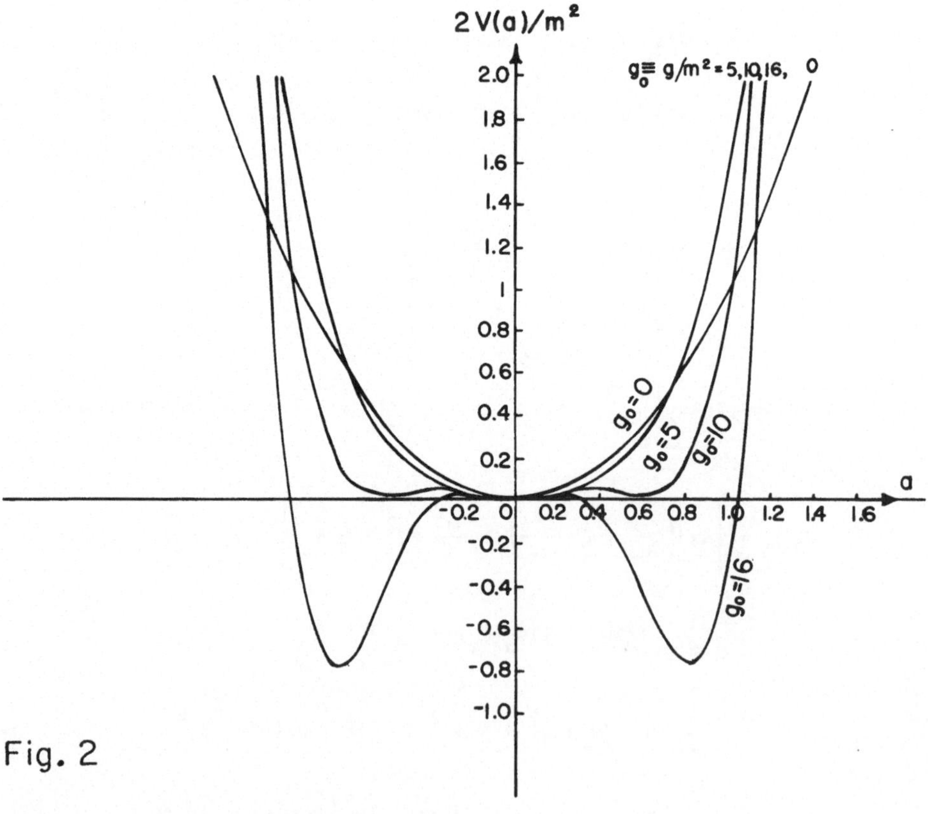

Fig. 2

The Hartree effective potential $V(\phi_c)$ as a function of ϕ_c = a.

VI. Effective Potential

We now apply the Hartree approximation to the calculation of the effective potential, $V(\phi_c)$. The effective potential is defined as the minimal energy density associated with a given constant ϕ_c. We can evaluate the effective potential in the Hartree approximation by

(1) Choosing a constant ϕ_c,

(2) Solving ϕ_q and $\Delta<\phi_q^2>$ self-consistently via (5.3) - (5.5), and

(3) Computing the energy density from (5.6).

We can solve (5.3) - (5.5) easily by noticing that

$$m'^2 \equiv m^2 + 3g\phi_c^2 + 3g\Delta<\phi_q^2> \tag{6.1}$$

is a constant, and that

$$\Delta<\phi_q^2> = \int \frac{d^2h}{(2\pi)^2} \left(\frac{i}{k^2 - m'^2 + iE} - \frac{i}{k^2 - m^2 + iE} \right) = \frac{1}{4\pi} \ln \frac{m^2}{m'^2} . \tag{6.2}$$

The effective potential can be written as

$$V(\phi_c) = \frac{1}{2} m^2 \phi_c^2 + \frac{1}{4} g\phi_c^4 + \frac{1}{8\pi} (m'^2 - m^2)$$
$$+ \frac{1}{2} (m^2 + 3g\phi_c^2) \frac{1}{4\pi} \ln \frac{m^2}{m'^2} + \frac{3g}{4} \left(\frac{1}{4\pi} \ln \frac{m^2}{m'^2} \right)^2 \tag{6.3}$$

with

$$m'^2 = m^2 + 3g\phi_c^2 + \frac{3g}{4\pi} \ln \frac{m^2}{m'^2} . \tag{6.4}$$

The Hartree effective potential in a $(\phi^4)_2$ theory is shown in Fig. 2. We can summarize the results as follows:

(1)   For a small value of g, $V(\phi_c)$ has a minimal at $\phi_c = 0$ which implies an unbroken ground state;

(2)   For a large value of g, $V(\phi_c)$ has two minima at $\phi_c \neq 0$ which implies a spontaneously broken symmetry; and

(3)   There is a first order phase transition at g = 10.211.

It turns out that the third conclusion concerning the first order phase transition is wrong.   The phase transition occurs here is actually a second order transition.[3/]

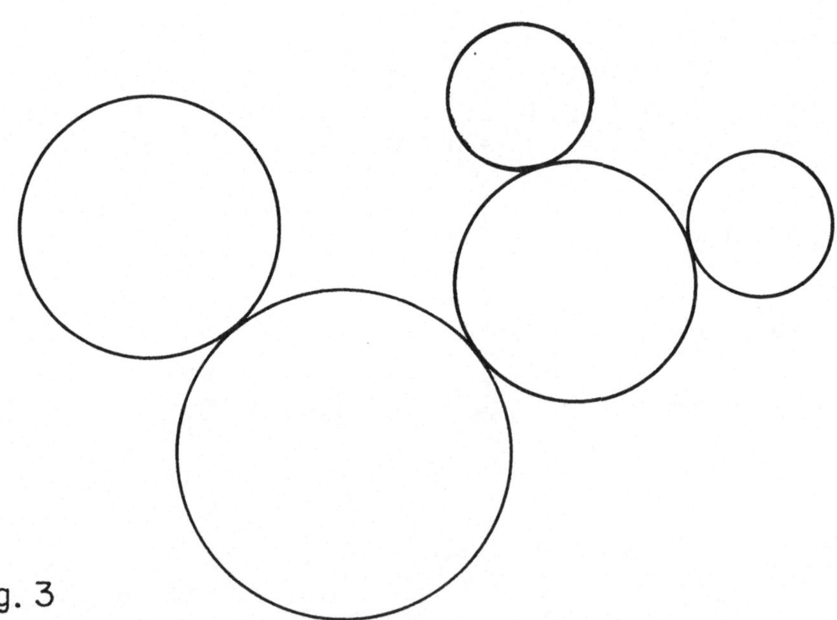

Fig. 3

A typical cactus diagram included in the Hartree calculation.

VII. Several Comments

I shall review to you several results without going into details:

(a) Graphical representation

Graphically, the Hartree approximation is equivalent to a sum over the Cactus-type diagrams as shown in Fig. 3. This is the same set of diagrams appeared in the leading $\frac{1}{N}$ - expansion.[4/] Thus, in the large N limit, the Hartree approximation becomes exact.

(b) Gaussian trial functions

In the variational language, the Hartree approximation is equivalent to using the gaussian trial functions. To give the gaussian trial function a precise meaning, we consider the ϕ-representation described by

$$<\phi|\hat{\phi}(x) = \phi(x) <\phi| , \tag{7.1}$$

$$<\phi|\hat{\dot{\phi}}(x) = \frac{1}{i} \frac{\delta}{\delta\phi(x)} <\phi| . \tag{7.2}$$

with the completeness condition

$$\int \mathcal{D}(\phi) \; |\phi><\phi| = I. \tag{7.3}$$

In Eqs. (7.1), (7.2), we put a hat over ϕ and $\dot{\phi}$ ($\hat{\phi}$ and $\hat{\dot{\phi}}$) to indicate that they are field operators. The $\phi(x)$ in (7.1) is a c-number realization of $\hat{\phi}(x)$. The Hamiltonian operator,

$$H = \frac{1}{2} \dot{\phi}^2 + \frac{1}{2} \left(\frac{\partial\phi}{\partial x}\right)^2 + \frac{1}{2} m^2\phi^2 + \frac{g}{4} \phi^4 , \tag{7.4}$$

takes the following form in the ϕ-representation:

$$H = - \frac{1}{2} \left(\frac{\delta}{\delta\phi(x)}\right)^2 + \frac{1}{2}\left(\frac{\partial\phi}{\partial x}\right)^2 + \frac{1}{2} m^2\phi^2 + \frac{g}{4} \phi^4 \tag{7.5}$$

A gaussian trial state $\langle g|$ is defined by

$$\langle \phi | g \rangle = N^{1/2} \exp \left[- \frac{1}{4} \int dxdy \ (\phi(x) - \phi_c) \right.$$
$$\left. \times \ g^{-1}(x, \ y) \ (\phi(y) - \phi_c) \right] \tag{7.6}$$

where $g^{-1}(x, \ y)$ and ϕ_c are variation parameters. The normalization factor $N^{1/2}$ is determined by

$$\langle g | g \rangle = \int \mathcal{D}(\phi) \ \langle g | \phi \rangle \langle \phi | g \rangle = 1. \tag{7.7}$$

This trial function implies immediately that

$$\langle g | \phi | g \rangle = \int \mathcal{D}(\phi) \langle g | \phi \rangle \phi(x) \langle \phi | g \rangle = \phi_c. \tag{7.8}$$

The condition that $\langle g | H | g \rangle$ be stationary with respect to the variation of $g(x, \ y)$ leads to the required Hartree equations.

(c) Reliability of Hartree approximations

Since the Hartree approximation is equivalent to the use of gaussian trial functions, we expect that the approximation is good when we are away from a phase transition. Hence, the Hartree approximation in a $(\phi^4)_2$ theory is reliable for both the weak and the strong coupling limits.[3/]

VIII. Renormalization Group Equation

One of the conclusions based on the Hartree approximation is that the strong coupling $(\phi^4)_2$ theory is equivalent to a weak coupling $(\phi^4)_2$ theory of a negative mass term.[3/] This result can be established rigorously by solving a renormalization group equation. Consider the Hamiltonian

$$H = H_o + H_I \tag{8.1}$$

with

$$H_o = \frac{1}{2} \dot{\phi}^2 + \frac{1}{2} \left(\frac{\partial \phi}{\partial x}\right)^2 + \frac{1}{2} m^2 \phi^2, \tag{8.2}$$

$$H_I = \frac{1}{2} \sigma \phi^2 + \frac{g}{4} \phi^4 . \tag{8.3}$$

We define a finite perturbation theory by removing the only divergent self-energy diagram as shown in Fig. 1. We shall see how the renormalized Green's function should change when we alter m^2 and σ. Let $G^{(n)}(x_1,\ldots,x_n)$ be an n-point Green's function. The contribution to G due to a variation of σ is equivalent to an arbitrary insertion of a mass term to an internal line, as indicated in Fig. 4. On the other hand, there are two types of contribution due to a variation of m^2. The first kind of contribution is a mass insertion as given in Fig. 4. The second kind of contribution is due to the change of mass counter term. Thus, the total contribution to $G^{(n)}$ due to a variation of m^2 is given graphically in Fig. 5, and obeys the following renormalization group equation,

$$\frac{\partial G^{(n)}}{\partial m^2} = \frac{\partial G^{(n)}}{\partial \sigma} - \frac{\partial}{\partial m^2} \text{(mass counter term)} \cdot \frac{\partial G}{\partial \sigma}$$

$$= \left(1 + \frac{3g}{4\pi m^2} \frac{\partial G^{(n)}}{\partial \sigma}\right) . \tag{8.4}$$

Fig. 4

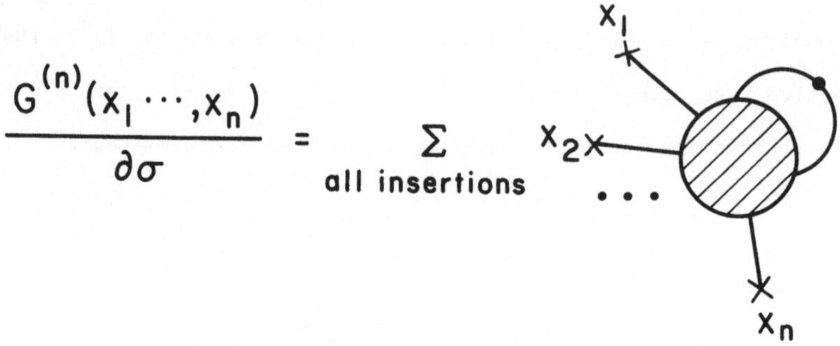

$$\frac{G^{(n)}(x_1\cdots,x_n)}{\partial\sigma} = \sum_{\text{all insertions}}$$

The effect of a mass insertion to a n-point function $G^{(n)}$.

Fig. 5

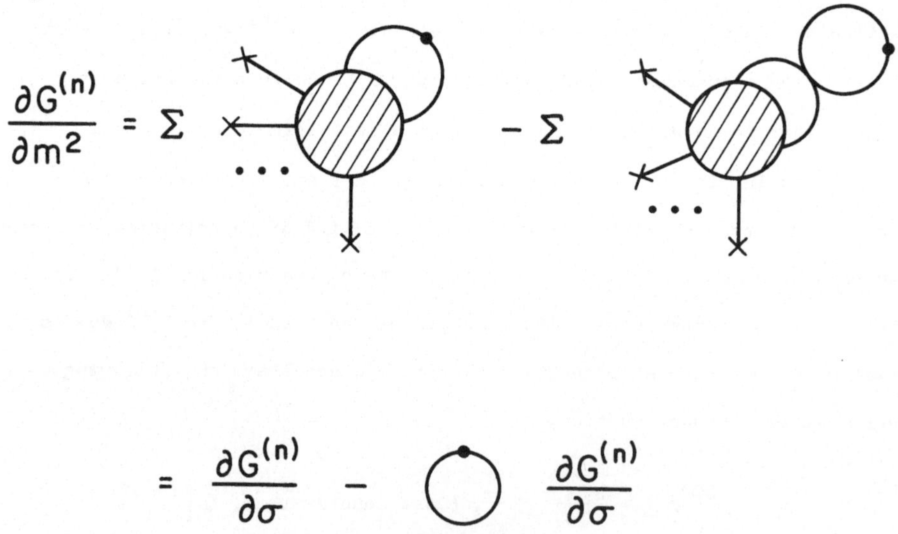

$$\frac{\partial G^{(n)}}{\partial m^2} = \sum \quad - \sum$$

$$= \frac{\partial G^{(n)}}{\partial\sigma} - \bigcirc \frac{\partial G^{(n)}}{\partial\sigma}$$

The effect of a change of renormalization point m^2 to a n-point function $G^{(n)}$.

It is important to point out that Eq. (8.4) is valid for both the exact and the Hartree approximate $(\phi^4)_2$ theory. Eq. (8.4) implies that $G^{(n)}$ is a function of

$$w = \sigma + m^2 + \frac{3g}{4\pi} \ln \frac{m^2}{g} \tag{8.5}$$

only. Equation (8.5) indicates that the Green functions $G^{(n)}$ obtained from the Hamiltonian (7.4),

$$H = \frac{1}{2} \dot{\phi}^2 + \frac{1}{2} \left(\frac{\partial \phi}{\partial x}\right)^2 + \frac{m^2}{2} \phi^2 + \frac{g}{4} \phi^4 \quad , \tag{7.4}$$

renormalized at m^2, and from the Hamiltonian

$$H' = \frac{1}{2} \dot{\phi}^2 + \frac{1}{2} \left(\frac{\partial \phi}{\partial x}\right)^2 - \frac{\mu^2}{4} \phi^2 + \frac{g}{4} \phi^4, \tag{8.6}$$

renormalized at μ^2 are identical to each other if

$$m^2 + \frac{3g}{4\pi} \ln \frac{m^2}{g} = - \frac{\mu^2}{2} + \frac{3g}{4\pi} \ln \frac{\mu^2}{g} \quad . \tag{8.7}$$

In particular, one finds that a strong coupling H is equivalent to a weak coupling H'. This fact implies the existence of a phase transition as we increase the coupling strength in H for both the exact $(\phi^4)_2$ theory as well as for the Hartree approximation. See Ref. (3) for detailed discussions.

IX. Generalization to $(\phi^4)_3$ Theories

The Hartree approximation can be generalized to other field theories as well. For instance, the unrenormalized Hartree field equations for $(\phi^4)_3$ and $(\phi^4)_4$ are the same. However, their renormalization counter terms are different. In a $(\phi^4)_3$ theory, we encounter two self-energy divergent diagrams as given in Fig. 6. This implies that the renormalized Hartree equation should be modified accordingly. To see the importance of the renormalization effect, we derive the renormalization group equation in the $(\phi^4)_3$ theory. As we have mentioned earlier, we expect that this renormalization group equation is valid for both the exact theory as well as for the Hartree approximate theory.

It is straightforward to see that the n-point, renormalized, Green's function $G^{(n)}(x_1,\ldots,x_n)$ associated with the $(\phi^4)_3$ Hamiltonian (8.1) - (8.3) obeys[6]

$$\frac{\partial G^{(n)}}{\partial m^2} = \frac{\partial G^{(n)}}{\partial \sigma} - \frac{\partial}{\partial m^2} \text{ (mass counter terms) } \frac{\partial G^{(n)}}{\partial \sigma}$$

$$= \left(1 + \frac{3g}{8\pi m} - \frac{3g^2}{16\pi^2 m^2} \right) \frac{\partial G^{(n)}}{\partial \sigma} \quad .$$

$$(9.1)$$

Eq. (9.1) implies that $G^{(n)}$ is a function of

$$w = \sigma + m^2 + \frac{3gm}{4\pi} - \frac{3g^2}{8\pi^2} \ln \frac{m}{g}$$

$$(9.2)$$

only. In Eq. (9.1), the $O(g)$ term originates from the subtraction term associated with the divergent diagram Fig. (6a), and the $O(g^2)$ term with the new divergent diagram Fig. (6b). At large g, the contribution due to the $O(g^2)$ term dominates. It determines the large g behavior of the Green's function.[6]

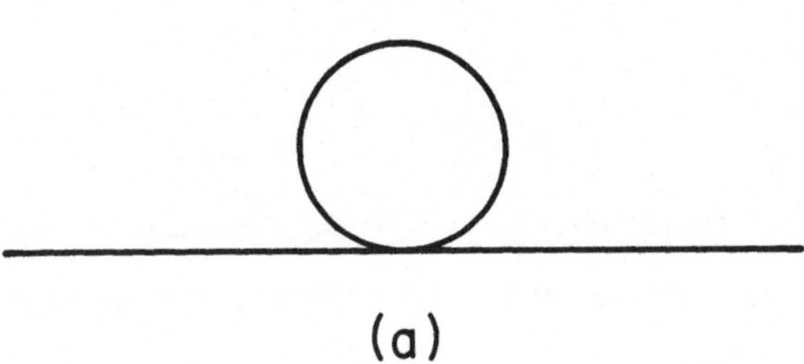

(a)

Fig. 6

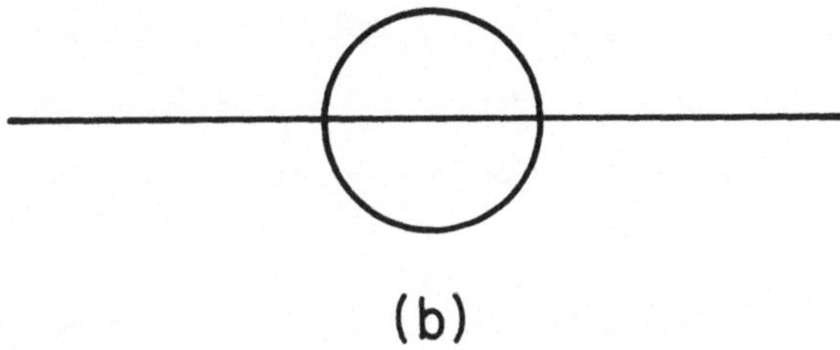

(b)

The divergent self-energy diagrams in a $(\phi^4)_3$ theory. Note that diagram (a) is $O(g)$, and diagram (b) is $O(g^2)$.

Acknowledgement

I wish to thank Professor and Mrs. Walter Dittrich and the organizers of the Symposium for their hospitality.

References

1. See, e.g. A. L. Fetter and J. D. Walecka, "Quantum Theory of Many-Particle Systems" (McGraw-Hill, N.Y., 1971); T. Kinoshita and Y. Nambu, Phys. Rev. $\underline{94}$, 598 (1954).

2. S. J. Chang, Phys. Rev. D $\underline{12}$, 1071 (1975); and S. J. Chang and J. A. Wright, Phys. Rev. D $\underline{12}$, 1595 (1975).

3. S. J. Chang, Phys. Rev. D $\underline{13}$, 2778 (1976).

4. The $\frac{1}{N}$ - expansion was known to statistical Mechamics applications for some time. The $\frac{1}{N}$ - expansion which is relevant to our problem can be found in, e.g., S. Coleman, R. Jackiw, and H. Politzer, Phys. Rev. $\underline{D}$ 10, 2491 (1974); H. J. Schnitzer, ibid, $\underline{10}$, 1800 (1974); $\underline{10}$, 2042 (1974).

5. See, e.g. J. M. Cornwall, R. Jackiw and E. Tomboulis, Phys. Rev. D. $\underline{10}$, 2428 (1974). These authors credited this trial function to J. Kuti (unpublished)

6. S. Magruder, Phys. Rev. D $\underline{14}$, 1602 (1976); S. J. Chang and S. Magruder, ibid $\underline{16}$, 983 (1977).

Two Topics in Eikonal Physics

H. M. Fried

Physique Théorique, Université de Nice and Physics Department, Brown University, Providence, R.I. 02912, U.S.A.

These remarks des cribe the beginning steps towards the resolution of two distinct and longstanding eikonal problems. Both employ "semi-classical" techniques; both produce "intuitive" results; and both suggest further refinements, closely related to the interpretation of high-energy experiments. The two topics treated here are

A. an Estimation of <u>all</u> eikonal pionization graphs (tower and non-planar) in a relevant Lagrangian field theory; and

B. a Construction of an eikonal representation, from first principles, when there exist isotopic constraints.

The reasons for the importance of these subjects may be most easily presented by enumerating the following sets of questions and remarks. For Topic A : What lies beyond the Cheng-Wu/Chang-Yan tower graphs [1]? Is a previous, crude estimate [2] of the effects of more complicated, non-planar (checkerboard) eikonal graphs correct - in which a typical hadronic σ_{TOT} vanishes as s^{-a} - or would a refined estimate continue to saturate the Froissart bound? How <u>does</u> one treat a strong-coupling problem, which is effectively the case at ultra-high energies, $Y = \ln\left(\frac{s}{\mu^2}\right) \gg 1$? Would the results have any resemblence to those recently found [3] in super-critical Pomeron theories?

Concerning Topic B : Essentially all of eikonal and Pomeron physics sidestep the restrictions that exist when virtual exchanges or real emissions can themselves carry quantum numbers. For example, in the simplest possible example of nucleon scattering by the exchange of many virtual, massive, neutral vector mesons (NVM),and including no other more complicated virtual structures, one obtains the old Lévy-Sucher relativistic eikonal model [4] in the appropriate limit of $|t|/s \ll 1$. The essence of that approximation is that every NVM emitted or absorbed by either

nucleon is treated independently of every other NVM; no correlations
between these NVMs are possible (with a few special exceptions [5,6]),
for then the simple combinatoric arguments used in the derivation would
break down. Thus the eikonal model can be formulated when the particles,
or entities, exchanged form the singlet representation of some relevant
symmetry group. In SU(2), for example, one can have multiple exchange of
ω s, but not of $\rho_\pm$, ρ_0 ; the physical reason is that if a proton emits
a ρ_+, the next charged particle it can emit must be a ρ_-, which arith-
metic restriction generates a correlation between that ρ_+ and that ρ_-,
and promptly ruins the combinatoric solubility.

It may be that ω –nucleon couplings are experimentally somewhat
larger than ρ –nucleon couplings, and models neglecting the latter [7] can
be in reasonable agreement with experiment; but as a matter of principle,
the description of such interchanges is incomplete. Because of the virtual
certainty that underlying quark fields exist, and the wide suspicion that
SU(3) Yang-Mills fields form the intervening glue which bind the quarks
into recognizable hadrons, it becomes a matter of interest to learn to
formulate eikonal approximations involving such non-Abelian fields; questions of quark binding and scattering _should_ be capable of dynamical explanation. Of course, phenomenological, and unitary, eikonal models can be
written [8] involving isotopics, and other quantum numbers; but starting
from a given field theory, this problem has never been solved.

Perhaps it would be appropriate to state what is "good" about eikonal models, why they continue to reappear, and why they always will :
Eikonal models automatically produce sums over all necessary permutations
of coordinates of the exchanged and emitted particles, in any reaction,
while permitting a simple formulation of s-channel unitarity. These are
non-trivial features of any approximation scheme, and will be used to advantage in the two computations described here.

At the outset, the eikonal idea assumes that all momenta can be divided into two classes, the "large" (e.g., incident $p_{1,2}$ and final $p'_{1,2}$ nucleon momenta) and the "small" (just about everything else, e.g., all exchanged meson momenta q_i); with the exception of the not-too-large diffractive disassociation processes, at current energies this seems to be an assumption in reasonable agreement with experiment. The result of such approximations will be the replacement of nucleon propagators G_c (x,y | A), defined in the presence of ficticious (but convenient !) c-number fields A_μ (z), by "no-recoil" or Bloch-Nordsieck (BN) propagators [9] G_c^{BN}(x,y | A), in effect an expression of the judicious neglect of any $q_i \cdot q_j$ combination in comparison to terms of form $p \cdot q_i$. And then, because the explicit source dependence of $G_c^{BN}[A]$ is usually simple, one may carry through almost all functional computations in a straight-forward way.

It will be useful to set the stage for what is to come by considering the eikonal forms relevant to a "hybrid" theory, one resembling both ϕ^3-theory and (massive photon) QED. Here, one treats nucleons (ψ) scattering by the exchange of many NVMs (A_μ), which may themselves exchange arbitrary numbers of "scalar pions" π ,

$$\mathcal{L}' = ig \, \bar{\psi} \gamma \cdot A \, \psi - \tfrac{\lambda}{2} \pi A^2 \, . \tag{1}$$

The "natural language" to use in representing the scattering amplitude

$$T_{EIK}(s,t) = i \frac{s}{2} \int d^2 b \; e^{i \vec{q} \cdot \vec{b}} \left[1 - e^{i \chi(s,b)} \right] , \tag{2}$$

with normalization given by $d\sigma_{EL}/dt = |T|^2 / \pi s^2$ and $\sigma_{TOT} = \frac{4}{s} \, \text{Im} \, T(s,0)$, is the language of functional methods, wherein all normalization factors, phases, etc..., are automatically included. Neglecting radiative corrections corresponding to NVM self-linkages on the same nucleon line, and neglecting closed nucleon and closed NVM loops (e.g., discarding A- and π-self-energies), it is easy to see that [2]

$$e^{i\chi} = \exp\left[-\tfrac{i}{2}\int \tfrac{\delta}{\delta\pi} D_c \tfrac{\delta}{\delta\pi}\right] \cdot \exp\left[ig^2\int \mathcal{F}_1 \bar{\Delta}_c(\pi) \mathcal{F}_2\right] \Big|_{\pi\to 0} , \qquad (3)$$

or

$$i\chi = \left(\exp\left[-\tfrac{i}{2}\int \tfrac{\delta}{\delta\pi} D_c \tfrac{\delta}{\delta\pi}\right] \cdot \exp\left[ig^2\int \mathcal{F}_1 \bar{\Delta}_c(\pi) \mathcal{F}_2\right]\right) \Big|_{\substack{\pi\to 0, \\ conn}} - 1 , \qquad (4)$$

where the $\mathcal{F}^\mu_{1,2}(u) = P^\mu_{1,2} \int_{-\infty}^{+\infty} d\xi \; \delta(u - z_{1,2} + \xi P_{1,2})$ denote classical currents of the interacting nucleons, with moment $p_{1,2}$ and spatial coordinates $z_{1,2}$; the impact parameter is $\vec{b} = (\vec{z}_1 - \vec{z}_2)_\perp$; D_c and $\delta_{\mu\nu}\Delta_c$ denote pion and NVM propagators of masses μ and m, respectively; $\bar{\Delta}_c(\pi) = \Delta_c(1 + \lambda\pi\Delta_c)^{-1}$ represents the NVM propagator in the presence of a ficticious, c-number field $\pi(x)$; and the subscript "conn" specifies the retention of connected graphs only, with at least one pion linkage between any NVM and the rest of that graph. S-channel unitarity then requires that $\sigma_{TOT} = \sigma_{EL} + \sigma_{IN}$, where

$$\sigma_{TOT} = 2\, Re \int d^2b \left[1 - e^{i\chi}\right], \qquad (5a)$$

$$\sigma_{EL} = \int d^2b \left|1 - e^{i\chi}\right|^2 , \qquad (5b)$$

$$\sigma_{IN} = \int d^2b \left[1 - e^{-2Im\chi}\right] . \qquad (5c)$$

Expansion of the g^2-dependent exponential of (4) provides a representation of the conventional eikonal approximations. Omitting all self-linkages (all pion radiative corrections to the same NVM line), to order g^2 one obtains just the Lévy-Sucher result [4], $i\chi_1(s,b) \simeq -\frac{g^2}{2\pi} \gamma(s) K_0(mb)$, where $\gamma(s) \to 1$ as $s/m^2 \to \infty$, in this NVM theory. (Were the NVM replaced by scalar mesons A, then $\gamma(s) \to -m^2/s$ as $s/m^2 \to \infty$). The g^4-term of this expansion contains all the ladder and crossed pion linkages between a pair of NVM lines; and upon extracting the leading rapidity dependence from the set of ladder graphs, one reproduces an absorptive eikonal $i\chi_2 \sim -a_1 s^{a_2}/\ell ns. \exp\left[-a_3 b^2/\ell ns\right]$, equivalent in physical content to the original Cheng-Wu/Chang Yan calculations[1]; here, the a_i are constants, and $a_2 \sim \lambda^2/m^2$. (In a scalar A-meson theory, $a_2 \sim \lambda^2/m^2 - 2$); and the phenomenological choice $a_2 = 0$ produces an effective, unitarized Pomeron, complete with the correct cuts). Estimates of the remaining $i\chi_n$, $n \geq 3$ are exceedingly difficult to obtain; and when obtained are difficult to believe [2]. The subject of Topic (A) represents one attempt to rectify this situation.

In order to discuss possible eikonal construction in the presence of isotopics, Topic (B), it will be simplest to neglect all pionization effects, setting $\lambda = 0$ everywhere above. Thus one is concerned only with generalizations to the simplest eikonal, $i\chi = i g^2 \int \mathcal{F}_1 \Delta_c \mathcal{F}_2$, expressing the result of multiple NVM exchanges between the scattering nucleons. It will be useful to mark exactly that point at which the exchange of charged mesons causes a combinatoric breakdown in the conventional eikonal analysis. For this, one may write the configuration space amplitude representing multiple NVM exchange in the form

$$M(x,\eta_1,x_2\eta_2) = i^2 \exp\left[-i\int \frac{\delta}{\delta A_1}\Delta_c \frac{\delta}{\delta A_2}\right]\cdot G_{c_1}(\eta_1,x_1|A_1)\cdot G_{c_2}(\eta_2,x_2|A_2)\Big|_{A_{1,2}=0}\ , \tag{6}$$

or better yet, the more convenient expression for the derivative of (6)

$$\frac{\partial M}{\partial g^2} = -i\,(2\pi)^{-4}\int d^4z_1\int d^4z_2\,\Delta_c(z_1-z_2)\cdot\exp\left[-i\int\frac{\delta}{\delta A_1}\Delta_c\frac{\delta}{\delta A_2}\right]\cdot \tag{7}$$

$$\cdot\left[G_{c_1}(\eta_1,z_1|A_1)\gamma_\mu\,G_{c_1}(z_1,x_1|A_1)\right]\cdot\left[G_{c_2}(\eta_2,z_2|A_2)\gamma_\mu\,G_{c_2}(z_2,x_2|A_2)\right]\Big|_{A_{1,2}=0}.$$

Subsequent mass-shell amputation of (7), and projection to appropriate
isotopic states will be necessary to extract the appropriate $\partial T/\partial g^2$;
but the forms (6) and (7) make clear the initial need for the construction
of $G_c^{(BN)}(y,\,x\,|\,A)$, the "no-recoil" approximation to each $G_c[A]$ of (7).
The mass-shell amputated form of each Green's function may be written as $\tag{9}$

$$\int d^4x\ e^{ip\cdot x}\,\overline{G}_c^{BN}(y,\bar{x}|A) = e^{ip\cdot y}\,\mathcal{F}(\infty;p) \tag{8a}$$

$$\int d^4y\ e^{-ip'\cdot y}\,\overline{G}_c^{BN}(\bar{y},x|A) = e^{-ip'\cdot x}\,\mathcal{F}(\infty;-p')\ , \tag{8b}$$

where the barred notation represents appropriate amputation, and where
$\mathcal{F}(\zeta;p)$ is the solution of a trivial equation,

$$\frac{\partial\mathcal{F}(\zeta;p)}{\partial\zeta} = i\,\pi(\zeta)\,\mathcal{F}(\zeta;p)\ , \tag{9}$$

with $\mathcal{F}(0)=1$ and $\pi(\zeta) \equiv g\,P_\mu\,A_\mu(z-\zeta p)$; thus the necessary solution is $\mathcal{F}(\zeta;p) = \exp\left[i\int_0^\zeta d\zeta'\,\pi(\zeta')\right]$, generating an exponential of linear A-dependence for the $G_c^{BN}[A]$ of (7).

When the exchanged quanta are allowed to carry quantum numbers,
the relevant generalizations of (9) become more complicated; and while
formal solutions can readily be written, their explicit use in the remainder of the analysis has always been prohibitively complicated. For the
simplest case of SU(2) isotopic dependence in a strong-coupling limit,
Topic (B) suggests a possible, approximate method of eikonal construction.

A. Semi-Classical Estimation of all Eikonal Pionization Graphs

In a recent paper [10], an attempt was made to incorporate the
effects of all the χ_n, in an approximate way, by performing a semi-classical calculation for an averaged, "self-consistent" pion field, whose properties would reflect the probable importance of the higher eikonal terms, in
the limit of ultra-high energies and in the region of small impact parameters. There, a formalism was set-up to perform such an "averaged" estimate, by first converting the functional differential operations of (3) and
(4) to an equivalent functional integral -- along with the corresponding
functional representations of all the pion inelastic cross sections --
and then approximating these functional integrals by a stationary phase
method, with the condition for stationary phase providing the "semiclassical" Euler equation for the "averaged" field. At very high energies,
one may expect that every pion exchanged between a pair of NVMs will, in a
leading-rapidity, ladder graph approximation, generate a factor of Y
essentially corresponding to the available phase space for the inelastic
production of a pion with limited transverse momentum; and hence one is
faced with an effective strong-coupling problem, it which it is not acceptable to neglect any particular sets of eikonal graphs. Mathematically, one
evaluates an integral in such a strong-coupling limit by a method such as

stationary phase; physically, one is shifting the description from one
of overly-many quanta to one of a single,averaged field.

This type of calculation has one distinct advantage over other,
more properly classical computations, whereing one merely hopes that quantum fluctuations are small, and will not distrurb the classical effects
obtained. Because one is here dealing with an eikonal representation of
quantum effects (which are then to be approximated by a semi- classical
technique), one has the added feature of s-channel unitarity, relating
inelastic pion emission at every impact parameter to the imaginary part
of the elastic eikonal. After the transition to an averaged field has been
made, this requirement of unitarity translates into a restriction on the
possible solutions of the Euler equations. Hence an averaged field that
satisfies this unitarity condition provides an approximate representation
of all the complicated, non-linear, internal dynamics specified by the
original Lagrangian interaction; and one which is also consistent with
the quantum mechanical restrictions of probability for scattering and all
production processes.

One may add that it is precisely because the Euler equations are
non-linear that the method works at all. That is, solutions to a particular non-linear Euler equation have no reason to satisfy _another_ non-linear
relation, that of unitarity; one cannot expect miracles. But because the
semi-classical equations are non-linear, they involve the magnitude, phase
and _branch_ of their solutions -- three parameters, or functions, rather
than just two. In essense, unitarity determines the third quantity, for
example, the branch; and in this sense, the non-linearity of the problem
is essential.

The equation for the averaged field $\varphi_0(x)$ was obtained in ref.(10) as

$$K \, \varphi_0(x) = -i \, \frac{\delta \mathcal{F}}{\delta \varphi(x)} \bigg|_{\varphi=\varphi_0} \, , \tag{10}$$

with

$$\mathcal{F}[\varphi] = i g^2 \int \mathcal{F}_1 \, \bar{\Delta}_c[\varphi] \, \mathcal{F}_2 \, , \tag{11}$$

and $K = \mu^2 - \partial^2 = D_c^{-1}$. Eq.(10) is "self-consistent" in the sense that each "quasi-pion" φ_0 emitted by a NVM is defined by the behavior of many (an infinite number of such) quasi-pions along that line. Inelastic emission then turns out to be described by the simplest of all unitary mechanisms, as if no more than a single particle is emitted by each multiperipheral chain. It is most convenient to introduce a specific form for $\bar{\Delta}_c[\varphi]$, the use of which is equivalent to the extraction of the leading rapidity dependence of all ladder-graph pions exchanged between any pair of NVMs,

$$\bar{\Delta}_c(z|\varphi) = \int \frac{d^4 g}{(2\pi)^4} \cdot \frac{e^{i g \cdot z}}{g^2 + m^2} \cdot \exp\left[\lambda \int d^4 k \, \frac{\vec{\varphi}(k)}{(g-k)_\perp^2 + m^2}\right] \, ,$$

together with the subsequent replacement

$$\int \frac{d^4 k}{k^2 + \mu^2} \rightarrow i\pi \, Y \cdot \int d^2 k_\perp \, , \quad Y \sim \ln\left(\tfrac{s}{\mu^2}\right) \, .$$

In terms of possible solutions φ_0 to (8), the eikonal is given by

$$\chi = -i \mathcal{F}[\varphi_0] - \tfrac{1}{2} \int \varphi_0 K \varphi_0 + \tfrac{i}{2} T_r \ln\left[1 + i D_c \frac{\delta^2 \mathcal{F}}{\delta \varphi \, \delta \varphi} \bigg|_{\varphi=\varphi_0}\right] \, . \tag{12}$$

We do take seriously the trace-log contribution of (12), which might
be thought of as a normalization correction to the eikonal, and physically corresponds to a sum of quasi-pion linkages between the NVMs such
that no more than two virtual particles are emitted or absorbed by any
NVM line. However, due to the non-linearity imposed by unitarity, inelastic emissions still have the form of just one particle ejected per chain.

One point concerning self-linkages along every NVM line should
be made, for the formalism of (3) and (4) will automatically include such
dependence, even though the propagator $\bar{\Delta}_c[\varphi]$ has really been designed
to reproduce those pionization effects delivered by cross-linkages between
two different NVMs. But one may expect that this difficulty is relatively
unimportant. In the context of the stationary phase approximation, such
unwanted self-linkages may be removed by subtracting from χ the first-
order expansion of the trace-log of (12),

$$\chi \to \chi' = \chi - \frac{i}{2} \, \mathrm{Tr} \left[i \, D_c \, \frac{\delta^2 \varphi}{\delta y \cdot \delta \varphi} \Big|_{y=y_o} \right] \, , \tag{13}$$

and this form will be used in what follows. For small b this has essentially no effect; but in the region of larger impact parameters close to
$b_{max} \sim \frac{1}{m} Y$, it ensures the Froissart bound in its usual form, rather than
$b_{max} \sim \frac{1}{m} \left[Y \, \ell n \, Y \right]^{\frac{1}{2}}$.

In momentum space, (10) reads

$$\tilde{\varphi}_o(k) = - \frac{g^2 \lambda}{(2\pi)^2} \cdot \frac{1}{k^2 + \mu^2} \cdot \int \frac{d^2q \, e^{i\vec{q} \cdot \vec{b}}}{q^2 + m^2} \, \frac{e^{R(\vec{q}/m)}}{(q - k_\perp)^2 + m^2} \, , \tag{14}$$

with

$$R(\vec{q}/m) = \lambda \int d^4p \, \vec{\varphi}_o(p) \left[m^2 + (q - p_\perp)^2 \right]^{-1} \, ,$$

where we reject, at the outset, any additional mass-shell contribution
to φ_0. It is immediately clear that finding solutions to (14) is a
difficult task. To simplify the analysis, we look for solutions under the
assumption mb $\gtrsim$ 1, which suggests that the only significant values of q,
in the integrand of (14), will be limited by q $\lesssim$ m. Hence the R(q/m)
factor may be approximated as

$$R(\varphi/m) \simeq R_0 + i\, \vec{q}\cdot\vec{R}_1 - q_i\, q_j\, R_2^{ij} \, ,\tag{15}$$

retaining only (the integrable) quadratic q dependence. If the $\left| R_2^{ij} \right|$ are
of size m^{-2}, this manipulation is equivalent to a cut-off of $\int d^2q$ at
$q_{max} \sim$ m; but non-linear effects can serve to significantly decrease q_{max}.
For example, if $m^2 \left| R_2^{ij} \right| \sim$ Y, or larger (and here we are assuming ultra-
high energies, where Y $\gg$ 1), then $q^2_{max} \ll m^2$. It is solutions of this
form, corresponding to sizable φ_0, that we have in mind.

The substitution of (15) into (14), together with the use of the
pionization form of $\bar{\Delta}_c[\varphi]$ and the assumption that $q^2_{max} \ll m^2$, generates

$$\vec{R}_1 = 0 \qquad \text{(by symmetry)} \, ,$$

$$R_2^{ij} = R_2\, \delta_{ij}, \quad R_2 = R_0/6m^2, \tag{16}$$

and

$$R_0^2 = -i\, \xi\, \exp\left[R_0 - \eta/R_0 \right] \, , \tag{17}$$

with $\xi = 3(g\lambda)^2 \pi Y / 2m^2$, $\eta = \frac{3}{2}(mb)^2$. By the use of these approximations, which follow naturally for large R_2, or R_0, all the non-linear complexity of the original eq.(14) has been transformed into a single relation

for a complex R_0, (17). If solutions to (17) can be found which have
the property

$$Re(R_0) > 0 ,\tag{18}$$

necessary for the convergence of the original q-integral of (14), then
$\widetilde{\varphi}_0$ may be written in the form

$$\widetilde{\varphi}_0(k) = -i \left(\frac{m^2 R_0}{\lambda \pi^2 y} \right) \frac{1}{(k^2 + \mu^2)(k_\perp^2 + m^2)} .\tag{19}$$

It is then straightforward to substitute (19) into (13), and obtain

$$i\chi = +\left(\frac{m}{\lambda\pi}\right)^2 \frac{1}{y} R_0 - \left(\frac{m}{\lambda\pi}\right)^2 \frac{1}{2y} R_0^2 - \frac{1}{2} \ell n (1-R_0) - \frac{1}{2} R_0 .\tag{20}$$

Evaluation of the trace-log terms is performed by an approximate evaluation of its <u>nth</u> iterate, using the same approximation methods followed in reaching (19); and then the sum over all iterates (each is finite !)
has been put into logarithmic form by a simple continuation argument.

With (20) one can now supply a specific input to the s-channel
unitarity relation of ref.10, which demands, for any such eikonal, that

$$Im \chi = \frac{1}{2} \int (k\varphi_0)^* \cdot D_{(+)} \cdot (k\varphi_0) ,\tag{21}$$

where $\widetilde{D}_{(+)}$ represents the (positive definite) pion's phase space function,
$\widetilde{D}_{(+)}(k) = (2\pi)^{-3} \cdot \theta(k_0) \cdot \delta(k^2 + \mu^2)$. In terms of the explicit solution (19), this becomes

$$\text{Im}\,\chi = \frac{1}{2y}\left(\frac{m}{\lambda\pi}\right)^2 |R_0|^2, \qquad (22)$$

where the LHS of (22) is given by the imaginary part of (20). We shall insist that any solution for R_0 to (17) and (18) must also satisfy (22).

Approximate solutions to the Euler equations-plus-unitarity have been found which depend on b and (very large !) Y in the following way: $R_0 = \rho \exp\left[i\left(\theta + 2\pi n\right)\right]$.

(i) For relatively small b, $mb \lesssim \left[Y \ln Y\right]^{\frac{1}{2}}$:

$$\rho \sim ay \ln Y, \qquad \theta \sim \pi/2 .$$

Here, the branch n is large, $n \sim \rho$, while the eikonal is properly absorptive,

$$i\chi \sim -\tfrac{1}{8}\ln Y - ia\left[\tfrac{1}{6}Y\ln Y\right]^{\frac{1}{2}}, \qquad a = \left(\frac{\pi\lambda}{2m}\right)^2 .$$

(ii) As the Froissart limit is approached, $\left[Y\ln Y\right]^{\frac{1}{2}} \lesssim mb \lesssim Y$:

$$\rho \sim \alpha Y + \beta \ln Y , \qquad \theta \to \theta_0 < \pi/2 ,$$

with $\alpha = 2a\cos\theta_0 \cdot \left(\sin\theta_0\right)^{-2}$, $\beta^{-1} = \cos\theta_0$. Here $n \sim \rho$ again; and hence we are on a higher branch. In this region, $\text{Re}\,(i\chi) \sim -Y$.

(iii) As b continues to increase, $\theta \to o$, and there occurs a sharp drop in ρ such that $\eta > \varsigma > \rho \sim O(\imath)$. Then, for larger b, $mb > Y$,

$$\rho \sim Y^{-1}\cos\theta \sim Y^{\frac{1}{2}} \exp\left[-\frac{a}{4\pi}\eta Y\right] ,$$

$$- 161 -$$

so that $\theta \to \frac{\pi}{2}$ and $n \sim \eta/\rho$. Here, ρ falls off very rapidly as Yb^2 increases, and is effectively zero for $b > b_{max} \sim \frac{1}{m} Y$. This behavior provides the Froissart bound for our inclusive, and total cross sections,

$$\sigma_{TOT} \sim Y^2 .$$

Inclusive cross sections are obtained by calculating the functional derivatives of

$$\sigma_{IN} = \int d^2b \; e^{-2\,Im\,\chi} \left[e^{\int K\psi_0^* \cdot D_{(+)} \cdot K\psi_0} - 1 \right]$$

with respect to $\tilde{D}_{(+)}(k)$, and produce non-correlated inclusive distributions. The one-particle inclusive cross section, for example, is given by

$$\omega \frac{d\sigma}{d^3k} = \pi \int d^2b \; \left| K\overset{\sim}{\psi}_0(k) \right|^2 ,$$

with an integrand that cuts off sharply at $b \gtrsim b_{max}$. Using the larger b forms throughout, this gives approximately

$$\omega \frac{d\sigma}{d^3k} \sim b_{max}^2 \cdot \left(\frac{m^2}{m^2 + k_\perp^2} \right)^2 . \tag{23}$$

The limited k-distribution of (23) is not significant, since it was effectively assumed at the beginning; but the growth of such an inclusive cross section with Y^2, and the flat plateau in the rapidity version of this result are specific predictions of the model. The other n-particle inclusive cross sections show a similar rise with Y^2, and contain no correlations between any such inclusive emissions. Except for the flat plateaux, these results are similar to those recently found in a strong-coupling, super-critical Pomeron computation [3].

Within the context of this averaged, or semi-classical calculation, the present computation thus provides an answer to the long-standing eikonal question : How important are <u>all</u> the $i\chi_n$, $n \geqslant 2$? We do not find the almost complete cancellation suggested in ref.2; rather there is an effective saturation of the Froissart bound, which would have been given by the pionization model of $i\chi_2$, except for a slight decrease in σ_{TOT} and the inclusive cross sections coming from the region of small b, $mb < [Y \ell n Y]^{\frac{1}{2}}$. One difference is that here the ratio of total one-particle inclusive cross section to σ_{TOT} produces a multiplicity $\langle n \rangle \sim Y$, in contrast to the Cheng-Wu/Chang-Yan behavior $\langle n \rangle \sim s^{a/1+2a}$. Except for the complete absence of correlations, these results are not in overt disagreement with existing data, and may be looked upon as a prediction of future, higher-energy experiments.

B. Eikonal Construction Including Isotopics

Consider the simplest possible extension of the NVM theory discussed above, the exchange of multiple $\rho_{\pm}$, ρ_0 mesons (represented by the VM fields A_μ^α) between a pair of scattering nucleons, according to the SU(2)-invariant interaction Lagrangian $\mathcal{L}' = i g_v \bar{\psi} \gamma \cdot \vec{A} \cdot \vec{\tau} \, \psi$
To this will later be added isoscalar exchanges, such as ω_s , specified by $\mathcal{L}'' = i g_s \bar{\psi} \gamma \cdot A \, \psi$. All other interactions are neglected. One is then faced with the necessity of constructing a solution to the appropriate generalization of (9)

$$\frac{\partial \mathcal{F}(s)}{\partial s} = i \, \vec{\tau} \cdot \vec{\pi}(s) \, \mathcal{F}(s),$$

$$(24)$$

with $\mathcal{F}(0) = 1$ and $\pi^\alpha(s) \equiv g_v \, p_\mu \, A_\mu^\alpha (z - s p)$. Until this is done, one cannot calculate the corresponding $\partial T / \partial g_v^2$, of (7); the lack of a useful solution to (24) has been the essential stumbling-block in this problem. Of course, one has always the formal solution to (24)

$$\mathcal{F}(\xi) = \left(\exp\left[i \int_0^\xi d\xi' \; \vec{\tau} \cdot \vec{\pi}(\xi') \right] \right)_+$$

(25)

in terms of an ordered exponential (in the variables ξ'); but for practical computations, (25) is worthless. In the absence of isospin, the
ordered exponential of (25) becomes an ordinary exponential, and the functional operations of (7) can be trivially obtained; with isotopics, however,
one has not been able to progress past this point.

One may, however, note that the unitarity property of the formal
solution (25), $\mathcal{F}^+ = \mathcal{F}^{-1}$, suggests a convenient representation for $\mathcal{F}$ in
the form

$$\mathcal{F}(\xi) = \exp\left[i\, G_0(\xi) + i\, \vec{\tau} \cdot \vec{G}(\xi) \right]$$

with real functions $\vec{G}(\xi)$ and $G_0(\xi)$ that obey $\vec{G}(0) = 0$, $G_0(0) = 0$.
Substitution into (9), with the aid of the relation

$$\frac{d}{d\xi}\, e^{Q(\xi)} = \int_0^1 d\lambda\; e^{\lambda Q(\xi)} \cdot \frac{dQ}{d\xi} \cdot e^{(1-\lambda)Q(\xi)}$$

immediately shows that $G_0(\xi) = 0$; and that $\vec{G}(\xi)$ must satisfy

$$\vec{\tau} \cdot \vec{\pi}(\xi) = \int_0^1 d\lambda\; e^{i\lambda \vec{\tau} \cdot \vec{G}}\; \vec{\tau} \cdot \frac{d\vec{G}}{d\xi}\; e^{-i\lambda \vec{\tau} \cdot \vec{G}} \quad ,$$

or

$$\vec{\pi} = \frac{d\vec{G}}{d\xi} - \left(\hat{G} \times \frac{d\vec{G}}{d\xi} \right) \cdot \frac{1}{2G} \left[1 - \cos(2G) \right]$$

$$\qquad - \hat{G} \times \left(\frac{d\vec{G}}{d\xi} \times \hat{G} \right) \left[1 - \frac{\sin(2G)}{2G} \right] \quad ,$$

(26)

$$- 164 -$$

where $\hat{G} = \vec{G}/G$ and $G = +\sqrt{\vec{G}^2}$. This exact differential equation is of course most difficult to solve; but it has a weak- and a strong-field limit which are curiously similar.

The weak-field limit, $G \ll 1$, must satisfy the equation

$$\vec{\pi} \simeq d\vec{G}/d\xi \quad ,$$

with solution

$$\vec{G}(\xi) \simeq \int_0^\xi d\xi' \; \vec{\pi}(\xi') \; . \tag{27}$$

It can be considered a weak-coupling solution in the sense that $\vec{\pi}$ is proportional to g, which will then enter "weakly" into all expressions built out of (27). The awkwardness of such a model lied here, for one is never sure of how much significance may be attached to the retention of all powers of g^2 in subsequent quantities, when one in fact defined the eikonal by a perturbation expansion in g.

The strong coupling limit of (26) follows from the assumption $G \gg 1$,

$$\vec{\pi} \simeq \frac{d\vec{G}}{d\xi} - \hat{G} \times \left(\frac{d\vec{G}}{d\xi} \times \hat{G} \right) = \hat{G} \left(\hat{G} \cdot \frac{d\vec{G}}{d\xi} \right) \; , \tag{28}$$

from which one immediately obtains the solution

$$\hat{G}(\xi) = \hat{\pi}(\xi), \quad G(\xi) = \int_0^\xi d\xi' \; \pi(\xi') \; . \tag{29}$$

The difference in these two cases is that, effectively, in (29) the direction of the vector $\vec{\pi}$ has been decoupled from its magnitude, as given by the weak-coupling solution (27). Presumably, the exact solution to (26) corresponds to an intermediate situation.

Because $\pi \sim |\vec{a}|$, (29) may be considered as a strong-coupling limit. As such, there is no question of the validity of perturbation expansions here, if only the necessary functional operations of (7) can be performed upon the functions constructed with (29). (There is, of course, no justification - other than simplicity - for the neglect of closed-nucleon loop structure, which approximation was used in reaching (7)). In fact, the functional operations of (7) will themselves require an approximation, but one that is at least intuitively reasonable for strongly coupled fields.

This strong-field solution has the form

$$\mathcal{F}(\varsigma) = \exp\left[i\, \vec{\tau} \cdot \hat{\vec{\pi}}(\varsigma) \int_0^\varsigma d\varsigma'\, \pi(\varsigma') \right] ,$$

and one must now decide what value to assign to the quantity

$$\hat{\vec{\pi}}(\varsigma) = p \cdot \vec{A}(z - \varsigma p)\left[\left(p \cdot \vec{A}(z - \varsigma p) \right)^2 \right]^{-\frac{1}{2}}$$

as $\varsigma \to \infty$, for this is the form needed in the mass-shell amputated form of (7). A suggestion for an answer follows from the correlation between the limit $\varsigma \to \infty$ in the different $\mathcal{F}(\infty ; \pm p)$, and the asymptotic procedure of specifying the mass-shell properties of a particular particle. Imagine, in (7), that the incoming particle p_1 represents a proton. The limit $\varsigma \to \infty$ in $\mathcal{F}(\varsigma; p_1)$ may be looked upon as the statement that, in principle, there was an infinite amount of time available to measure the 4-momentum p_1 with perfect accuracy. But if that incident particle is prepared as a proton, its Green's function during that time of preparation can only emit or absorb neutral mesons; otherwise, it would be able to change its charge, and no longer be a certified proton . That is, any measurement of that initial state by an external electromagnetic field during the course of its preparation, must yield a charge +1; and hence during that time, as $\xi \rightarrow \infty$, only neutral meson emission or absorption should be permitted. This property is guaranteed if $\hat{\pi}(\infty)$ is chosen to point in the $\pm \hat{e}_3$ direction. A similar argument suggests that the corresponding $\hat{\pi}(\infty)$ for a neutron should be $\mp \hat{e}_3$, for this double choice will then satisfy charge independence (of the forces generated by the exchange of neutral mesons between either proton or neutron and an external nucleonic "testing field") during those long asymptotic times of preparation. In fact, as is easily seen upon taking subsequent isotopic projections into states of $I = 0,1$, the restriction $\hat{\pi}_p(\infty) = -\hat{\pi}_n(\infty)$ guarantees the necessary degeneracy of the triplet state.

If, for definiteness, one chooses $\hat{\pi}_p(\infty) = -\hat{\pi}_n(\infty) = -\hat{e}_3$, upon appropriate projection of n or p states in (7), one will always find similar (negative) phase factors for both asymptotic proton and neutron Green's functions, simply because the nucleon projection operators have the property $\frac{1}{2}(1 \pm \tau_3) F\{\pm \tau_3\} = \frac{1}{2}(1 \pm \tau_3) F\{1\}$. The complete phase factors for each nucleon may then be written as $\exp\left[-i \int d^4 u \, \pi'(u) \, f(u)\right]$, where

$$f(u) = |g| \int_0^\infty d\xi \left\{ \delta(u - z + \xi p) + \delta(u - z - \xi p') \right\} ,$$

or

$$f_{1,2}(u) \rightarrow |g| \int_{-\infty}^{+\infty} d\xi \; \delta(u - z_{1,2} + \xi \, p_{1,2}) \;, \tag{30}$$

initially neglecting, as is usual in small-angle eikonal models, the q-
dependence of all factors other than the phase explicitly exhibited in
(2). One then obtains

$$\frac{\partial T_{EIK}}{\partial g^2} = -\left(\frac{s}{2m^2}\right) \int d^4z \; e^{i q \cdot z} \, \Delta_c(z) \cdot \langle N_1' N_2' | \; \vec{\tau}_1 \cdot \vec{\tau}_2 \, | N_1 N_2 \rangle \cdot$$

$$\exp\left[-i \int \frac{\delta}{\delta A_1} \Delta_c \frac{\delta}{\delta A_2}\right] \cdot \exp\left[-i \int f_1 \pi_1, -i \int f_2 \pi_2\right]\Big|_{A_{1,2} \rightarrow 0}, \tag{31}$$

with $\pi(u) = \left[\sum_{\alpha=1}^{3} \left(\sum_\mu p_\mu A_\mu^\alpha(u)\right)^2\right]^{\frac{1}{2}}$. An obvious notation has been used
for the isotopic matrix elements, which here have the form expected in
a Born Approximation.

The next question one must face is the evaluation of the functional operations of (31), a decidedly non-trivial matter because of the
appearance of the magnitudes, rather than the components, of the isotopic
fields A_μ^α . Another complication is that it is not clear, using the
form (31), just how one may comply with the instructions A $\rightarrow$ 0, at the
end of the computation, since this approximation has been defined for
large field strengths, or large couplings.

However, there exists an alternate and formally equivalent procedure, defined by a functional integration representation for the differential operator of (31),

$$\exp\left[-i\int \frac{\delta}{\delta A_{1\mu}^{\alpha}}\,\Delta_c\,\frac{\delta}{\delta A_{2\mu}^{\alpha}}\right] = C^{-1}\cdot\exp\left[-\text{Tr}\ln\Delta_c\right]\cdot$$

$$\cdot\int d[\varphi]\int d[\psi]\,\exp\left[-i\sum_{\alpha,\mu}\int\varphi_{\mu}^{\alpha}K\,\psi_{\mu}^{\alpha}+\int\varphi_{\mu}^{\alpha}\frac{\delta}{\delta A_{1\mu}^{\alpha}}+\int\psi_{\mu}^{\alpha}\frac{\delta}{\delta A_{2\mu}^{\nu}}\right], \tag{32}$$

where $K\,\Delta_c = 1$, and $C = (2\pi)^N$ is a typical functional integration constant for N degrees of freedom (as $N\to\infty$), which will cancel out of the
final result. Here, $\int d[\varphi] = \prod_{\ell=1}^{N}\int_{-\infty}^{+\infty}d\tilde{\varphi}_{\ell}$, where $\tilde{\varphi}_{\ell}\equiv\tilde{\varphi}(\ell_{\ell})$ is the $\ell^{\underline{th}}$ Fourier
mode of $\varphi(x)$, with isotopic and 4-momentum coordinates suppressed.

With the representation (32), one may write in place of (31),

$$\frac{\partial T_{EIK}}{\partial g_v^2} = -\left(\frac{s}{2m_2}\right)\cdot C^{-1}\cdot\exp\left[-\text{Tr}\ln\Delta_c\right]\cdot\langle N_1' N_2'\,|\,\vec{\tau}_1\cdot\vec{\tau}_2\,|\,N_1 N_2\rangle\cdot$$

$$\cdot\int d[\varphi]\int d[\psi]\,\exp\left[-i\int\varphi K\psi -i\int f_1\,\pi_1[\varphi] -i\int f_2\,\pi_2[\psi]\right], \tag{33}$$

where $\pi_{1,2}[x] \equiv \left[\sum_{\alpha}\left(\sum_{\mu}P_{1,2}^{\mu}\,\chi_{\mu}^{\alpha}{}_{(\mu)}\right)^2\right]^{\frac{1}{2}}$; and one may now look for alternate
methods of evaluation which maintain large values of $\pi_1[\varphi]$ and $\pi_2[\psi]$
The method that comes to mind immediately, in this strong-field approximation, is that of stationary phase. Here, one imagines that the coupling is
so strong, and the fields in question so large, that virtual processes
must contain very large numbers of quanta; and that a simpler way of pic-
turing the effects of so many quanta is obtained by replacing them by a
"semi-classical" field, here defined in some self-consistent, non-linear
way. It is understood that, by the imposition of such an "averaging" approximation, some of the fine details of the theory may be lost; but the intuitive hope remains that the important qualitative features of the strong-
field limit will be preserved.

The stationary phase approximation now adopted to evaluate (33)
treats the φ and ψ coordinates as independent variables; that is,
one considers the functional integral

$$\int d[\varphi] \int d[\psi] \ \exp f[\varphi, \psi] \tag{34}$$

and expands $f[\varphi, \psi]$ about non-zero (and large) φ_0 , ψ_0 values, which are determined by the simultaneous conditions $\left.\frac{\delta f}{\delta \varphi}\right|_0 = \left.\frac{\delta f}{\delta \psi}\right|_0 = 0$. One retains in $f[\varphi, \psi]$ only quadratic dependence in the variables $\varphi - \varphi_0$ and $\psi - \psi_0$, so that (34) is replaced by

$$\exp f[\varphi_0, \psi_0] \cdot \int d[\bar\varphi] \int d[\bar\psi] \cdot \exp\left[\tfrac{1}{2}\int \bar\varphi \ \frac{\delta^2 f}{\delta\varphi_0 \delta\varphi_0} \ \bar\varphi + \tfrac{1}{2}\int \bar\psi \ \frac{\delta^2 f}{\delta\psi_0 \delta\psi_0} \ \bar\psi + \right.$$
$$\left. + \int \bar\varphi \ \frac{\delta^2 f}{\delta\varphi_0 \delta\psi_0} \ \bar\psi \ \right] \tag{35}$$

In the variables $\bar\varphi = \varphi - \varphi_0, \ \bar\psi = \psi - \psi_0,$ (35) now corresponds to a pair of Gaussian functional integrals, and can be immediately evaluated in terms of the functions φ_0 , ψ_0 , or more properly $\varphi_{0\mu}^{\alpha}$, $\psi_{0\mu}^{\alpha}$. For notational ease, the subscript 0 is henceforth omitted.

The Euler equations for φ_μ^{α} , ψ_μ^{α} obtained in this way are

$$K \ \psi_\mu^{\alpha}(x) = - \ P_{1\mu} \ f_1(x) \ \frac{\left(P_{1\nu} \ \varphi_\nu^{\alpha}(x)\right)}{\pi_1[\varphi(x)]} \quad,$$

and

$$K \ \varphi_\mu^{\alpha}(x) = - \ P_{2\mu} \ f_2(x) \ \frac{\left(P_{2\nu} \ \psi_\nu^{\alpha}(x)\right)}{\pi_2[\psi(x)]} \quad.$$

Solutions have been found - self-consistent for the triplet state, but (unfortunately) not for the singlet - by two different methods. In each case, the unit vectors of φ_μ^{α} , ψ_μ^{α} in iso-space have been chosen (for simplicity) as constants, after which the remainder of the analysis is straight-forward.

If one also includes isoscalar exchanges, the isotopic projections into states of $I = 0,1$ of $\partial T / \partial g_v^2$ and $\partial T / \partial g_s^2$ can be shown to satisfy a restrictive integrability condition. The result of this analysis produces

$$T_{Eik}^{(I)}(s,t) = i \frac{s}{2} \int d^2b \, e^{i\vec{q}\cdot\vec{b}} \left[1 - e^{i\chi_1^{(s)} + i \langle \vec{\tau}_1 \cdot \vec{\tau}_2 \rangle \chi_1^{(v)}} \right] \cdot e^{Q^{(I)}(b,\mu,g,s)}$$

$$, \qquad (36)$$

where $\chi_1^{(s)}$ is the typical eikonal function of isoscalar exchange, $\chi_1^{(v)}$ is the <u>same</u> function with coupling and mass parameters of the isovector exchange, and $\langle \vec{\tau}_1 \cdot \vec{\tau}_2 \rangle = 2\, I\,(I + 1) - 3$. The quantities $Q(I)$ are functionals of the solutions φ_μ^α, ψ_μ^α, defined as

$$Q = -\frac{1}{2} \operatorname{Tr} \ell n \left[1 - S \Delta_c T \Delta_c \right] ,$$

with

$$S_{\mu\nu}^{\alpha\beta}(u) = \left(\frac{P_{1\mu} P_{1\nu}}{P_1 \cdot P_2} \right) \frac{f_1(u)}{\varphi(u)} \left[\delta_{\alpha\beta} - \hat{\chi}_\alpha \hat{\chi}_\beta \right] , \qquad \varphi_\mu^\alpha = P_2^\mu x^\alpha \varphi ,$$

and

$$T_{\mu\nu}^{\alpha\beta}(u) = \left(\frac{P_{2\mu} P_{2\nu}}{P_1 \cdot P_2} \right) \frac{f_2(u)}{\psi(u)} \left[\delta_{\alpha\beta} - \hat{\rho}_\alpha \hat{\rho}_\beta \right] , \qquad \psi_\mu^\alpha = P_1^\mu \rho^\alpha \psi .$$

Very crude estimates, especially for the isosinglet amplitude, suggest that $Q^{(1)} \sim -\ell n \left[1 + \phi^2 \right]$, $Q^{(0)} \sim -\frac{1}{2} \ell n \left[1 + (4\phi)^2 \right]$, where $\phi \sim \frac{1}{\pi} \ell n\,(bm)$, $1 > \frac{g}{\sqrt{s}} > mb$ and $\phi \sim g / mb \sqrt{2\pi s}$, $1 > mb > g/\sqrt{s}$. One may expect better estimates of the Q to show some fine details missed by the rough arguments of this analysis.

Perhaps the most novel part of this construction arises in the evaluation of the $Q^{(I)}$. Although crude, these estimates point to the first situation forbidding the ordinary eikonal approximation of neglecting q-dependence everywhere except in the $\exp\left[i\vec{q}.\vec{b}\right]$ phase of (36). The iterates involved in the construction of the Q are extraordinarily sensitive to small q, and the approximation of (30) cannot be retained; this momentum transfer, and in particular the combination $q/\sqrt{s}$, must be held fixed and non-zero, until the end of the calculation.

An analogous computation of wide-angle effects, involving the multiple exchange of SU(2) isovector mesons in a strong-coupling limit, has also been carried through in a similarly crude way. The results indicate that, in part because of relative cancellations, and in part because ρ_s couple less strongly to nucleons than do ω_s , form factor damping due to $\rho_{\pm},\rho_0$ eschange, as well as the contributions to large p_t scattering and production amplitudes, may be neglected as a first approximation in the construction of massive-gluon bremsstrahlung models. One assumes, of course, that the strong-coupling limits and approximations used are actually applicable to ρ- exchange.

Finally, there are some interesting speculations that follow from this analysis. As m decreases, at fixed q and s, the damping of the exp(Q) factor can be become appreciable. In the limit as $m \rightarrow 0$, one must consider suitably inclusive cross sections, combining scattering with soft meson emission; and because of existing theorems [11], one expects that all logarithmic mass dependence in every order of perturbation theory will cancel. In the present case of strongly coupled fields, where the limit of large coupling is taken before m vanishes, it would be interesting to see if one can have an SU(2) version of the Yang-Mills behavior suggested in reference 12; The graphical origins of that effect, if it exists, would be quite different in each case; but it would be interesting to find an explicit case of damping by logarithmic mass dependence in a strongly coupled, non-Abelian theory.

If the strong-coupling analysis could be extended to SU(3), and the self-interactions of Yang-Mills included along with gluon exchange in a generalization of the stationary phase procedure, one would have a rough model for quark-quark scattering. It would then be most interesting to see if the Cornwall-Tiktopoulos damping occurs, as $m \rightarrow 0$. For $q\bar{q}$ scattering, one would expect a generalization of the Lévy-Sucher calculation of the bound-states of positronium, that could well be useful in QCD estimates; and one might even look forward to analogous estimates of the Yang-Mills binding of three quarks. Once the general framework exists, many possibilities are opened up; the present calculation sets the stage for the simplest such non-Abelian eikonal calculation.

References

(+)   H.MORENO and H.M.FRIED, Nice Preprint NTH 77/5.

(++)   H.M.FRIED, Nice Preprint NTH 77/3

(1)   H.CHANG and T.T.WU, Phys. Rev. Letters $\underline{24}$, 1455 (1970);
         S-J CHANG and T-M YAN, Phys. Rev. $\underline{D4}$, 537 (1971).

(2)   R.BLANKENBECLER and H.M.FRIED, Phys. Rev. $\underline{D8}$, 678 (1973).

(3)   D.AMATI, M.LE BELLAC, G.MARCHESINI and M.CIAFALONI, Nucl. Phys.
         $\underline{B112}$ (1976) 107.

(4)   M. LEVY and J.SUCHER, Phys. Rev. $\underline{D2}$, 1716 (1970).

(5)   S.WEINBERG, Phys. Rev. $\underline{D2}$, 674 (1970) and 3085 (1970);
         L.S.BROWN, Phys. Rev. $\underline{D2}$, 3083 (1970).

(6)   H.M.FRIED, Phys. Rev. $\underline{D2}$, 3035 (1970); C.E.CARLSON and T.L.NEFF,
         Phys. Rev. $\underline{D4}$, 532 (1971).

(7)   H.M.FRIED, T.GAISSER and B.KIRBY, Phys. Rev. $\underline{D8}$, 3210 (1973);
         H.M.FRIED, Phys. Letters, $\underline{51B}$, 90 (1974). For studies of corre-
         lations, see A.P.CONTOGOURIS and D.SCHIFF, Phys. Rev. $\underline{D13}$, 1914 (1976).

(8)   C.T.SACHRAJDA and R.BLANKENBECLER, Phys. Rev. $\underline{D12}$, 1754 (1975);
         L.N.CHANG and G.SEGRE, Phys. Rev. $\underline{D6}$, 2231 (1972);
         R.SUGAR, Phys. Rev. $\underline{D9}$, 2474 (1974).

(9)   Rather extensive discussions and applications of the fundamental
         Bloch-Nordsieck approximation may be found in : H.M.FRIED, "Functio-
         nal Methods and Models in Quantum Field Theory", M.I.T. Press,
         Cambridge (1972).

(10)   H.MORENO and H.M.FRIED, Phys. Rev. $\underline{D12}$, 2031 (1975).

(11)   T.KINOSHITA, J. Math. Phys. $\underline{3}$, 650 (1962); T.D.LEE and M.NAVENBERG,
         Phys. Rev. $\underline{133}$, B 1549 (1964).

(12)   J.M.CORNWALL and G.TIKTOPOULOS, Phys. Rev. $\underline{D13}$, 3370 (1976).

The Static Potential Energy of a Heavy Quark and Anti-Quark

K. Johnson

Center for Theoretical Physics
Laboratory for Nuclear Science and Department of Physics
Massachusetts Institute of Technology Cambridge, Massachusetts 02139

We have heard quite a bit about "linear" quark confining potentials in the past few years. This has been
stimulated by the great phenomenological success[1] for
charmonium spectroscopy obtained using a model based upon
heavy quarks which move with such a potential energy. It
is believed by many that the likely place to find a potential of this type is in the currently most fashionable
field theory, quantum chromodynamics. This is because a
potential which is proportional to the distance between
field sources is familiar in ordinary electrodynamics
when the electric field lines which connect the sources
are lined up parallel to each other, as for example, when
the sources are opposite charges located on parallel con-
ducting plates.

Of course, nobody thinks that the quark and antiquark
colors in charmonium are located on parallel plates, the
belief (or hope) is that non-linear effects associated
with the large color charge in QCD will lead to a focusing
of the color fields into a tube of parallel "chromoelec-
tric" flux. Now, QCD is a complicated theory so no one
actually knows whether or not this happens. One might
ask if there is a local way of describing an effect which
will cause the color electric field to become focused
short of simply hoping that some complicated non-linear
consequences of the QCD interaction achieve it. Suppose
there is a region of space which contains a chromo-static
field with energy density $\frac{1}{2}E^2$. The Maxwell stress corresponds to a tension per unit area ($\leftrightarrow$) along the lines
of flux equal to $\frac{1}{2}E^2$, and the same pressure ($\updownarrow$) between
the lines. If the field falls to zero at some surface, as
it must do if the lines are in a flux tube, then at the
surface the outward pressure of the lines, $\frac{1}{2}E^2$, must be
balanced by whatever effect causes them to focus. Let us
<u>assume</u> that color fields can only be supported in the va-
cuum when they have a strength which exceeds a critical
value, i.e.,

$$\frac{1}{2}E^2 \geq B$$

where B measures the critical strength. Since this refers
to a property of the vacuum, it must be a Lorentz invariant
condition, i.e.

$$-\tfrac{1}{4}F^{\mu\nu}F_{\mu\nu} \geq B \tag{1}$$

(or the non-Abelian fields $-\tfrac{1}{4}[F_a^{\mu\nu}F_{\mu\nu}^a] \geq B$.) In this case, at the edge of the flux tube, the pressure in the field lines will be equal to B, and hence to balance the Maxwell stress there must be a counter pressure equal to B. Since this is a property of the vacuum it will correspond to a stress tensor in the region exterior to the color fields equal to

$$g^{\mu\nu}B$$

where $g^{\mu\nu}$ is the Lorentz metric. This is equivalent to a stress associated with the interior region equal to $-g^{\mu\nu}B$, since a constant stress through space can have no observable consequence. The presence of such a stress is the basic postulate of the M.I.T. Bag Model.[2] We call B the "bag constant". Now we can see that a linear flux tube containing a constant field, E, can be in equilibrium. For with the addition of $-g^{\mu\nu}B$, there is a net tension per unit area along the field lines equal to $\tfrac{1}{2}E^2+B=2B$, (to be balanced at the sources of the lines) and a pressure between the lines equal to $+\tfrac{1}{2}E^2-B=0$. There is an energy density, $\tfrac{1}{2}E^2+B=2B$, equal to the tension per unit area. If the flux tube spans a cross section A, the net tension, T, equals 2BA. By Gausses law, $EA=e=A\sqrt{2B}$, so $T=e\sqrt{2B}$. e measures the magnitude of the color charge on which the chromo-static flux begins and terminates. In the case of non-Abelian QCD when, for example, the flux tube connects a color 3 to $\bar{3}$, $e^2=4/3\,e_s^2$. Here e_s is the rationalized unit of charge, so $e_s^2/4\pi=\alpha_s$ is the strong coupling constant.[3] Thus, the tension is $T=\sqrt{\alpha_s}4\sqrt{2\pi/3}\sqrt{B}$. An ordinary meson corresponds to a relativistically spinning flux tube,[4] and the slope parameter of the Regge mass spectrum α' is related to T by $1/\alpha'=2\pi T=8\pi\sqrt{2\pi/3}\sqrt{\alpha_s}\sqrt{B}=36.372\sqrt{\alpha_s}\sqrt{B}$. The observed slope α' is .9 GeV^{-2} so we find $\sqrt{\alpha_s}\sqrt{B} \simeq [.175$ GeV$]^2$. $B^{\tfrac{1}{4}}$ which has the dimensions of mass sets the scale of masses for ordinary spectroscopy of the hadrons and was determined to be $B^{\tfrac{1}{4}}\sim.145$ GeV,[5] so we find $\alpha_s\sim2.1$. That is, the strong interaction coupling parameter is about 100 times larger than the electromagnetic parameter.

Now the flux tube in low mass mesons terminates on light quarks which are kept separated by an angular momentum barrier. Imagine instead that it connects heavy, slowly moving quarks and acts as a static potential.[6] This potential will be proportional to the distance between the quarks, with a slope equal to $T=1/2\pi\alpha'=.18$ GeV^2, which is just about equal to the slope parameter determined from charomonium phenomenology.[1] However, we would seem to be unjustified to simply assume that the flux tube forms so quickly as the quark and antiquark begin to separate. We might only expect a potential of that sort asymptotically as the distance between the quarks gets large. The relevant length scale should be the radius of the flux tube, i.e. $R=\sqrt{A/\pi}=(e/\sqrt{2B}\ 1/\pi)^{\frac{1}{2}}=\sqrt{\alpha}'\sqrt{\alpha}_s(32\pi/3)^{\frac{1}{2}}\sim 8$ GeV^{-1}, that is, about 1.5 fermi. However, in the low lying states of the charmonium system, the quarks are separated by a distance less than a fermi.

We would like to show that the inequality (1) leads to the effective interaction developing the linear part as soon as the charges begin to separate.[6] For convenience, we shall carry out the discussion in the language of an Abelian analogue to quantum chromodynamics. The results are equivalent to the color theory with the replacement $e^2=e_s^2\ \frac{4}{3}$, since we will only require the lowest order of QCD perturbation theory to obtain our effective potential.

Then the analogue fields[7] can be obtained from the variational principle $\delta W=0$, with

$$W = \int (d^3x)(\tfrac{1}{2}E^2-B)\,\theta(\tfrac{1}{2}E^2-B) \tag{2}$$

subject to the constraint

$$\nabla\cdot E = e(\delta^{(3)}(\vec{x}-\tfrac{\vec{r}}{2}) - \delta^{(3)}(\vec{x}+\tfrac{\vec{r}}{2})) \tag{3}$$

so that the quarks, located at the positions $\pm\tfrac{\vec{r}}{2}$, are the sources of the "analogue" color electric field. We have inserted into (2), $\theta(x)=\begin{cases} 1 & x>0 \\ 0 & x<0 \end{cases}$ to enforce the constraint $\tfrac{1}{2}E^2>B$. For consistency, we have added the term $-B$ to $\tfrac{1}{2}E^2$ to make the integrand a continuous function of the field strength. We shall see that this simply corresponds to the addition of a uniform pressure exterior to the domain containing the fields. We have noted already the necessity of this term. If we require that (2) be stationary on arbitrary variation of E, we find,

$$\vec{E}\theta = -\nabla\phi \tag{4}$$

where ϕ is a Lagrange multiplier introduced to maintain the constraint (3) on $\vec{\nabla}\cdot E$. Then from (3) we obtain

$$-\nabla^2\phi=\vec{\nabla}\cdot(\vec{E}\theta)=\vec{\nabla}\cdot\vec{E}+\vec{E}\cdot\nabla(\theta)=e(\delta^{(3)}(\vec{x}-\tfrac{\vec{r}}{2})-\delta^{(3)}(\vec{x}+\tfrac{\vec{r}}{2}))+\hat{n}\cdot\vec{E}\delta(\tfrac{1}{2}E^2-B)\lambda$$

$$\tag{5}$$

where $\hat{n}$ is normal to the surface whose equation is $\frac{1}{2}E^2-B=0$, and beyond which the fields must vanish. That is $\nabla(\frac{1}{2}E^2-B)=\hat{n}\lambda$ on the surface. We see ϕ, where

$$-\nabla^2\phi = e(\delta^{(3)}(\vec{x}-\frac{\vec{r}}{2})-\delta^{(3)}(\vec{x}+\frac{\vec{r}}{2})) \tag{6}$$

with $\hat{n}\cdot\nabla\phi=0$ on the surface, gives a unique solution to the problem. As already noted,

$$\frac{1}{2}E^2 =\frac{1}{2}(\nabla\phi)^2 = B \tag{7}$$

is the equation for the surface which encloses the domain where the fields are non-vanishing. (This is the "bag" of the M.I.T. Bag Model.) We see that (7) corresponds to locally balancing the pressure of the static field by the bag on the surface confining pressure B. It is also possible to show with a little computation that the surface which encloses the fields and on which the potential obeys (7) can be gotten from an energy variational principle. Consider,

$$U = \int_{Bag} d^3x \ (\frac{1}{2}(\nabla\phi)^2+B) \tag{8}$$

where

$$-\nabla^2\phi = e(\delta^{(3)}(\vec{x}-\frac{\vec{r}}{2}-\delta^{(3)}(\vec{x}+\frac{\vec{r}}{2})) \tag{9}$$

and where

$$\hat{n}\cdot\nabla\phi = 0 \tag{10}$$

on the surface which encloses the "bag", the region of space included in the integral (8). Since there is a unique ϕ associated with each surface, (8) may be regarded as a surface functional. On the surface which minimizes (8),

$$\frac{1}{2}(\nabla\phi)^2 = B,$$

that is, we have the solution for the boundary condition (7). If we evaluate U for this case we obtain the desired static potential, $U_o(r)$. We shall show that $U_o(r)$ is very accurately (but not exactly) represented by the function

$$U_o(r) = -\frac{e^2}{4\pi r} + r\sqrt{2Be^2}. \tag{11}$$

That is, the effect of imposing the constraint

$$\frac{1}{2}E^2 \geq B$$

is simply to add to the Coulomb chromo-static energy a linear function of r. We have already indicated how this form obtains in the limit r→∞, when the flux arranges itself in a tube. But we shall now see that (11) works over the entire range.

Suppose the surface which minimizes (8) is given by the equation

$$|\vec{x}| = R(\hat{x};r) \tag{12}$$

when the quarks are separated by the distance r. Further let the field potential be given by the function

$$\phi_o(\vec{x};r), \tag{13}$$

so

$$-\nabla^2\phi_o = e(\delta^{(3)}(\vec{x}-\tfrac{\vec{r}}{2})-\delta^{(3)}(\vec{x}+\tfrac{\vec{r}}{2}))$$

and $\hat{n}\cdot\nabla\phi_o$=0 on the surface of (12). Now consider the scaled surface,

$$|\vec{x}| = \tfrac{1}{\lambda} R(\hat{x};\lambda r). \tag{14}$$

If we use (14), to calculate (8) for any $\lambda\neq 1$ it will give a larger energy than U_o. It is easy to check that the potential appropriate to (14) is,

$$\phi_\lambda(x;r) = \lambda\phi_o(\lambda\vec{x};\lambda r).$$

That is, ϕ_λ obeys (9), and (10) on the surface (14). If we evaluate (8) using the scaled surface, we find

$$U_\lambda = \lambda S(\lambda r) + \tfrac{1}{\lambda^3} BV(\lambda r)$$

where

$$S(r) = \int_{B_o} d^3\vec{x}\ \tfrac{1}{2}(\nabla\phi_o)^2$$

and

$$V(r) = \int_{B_o} d^3\vec{x}.$$

B_o is the region of space enclosed by the surface (12). That is, S and BV are the field and volume energies corresponding to the minimum energy surface. Since $\partial U/\partial\lambda$=0 at λ=1, we find

$$0 = \frac{\partial}{\partial r} (r[S(r)+BV(r)])-4BV(r) \tag{15}$$

and thus by integration of (15)

$$U_o(r) = S(r) + BV(r) = -e^2/4\pi r + 4B\frac{1}{r}\int_o^r dr' V(r') \qquad (16)$$

The constant of integration $-e^2/4\pi$ is obtained from the requirement that the interaction be Coulombic as $r \to 0$. (The divergent self energies have been subtracted from S.)

The second term in (16) represents the consequence of the condition $\frac{1}{2}E^2 > B$. As $r \to \infty$, we have already noted that $V(r) \to Ar = e/\sqrt{2B}\ r$, and (16) of course is consistent with the linear flux tube since $U_o(r) \to e\sqrt{2B}\ r$. However, $V(r)$ is also linear as $r \to 0$. We can see why intuitively. As $r \to 0$ we expect that the field is approximately of the dipole form, so on the surface

$$E \simeq \frac{re}{R^3 4\pi}$$

where R is the distance to the surface. By (7), on the surface

$$\sqrt{2B} = E \simeq \frac{r}{R^3}\ e\ \frac{1}{4\pi}\ ,$$

and so we find the volume $\sim 4\pi/3\ R^3$ is of order $r\ e/\sqrt{2B}\ \frac{1}{3}$. Thus, $V(r)$ is also a linear function near $r=0$, with perhaps a different numerical coefficient than that given by our guess of the dipole field. We have computed $U(r)$ + $e^2/4\pi r$ for a sphere and a cylinder as trial surfaces. In both cases we use the exactly calculable potential which is appropriate. We obtain the best sphere by minimizing (7) with respect to the radius, and the best cylinder by minimizing with respect to length and diameter. For the sphere $U(r) + e^2/4\pi r \to \sqrt{2/3}\sqrt{2B}e^2 r$ as $r \to 0$, ($\sqrt{2/3} = .816$) and for the cylinder, $\to (.840)\sqrt{2B}e^2 r$. Thus we anticipate that $V(r)$ in (16) is of the form $V(r) = e/\sqrt{2B}\ rk(r)$ where as $r \to \infty$, $k(r) \to 1$ and where as $r \to 0$, $k(r) \sim .8$. Thus the volume energy is almost exactly linear and hence we can understand the accuracy of the form (8). These simple considerations are supported by an exact calculation which can be done for the equivalent problem in two space dimensions,[8] and by detailed computer calculations in three space dimensions.[6] We may remark that a sphere doesn't look very much like a linear tube of flux. In the case of the cylindrical approximation near $r=0$, the cylinder is shorter than it is thick ($L/D \sim .74$ near $r=0$).

Now what should we make of this? We have found in the case of heavy quarks where a static potential makes

sense, that its form will be quite accurately linear plus
Coulombic if the particles interact through a gauge field
which obeys the restriction $\frac{1}{2}E^2 > B$. Thus, B determines
a critical field strength. The M.I.T. Bag Model is closely
related to a model studied some time ago by Dirac.[9] The
term in the Lagrangian which makes the fields "lump" into a
finite volume around the static charges corresponds to a
term in the action equal to

$$-B \int_{Bag} d^4x \qquad (17)$$

where the integral (17) extends over the tube in space-
time which is swept out by the volume occupied by the
fields belonging to the extended particle. The action
(17) is the three dimensional version of an action for an
extended particle considered some time ago by Dirac, who
associated a geometrical action with the surface of an extended particle. In Dirac's case, the geometrical term
has the form,

$$-\sigma \int dt \int d^2 S \sqrt{1-V_T^2} \qquad (18)$$

where V_T is the component of velocity of a point on the
surface along the space normal to the surface element d^2S.
The factor $\sqrt{1-V_T^2}$ is needed to make the above action relativistically invariant. Thus, by introducing a term into
the action associated with a geometrical extension for the
particle, in the case of "surface tension" one must make
the geometrical variables dynamical. However, in (18)
there is kinetic energy associated only with motion transverse to the extension. Such geometrical actions have also
played a role in the dual string model, where the geometrical extension is one dimensional.[10]
        In the M.I.T. model for hadrons the geometrical action,
by itself, cannot have any dynamics associated with it
since there is no kinetic energy which belongs to the internal spatial points. We made the model physical by fixing on these points field operators for creating and des-
troying elementary Dirac particles, in this case, quarks.
Thus, hadrons, the particles which have this structure, are
extended blobs carrying quark fields. As we originally con-
ceived our model[2] we assumed that hadrons were extended
"particles" carrying the geometrical term (17) in their action, along with Dirac fields attached to the geometry.
However, as we have seen here, an alternative formulation
can be achieved within the context of ordinary field
theory, if we assume[11] that the color gauge fields obey the
restriction (1).[12] In this case the fields must necessarily

lump around their sources which are then forced to be in colorless combinations.

References

1. E. Eichten, K. Gottfried, K. Lane, T. Kinoshita, T.M. Yan, Cornell Preprint #375, to appear in Phys. Rev. D.
2. A. Chodos, R.L. Jaffe, K. Johnson, C.B. Thorn, V.F. Weisskopf, Phys. Rev. D$\underline{9}$, 3471 (1974).
3. We have adopted the definition of the coupling constant favored by the majority, α_s. It is related to that used in our earlier work, α_c, by $\alpha_c = \alpha_s/4$, or $e_c = \frac{1}{2}e_s$.
4. K. Johnson and C.B. Thorn, Phys. Rev. D$\underline{13}$, 1934 (1976).
5. T. DeGrand, R.L. Jaffe, K. Johnson, J. Kiskis, Phys. Rev. D$\underline{12}$, 2060 (1975).
6. P. Gnädig, P. Hasenfratz, J. Kuti, A.S. Szalay, Phys. Letters, $\underline{64B}$, 62 (1976).
7. These considerations will be treated in more detail in a subsequent publication.
8. R. Giles, M.I.T. Preprint, CTP #706 (1978).
9. P.A.M. Dirac, Proc. R. Soc. A268, 57 (1962).
10. Y. Nambu, Lectures at the Copenhagen Summer Symposium (1970). (Unpublished).
11. K. Johnson, M.I.T. Preprint, CTP #719 (1978).
12. Recently, C. Callan, R. Dashen and D. Gross have argued that the free field vacuum of QCD is unstable against a first order phase change to a state in which static weak colored electric fields are excluded and which would phenomenologically correspond to an action of the form (2). B is then related to the scaling parameter associated with the running coupling constant of Quantum Chromodynamics. C. Callan, R. Dashen, D. Gross, Institute of Advanced Study Preprint, 1978.

Semiclassical Methods in Field Theory

A. Neveu

Laboratoire de Physique Théorique de l'Ecole Normale Supérieure,
Paris, France

The initial motivation for the introduction of semiclassical methods in field theory comes from facts observed in the dynamics of hadrons. At energies of a few hundred MeV to a few GeV, they behave as though they had a rich extended or composite nature. For example, primitive bound-quark models give a surprisingly good description of static properties. But the notion of potential is incompatible with relativistic kinematics needed at those energies.

Historically, the first consistent relativistic extended system was the dual resonance model[1]. In that model, hadrons are obtained by canonical quantization of a relativistic string. Dual models enjoy great formal beauty, but seem to lack the flexibility of ordinary quantum field theory. The MIT "bag" model[2] is another example of a consistent treatment of a relativistic extended object. More flexible than dual models, it gives a good description of low-lying hadrons, but the complexity of oscillation modes of the bag makes it essentially intractable for the description of collision processes.

Because of these problems, Nielsen and Olesen [3] suggested a study of extended models in field theory. More precisely, they consider the vortex solution of the relativistic extension of the Landau-Ginsburg Lagrangian for superconducting metals as an approximation of the dual string. Semiclassical methods were introduced in field theory to investigate the quantum meaning of such solutions. Although many elegant results have been obtained by those methods, none is yet directly applicable to hadron physics, presumably because of the poor understanding of the appropriate classical field theories.

The interest of classical equations of motion in quantum mechanics is already manifest for one degree of freedom: the WKB approximation for bound states is in general excellent, even for the ground state in the extreme strong coupling limit (for numerical results in the anharmonic oscillator, see [4]). When there are many degrees of freedom, one tries to separate them. If one can, the WKB method is applied to each of them separately. If one cannot, whether for fundamental or practical reasons, the problem is more complicated, and, the knowledge of the classical system being more restricted, so will be the validity of semi-classical quantization. In practice, for such non-separable systems, one knows only a very special set of simple classical motions (rather than all the motions). It is still possible to extract useful quantum mechanical information. Examination of simple systems leads to the general heuristic feeling that the results of semiclassical quantization will be good at least when the quantum fluctuations (around the classical motion which is quantized) remain small enough that the effect of nonlinearities is small. A general quantitative statement is difficult to form-ulate. Each system has to be examined separately. For field theory, it turns out that the region of validity includes the weak coupling region. This has the advantage that one can compare with the results of ordinary perturbation theory. Of course, in field theory, one has to deal with divergences and renormalization. This is rather straightforward, and the result is that all

divergences are handled by the ordinary one-loop counterterms.
All this fits with the widespread and rather vague belief that
"WKB = trees (classical) + one loop". Actually, semiclassical
results cut across the whole perturbations expansion, picking
in each order diagrams or pieces of diagrams that cannot be identified in any simple fashion.

The approach to semiclassical quantization used in ref [5]
involves the trace of the resolvent operator:

$$G(E) = \text{tr } \frac{1}{H-E} = \sum_n \frac{1}{E_n-E} \tag{1}$$

$G(E)$ has poles for $E = E_n$, nth eigenvalue of the quantum-mechanical
hamiltonian H. We do not worry here about the convergence of the
series in (1); subtractions could be necessary, but would not
affect the location of the poles. The next step involves writing
$G(E)$ as

$$G(E) = i \text{ tr } \int_0^\infty \frac{dT}{\hbar} \exp \left[i (E-H) \frac{T}{\hbar} \right] \tag{2}$$

Here again, we do not worry about possible divergences in the T
integration. In principle, an $i\varepsilon$ provides convergence. Let us
only remark that one is not restricted to an integration along
the real T axis, but that complex values of T can be considered,
as long as they are compatible with the $i\varepsilon$ prescription. Allowing
T to be complex would correspond to including possible tunnelling
phenomena, for which we do not yet know the general formalism.
We now outline the general method for the calculation of the right-
hand side of eq.(2), referring the reader to the original papers
for the details. The strategy by which classical solutions appear
naturally involves using Feynman path integrals[7]:

$$\text{tr } e^{-iHT/\hbar} = \int \mathscr{D}x(\tau) e^{iS/\hbar} \tag{3}$$

$$S = \int_0^T [\tfrac{1}{2} \dot{x}^2 - V(x)] \, d\tau \tag{4}$$

where S is the action computed along the path $x(\tau)$. Here $x(\tau)$ is
only a generic name for all the degrees of freedom of the theory:
in the simplest case, it is just the position of a particle in a
one-dimensional potential $V(x)$, but it can have many components,
for a motion in a multidimensional potential; and for field theory

it has an infinite number of components, namely the values of the
fields at each point in space. In this case, $V(x)$ contains both
space derivatives of the field and interaction terms. Finally,
the functional integration in (3) is to be done on periodic paths
only: $x(0) = x(T)$. This condition is just the translation in
position space language of the trace operation of the left-hand-
side.

The connection with classical mechanics is now evident: the
semiclassical (= small h) approximation consists in computing
the functional integral of eq. (3) by stationary phase around
classical periodic orbits of period T.

By examining separable systems, one can see that there are in
general many periodic orbits with the same period. In such a case,
it can be shown [5] that including all of them in the stationary
phase calculation of (3) leads to the same result as first separating the variables and then quantizing each of them a la WKB.
In field theory, one cannot hope to retain all possible periodic
classical motions, but only a limited set. The larger the set,
the more information one will get on the quantum system. Here,
we will restrict the discussion to one set of period orbits with
the period being the only varying parameter; a typical example of
such a set of motions is the doublet of the sine-Gordon theory:
the classical energy varies continuously with the period, and
quantization will restrict it to discrete values.

The stationary phase calculation of eq. (3) is no different
in principle from ordinary stationary phase calculation of a simple
definite integral: shifting the integration variable to the stationary phase point and expanding the exponent to second order makes the
integration gaussian:

$$\operatorname{tr} e^{-iHT/\hbar} \simeq e^{iS_{c\ell}/\hbar} \int \mathcal{D} x(\tau) e^{i\tilde{S}/\hbar} \tag{5}$$

where $S_{c\ell}$ is the classical action around one orbit, and

$$\tilde{S} = \int_0^T [\tfrac{1}{2}\dot{x}^2 - \tfrac{1}{2}x\, V''(x_{c\ell}(\tau))x]\,dt \tag{6}$$

$x_{c\ell}$ being the classical trajectory.

The x integration is now gaussian. Performing this integration gives the inverse of the square root of the product of the
eigenvalues of the differential operator $\partial^2 + V''(x_{c\ell}(\tau))$. This eigen
product is computed in ref.[8] in terms of the stability angles
of the classical trajectory $x_{c\ell}(\tau)$. The stability angles ν_α are

defined by the solutions of

$$[\partial_\tau^2 + V''(x_{c\ell}(\tau))]\, y_\alpha(\tau) = 0 \tag{7}$$

such that

$$y_\alpha'(\tau+T) = e^{-i\nu_\alpha}\, y_\alpha(\tau) \tag{8}$$

The classical trajectory is stable if all the ν's are real. We will assume that this is the case: quantum fluctuations remain small, and do not take the system to regions of phase space far from the classical trajectory.

Zero stability angles require a special treatment: any continuous symmetry of the classical system generates a corresponding zero stability angle. In practice, for field theory, it will be space-time translation and internal symmetries. The integration over the corresponding modes is not gaussian, and formula (5) has to be modified to take this into account. We refer the reader to the literature on this delicate subject (ref.[9]).

Finally, one should take into account the fact that if a trajectory with period T is known, traversing it n times trivially defines a trajectory with period n T: one has to sum over n. After which, the approximate form of G(E) turns out to be

$$G(E) = \int_0^\infty dT \sum_{n=1}^\infty \sum_{\{q_\alpha\}} e^{\,i\frac{n}{\hbar}\{(S_{c\ell}+ET) - \sum_\alpha (q_\alpha + \frac{1}{2})\nu_\alpha \hbar\}} \tag{9}$$

$\{q_\alpha\}$ being any set of positive integers (or zero).

The remaining integration over T is also done by stationary phase. The stationary phase point is

$$E = E_{c\ell} + \hbar \sum_\alpha (q_\alpha + \tfrac{1}{2})\frac{d\nu_\alpha}{dT} \tag{10}$$

and the summation over n in eq. (9) gives poles at

$$\oint p\, dq + \hbar \sum_\alpha (q_\alpha + \tfrac{1}{2})\,(T\frac{d\nu_\alpha}{dT} - \nu_\alpha) = 2n\pi\hbar \tag{11}$$

This is the generalization of the ordinary WKB formula

to many degrees of freedom. The "main" quantum number n quantizes
motions along the classical trajectory. The integers q_α (in number
equal to the number of degrees of freedom minus one, in general)
quantize the oscillations around that classical orbit. Since
these have been treated in the linear approximation, the validity
range of (11) is large n but small q_α. In practice, in field
theory, one will be looking for bounds states, and most of the
$v_\alpha's$ correspond to travelling waves, for which q_α will be taken
equal to zero. In field theory, because of the infinite number
of degrees of freedom, the sums in (10) and (11) diverge and have
to be renormalized: it turns out that for a renormalizable theory,
(10) and (11) are made finite by simply subtracting from E the
vacuum energy in the one loop approximation and using in the functional integral (3) the Lagrangian with the ordinary one-loop counterterms included.

A time independent solution is a particular case to which
formula (10) can apply. In that case, $v_\alpha = \omega_\alpha T$, where ω_α's are
the frequencies of the small oscillations around that solution,
treated in the harmonic approximation.

Examples of these methods have been worked out in some detail
for two-dimensional model field theories where calculations could
be done analytically, and compared with exact results. Both static
and time-dependent solutions have been considered.

The first and most remarkable example is the sine-Gordon
theory, defined by the Lagrangian

$$\mathcal{L} = - \ \tfrac{1}{2}(\partial_\mu \phi)^2 + \frac{m^4}{\lambda} \ [\cos \frac{\sqrt{\lambda}}{m} \phi - 1] \tag{12}$$

In (12), the variables have been defined so that the small ϕ
expansion

$$\mathcal{L} = - \ \tfrac{1}{2}(\partial_\mu \phi)^2 - \tfrac{1}{2} m^2 \phi^2 + \frac{\lambda}{4!} \phi^4 + \dots \tag{13}$$

corresponds to a field theory of a boson of mass m with a weakly
attractive contact interaction. On the other hand, the rescaling

$$\phi \to \frac{m}{\sqrt{\lambda}} \phi \qquad , \ x \to \frac{x}{m} \tag{14}$$

brings the Lagrangian (12) under the form

$$\mathcal{L} = \frac{m^4}{\lambda} \ [- \ \tfrac{1}{2}(\partial_\mu \phi)^2 + \cos \phi - 1] \tag{15}$$

The classical behavior of the sine-Gordon theory is completely
known. See ref. [10]. From (15), we see that the dimensionless
coupling constant λ/m^2 will play in the functional integral (3)
the role of $\hbar$. Hence, the results of semiclassical calculations
overlap with the validity range of perturbation theory. This is
a completely general feature, independant of the space-time dimension and of the specific lagrangian.

From the work of ref. [10] on the classical sine-Gordon theory,
we learn that there are two types of solutions that, when quantized,
will give particles and bound states. In their rest frame, these
are the soliton:

$$\phi = 4 \text{ Arctan } e^x \tag{16}$$

and the doublet (or breather), which is a soliton-anti-soliton
bound state:

$$\phi = 4 \text{ Arctan } \frac{\varepsilon \sin[t(1+\varepsilon^2)^{-\frac{1}{2}}]}{\cosh[\varepsilon x(1+\varepsilon^2)^{-\frac{1}{2}}]} \tag{17}$$

(ε is any real positive number).

The theory being Lorentz invariant, solutions (16) and (17)
can be boosted to arbitrary velocities. By putting the system
in a periodic box, there results further periodic motions.
Quantization of these motions can be done [8]: this is actually
a case of separation of variables, with the center of mass position
being one of the variables. One then just gets the expected quantization of momentum in the box.

Application of eq. (10-11) to the solutions (16-17) is described
in ref. [8]. All calculations can be done analytically, thanks to
the fact that the classical sine-Gordon system is integrable: in
particular, one can find analytically the solutions of eq. (7-8).
The final results are: the mass of the soliton is

$$M(\text{soliton}) = \frac{8m^3}{\lambda} - \frac{m}{\pi} \tag{18}$$

The first term on the right hand side of eq. (18) is the classical
mass. The second is the contribution of small oscillations around
the static solution. It turns out in the course of the calculation
of the first quantum correction that the only non-zero stability
angles are part of the continuum: there is no soliton-meson bound
state. This is in contrast with $\lambda\phi^4$ (see below). For weak

coupling ($\lambda/m^2 \ll 1$), the soliton is a very heavy particle. It cannot decay into ordinary particles because of its topological properties. The topological conservation laws of static field theory classical solutions is discussed in ref. [11].

The doublet (17) produces the remaining series of states at masses

$$M_n = \frac{16m}{\gamma'} \sin \frac{n\gamma'}{16} \qquad n = 1,2,3,--- \quad <8\pi/\gamma' \qquad (19)$$

with $\gamma' = (\lambda/m^2)(1 - \lambda/8\pi m^2)^{-1}$

The original "elementary particle" of the theory is the $n = 1$ state in eq. (19), as can be seen in the small coupling limit $\lambda \ll m^2$. The other n>1 states can be considered as bound states (at least in weak coupling) of the elementary particle. It is interesting that one can either consider ϕ as the fundamental field, of which the soliton is a complicated collective excitation or the soliton as the fundamental object, other states, including the elementary particle, being soliton-antisoliton bound states. Indeed, Coleman has shown that the sine-Gordon theory is equivalent to the massive Thirring model, the fermion field being identified as the field of the soliton. Thanks to that equivalence, one can show that the mass ratios as given by eq. (19) are exact to all orders in λ/m^2: see ref. 12 . This is of course an accident, analogous to the non-relativistic hydrogen atom.

The other model field theory considered in ref. [8] is $\lambda\phi^4$ in the two-phase region:

$$\mathcal{L} = - \tfrac{1}{2}(\partial_\mu\phi)^2 + \tfrac{1}{2}m^2\phi^2 - \tfrac{1}{4}\lambda\phi^4 \qquad (20)$$

which, after the rescaling (14) becomes

$$\mathcal{L} = \frac{m^4}{\lambda} [- \tfrac{1}{2}(\partial_\mu\phi)^2 + \tfrac{1}{2}\phi^2 - \tfrac{1}{4}\phi^4] \qquad (21)$$

Ordinary perturbation theory of (21) involves first shifting the field to its vacuum value + 1 (it could also be -1), thus spontaneously breaking the $\phi \to -\phi$ discrete symmetry. The first state above the vacuum is a particle state, with mass $m\sqrt{2}$ in lowest order of perturbation theory. Here again, perturbation theory is made in powers of λ/m^2.

There is a static space dependent solution of (21), the

kink, analogous to the sine-Gordon soliton:

$$\phi = \tanh \frac{1}{\sqrt{2}} x \qquad (22)$$

This solution connects the two vacua ± 1. It is stable, both classically and quantum-mechanically [11]. The computation of the first quantum mechanical correction to the mass of the solution (22) is performed in ref. [13]. Contrary to the case of sine-Gordon, there is one isolated non-zero stability angle, which means that there can be bound states of a kink and an ordinary particle, labeled by an integer $q \geq 0$.

The mass of such states is

$$M_q = \frac{2}{3} \sqrt{2} \; \frac{m^3}{\lambda} + m \left(- \frac{3}{\pi\sqrt{2}} + \frac{1}{2\sqrt{6}}\right) + m \sqrt{\frac{3}{2}} \; q \qquad (23)$$

where the first term on the right hand side is the classical mass of the kink, the second the first quantum mechanical correction. The state $q = 0$ is the unexcited kink. The state $q = 1$ is a kink-meson bound state, and is stable because of energy conservation. States with $q > 1$ can decay into an unexcited kink and a meson; they would gain a width in higher order of perturbation theory.

Non-trivial time-dependant solutions of the ϕ^4 theory are not known analytically. However, one can find them in perturbation by analogy with eq. (17), in which ϵ is considered a small parameter. The strategy is to expand simultaneously in harmonics of the fundamental frequency and in powers of ϵ. This is explained in detail in ref. [8]. The result is a classical motion qualitatively analogous to the doublet (17), which, when quantized, gives a set of bound states built out of $n(n \geq 1)$ elementary particles. This solution also seems to be a kink-antikink bound state [8]. States with $n > 2$ are not expected to retain their stability in higher orders: ϕ^4 does not have all the higher conservation laws of sine-Gordon which stabilize all the states of eq. (19).

The non-linear Schrödinger equation [14] , derived from the classical Lagrangian

$$\mathcal{L} = i(\varphi^* \, \partial_t \varphi - \varphi \, \partial_t \varphi^*) - \partial_x \varphi^* \, \partial_x \varphi + \frac{1}{2}(\varphi^* \varphi)^2 \qquad (24)$$

is another completely integrable classical field theory [15] . It can be considered as the non-relativistic limit of the sine-Gordon equation. The soliton

$$\varphi = \sqrt{2\omega} \; \frac{e^{i\omega t}}{\cosh^2 x \sqrt{\omega}} \qquad (25)$$

can be quantized by the semiclassical method 16 . One obtains the binding energy of the bound state of a non-relativistic n-body system with δ-function interactions. Here again, the semiclassical answer coincides with the exact result [17] .

The introduction of fermions in semiclassical methods is delicate : the only way one can reach a classical limit (= large quantum numbers) is by the introduction of a large number of fermion species, so that there can be many fermions in the same state. A two-dimensional model of this type has been considered in ref. [18] and gives a remarkably rich spectrum of bound states. It has recently been shown to be exactly soluble and its exact S-matrix can be found in ref. [19] .

In higher space-time dimensions, all known classical solutions are static or have a simple time dependence like in eq. (25). Few quantum mechanical calculations have been done with them.

A whole class of quantum mechanical phenomena have no classical analogue : their amplitude goes exponentially to zero when h goes to zero, the classical limit. These phenomena [20] include barrier penetration by tunnelling, reflection in an attractive potential, level splitting in a potential with many minima, among others. Such phenomena nevertheless can be analyzed semiclassically. They always involve regions of phase-space where the wave function decreases exponentially, which are classically forbidden regions. In these regions, the classical trajectory either runs with a complex time parameter [21] or the trajectory itself is complex.

The extension of semiclassical methods to deal with such problems in field theory is being investigated vigorously since the discovery of imaginary time (i.e. Euclidean) solutions of four-dimensional Yang-Mills equations with finite action [22] . These solutions correspond to tunnelling between topologically inequivalent classical vacua of the theory : the classical vacua are labelled by an integer n (running from − ∞ to + ∞), exactly like the classical ground states of a one-dimensional periodic potential. Quantum mechanically, there appears a similar band structure [23] . The physical consequences of such tunnel-ling in unified theories of weak and electromagnetic interactions are being investigated [24] : it leads to the violation of certain conservation laws.

Some years ago, Bender and Wu [25] had already noticed that the essential singularity of perturbation theory at the origin in the coupling constant plane has a semiclassical interpretation. In the anharmonic oscillator, for example, reversing the sign of the quartic coupling constant leads to instability : the potential has no lower bound and the "ground state" can actually tunnel to regions where the potential is arbitrarily negative. This means that perturbation theory cannot have any finite radius of convergence. Bender and Wu [25] have shown that quantitative information on the very high order behavior of perturbation theory for the anharmonic oscillator can be extracted from this semiclassical tunnelling instability. These methods have been extended to field theory in refs. [26,27] . Besides information about the nature of perturbation theory, these methods can be applied for a practical and accurate resummation of the asymptotic series of the ε-expansion in statistical mechanics [27] .

References

[1] J.H. Schwarz, Phys. Rep. 8C (1973) 269 ;
S. Mandelstam, Phys. Rep. 13C (1974) 259 ;
C. Rebbi, Phys. Rep. 12C (1974) 1 ;
J. Scherk, Rev. Mod. Phys. 47 (1975) 123 ;
P. Frampton, Dual Resonance Models (New York, Benjamin, 1976) ;
G. Veneziano, Phys. Rep. 9C (1974) 199.

[2] A. Chodos, R.L. Jaffe, K. Johnson, C.B. Thorn and V.F. Weisskopf,
Phys. Rev. D9 (1974) 3471 ;
K. Johnson and C.B. Thorn, Proc. 17th Int. Conf. on High Energy
Physics, London 1974.

[3] H.B. Nielsen and P. Olesen, Nucl. Phys. B61 (1973) 45.

[4] J. Kilpatrick and M. Kilpatrick, J. Chem. Phys. 16 (1948) 781.

[5] R. Dashen, B. Hasslacher and A. Neveu, Phys. Rev. D10 (1974) 4114 ;
see also ref. 6 .

[6] M. Gutzwiller, J. Math. Phys. 12 (1971) 343, 11 (1970) 1791,
10 (1969) 1004, 8 (1967) 1979 ;
J.B. Keller, Ann. Phys. 4 (1958) 180 ;
V. Maslov, Theor. Math. Phys. 2 (1970) 21.

[7] R.P. Feynman and A.R. Hibbs, Quantum Mechanics and Path Integrals
(New York, McGraw Hill, 1965).

[8] R. Dashen, B. Hasslacher and A. Neveu, Phys. Rev. D11 (1975) 3424.

[9] J.L. Gervais, A. Jevicki and B. Sakita, Proc. of the Conf. on
Extended Systems in Field Theory, Paris, June 1975, Phys. Rep.23C
(1976) 281.

[10] L.A. Takhtadzhyan and L.D. Faddeev, Theor. Math. Phys. 21 (1974) 160.

[11] S. Coleman, Lectures delivered at the 1975 Int. School of Subnuclear
Physics "Ettore Majorana", Erice, Italy.

[12] A. Luther, Phys. Rev. B14 (1976) 2153.

[13] R. Dashen, B. Hasslacher and A. Neveu, Phys. Rev. D10 (1974) 4130.

[14] A. Scott, F. Chu and D. McLaughlin, Proc. IEEE 61 (1973) 1443.

[15] V. Zakharov and A. Shabat, Sov. Phys. JETP 34 (1971) 62.

[16] C. Nohl, Ph. D. Thesis, Princeton University, 1975.

[17] C.N. Yang, Phys. Rev. 168 (1968) 1920.

[18] R. Dashen, B. Hasslacher and A. Neveu, Phys. Rev. D12 (1975) 2443.

[19] A. Zamolodchikov and A. Zamolodchikov, Moscow ITEP preprint 112 (1977).

[20] L. Landau and E. Lifshitz, Quantum Mechanics, Non-Relativistic Theory (Pergamon, London), p. 171-182.

[21] D. McLaughlin, J. Math. Phys. 13 (1972) 1099.

[22] A. Belavin, A. Polyakov, A. Schwartz and Y. Tyupkin, Phys. Lett. 59B (1975) 85.

[23] C. Callan, R. Dashen and D. Gross, Phys. Lett. 63B (1976) 334 ;
R. Jackiw and C. Rebbi, Phys. Rev. Lett. 37 (1976) 172.

[24] G. 't Hooft, Phys. Rev. Lett. 37 (1976) 8.

[25] C. Bender and T.T. Wu, Phys. Rev. D7 (1973) 1620.

[26] L. Lipatov, JETP Letters 25 (1977) 116.

[27] E. Brézin, J.C. Le Guillou and J. Zinn-Justin, Phys. Rev. D15 (1977) 1544, 1558, D16 (1977) 408.

Lectures on the Relativistic String

F. Rohrlich

Syracuse University

The purpose of these lectures is to present a survey of some recent work on the relativistic string, especially as carried out at Syracuse University.[1-9] There, the primary objective has been to provide a consistent treatment of this apparently very simple physical system, devoid of invariance difficulties, occurrences of tachyonic states, etc. that have been encountered by other research groups in the past. It was felt that only after these problems are resolved can one seriously consider hadronic models of quarks bound together by strings. The latter goal is encouraged by the strong-coupling limits of various field theories which lead to geometrical confinement of quarks, i.e. to their restriction into strings, bubbles, or bags.

The primary objective is thus a study of a relativistic Hamiltonian system with constraints which is simple enough to be solved explicitly, and which permits canonical quantization. Apart from its interest for hadron models, this study provides a sample of such systems which, after all, occur in many other areas of theoretical physics.

These lectures will consist of three parts. The first and longest part will deal with the classical free string as defined by Nambu, Susskind, and Goto[10]. As a free physical system of total energy $M > o$ it follows the laws of relativistic dynamics with a total momentum fourvector P^μ which is time-like with $P^o > o$. We show how the trivial uniform motion of the system as a whole can be separated covariantly from its internal motion taking proper account of all constraints. The internal motion is then discussed.

The second part deals with the covariant quantization of the relativistic string. This will be an easy task after the careful classical preparation.

Finally, in the third part a brief comparison is made between the older literature on the string (1968-1975) and the recent work by the Syracuse group. Time will not permit a discussion of string models of hadrons, so that we shall only give some important references to papers published in the last few years.

I. The Classical Free String

<u>(1a)   Definitions</u>

One distinguished open strings and closed strings (loops). A par-
amater u^1 is chosen to vary monotonically from 0 to π for the open, from
0 to 2π for the closed string, characterizing the points along the string
at a particular "time" u^0 . The string sweeps out a two-dimensional surface
in Minkowski space as time progresses. This surface is a simply connected
"rectangular" surface for the open string, a doubly connected "cylindrical"
surface for the closed string, as shown in Fig. 1.

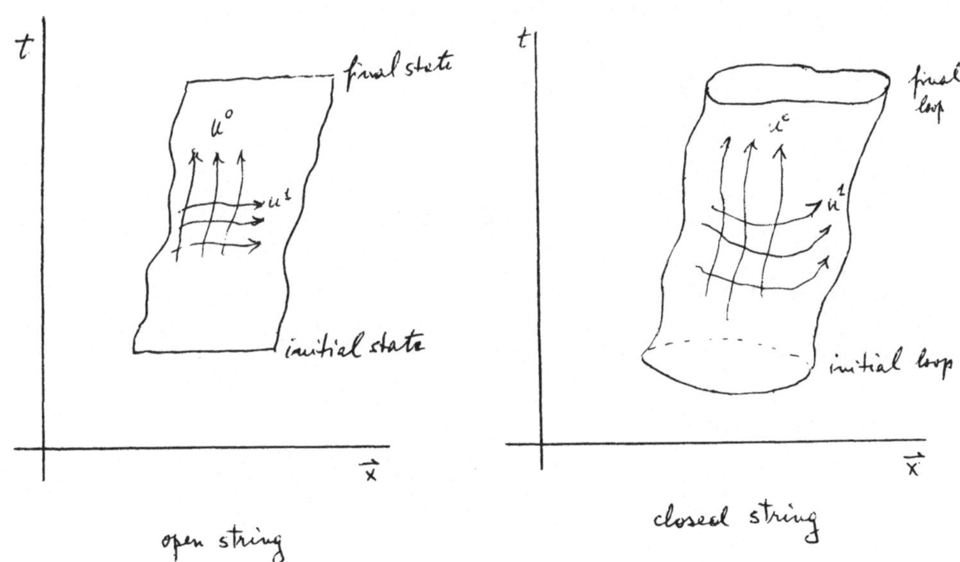

Fig. 1 The two-surface Σ_2 embedded in M_4.

Mathematically, one is dealing with a two-dimensional pseudo-Riemannian space Σ_2 , with metric $g_{\alpha\beta}$ $(\alpha, \beta = 0, 1)$ embedded in flat Minkowski space M_4 with metric $\eta_{\mu\nu}$; $\mu, \nu = 0, 1, 2, 3$; $\eta_{00} = -1$, $\eta_{kk} = +1$ ($k = 1, 2, 3$). The relation between $g_{\alpha\beta}$ and $\eta_{\mu\nu}$ is expressed by the tangent vectors to Σ_2 , $\partial x^\mu / \partial u^\alpha$,

$$g_{\alpha\beta} = \frac{\partial x^\mu}{\partial u^\alpha} \frac{\partial x^\nu}{\partial u^\beta} \eta_{\mu\nu}. \tag{1.1}$$

The quantity $x^\mu(u)$ is a fourvector in M_4 whose u-dependence defines the surface Σ_2 .

The string is defined by the action integral which yields an extremum for Σ_2 given the initial and final states of the string.

$$A = \iint \mathcal{L} \, d^2u = - \iint \sqrt{-g} \; d^2u \tag{1.2}$$

with

$$g \equiv \det g_{\alpha\beta} \tag{1.3}$$

If one uses the abbreviations

$$\dot{x} \equiv \partial x / \partial u^0, \qquad x' \equiv \partial x / \partial u^1, \qquad x \cdot y \equiv x^\mu y^\nu \eta_{\mu\nu} , \qquad x^2 \equiv x^\mu x^\nu \eta_{\mu\nu}$$

then A can be written

$$A = - \iint \sqrt{(\dot{x} \cdot x')^2 - \dot{x}^2 x'^2} \; du^0 \, du^1 \tag{1.4}$$

When all coordinates are measured in some suitable unit of length, A has a numerical factor of dimension of reciprocal length (to make it dimensionless) which we put here equal to one. For later comparison with experiment this unit of length is chosen to be $\sqrt{2\pi\alpha'}$ where α' is the slope of the Regge trajectory (spin vs. square of the mass) in units $\hbar = c = 1$.

(1b) Constraints and gauge

The action (1.4) exhibits an invariance under reparametrization which is usually called "gauge invariance" in this context,

$$u^\alpha \longrightarrow u'^\alpha = f^\alpha(u^0, u^1).$$

(1.5)

This invariance entails the existence of identities between the $\partial x^\mu / \partial u^\alpha$ and the canonical momenta

$$p_\mu^\alpha \equiv \frac{\partial \mathcal{L}}{\partial x_\alpha^\mu} \qquad\qquad X_\alpha^\mu \equiv \partial x^\mu / \partial u^\alpha.$$

(1.6)

This fact follows from Noether's second theorem[11]. In this case these identities are

$$\left.\begin{array}{r} p^\alpha \cdot X_\beta = 0 \\[2mm] (p^\alpha)^2 + X_\beta^2 = 0 \end{array}\right\} \quad (\alpha \neq \beta)$$

(1.7)

When the first variation of A is equated to zero one obtains the Lagrangian equation of motion and a boundary condition. Both of these are most conveniently expressed in terms of the p_μ^α . The equations of motion are

$$\partial p_\mu^0 / \partial u^0 + \partial p_\mu^1 / \partial u^1 = 0$$

(1.8)

and the boundary conditions are: for the open string

$$p_\mu^1(u^0, 0) = 0 = p_\mu^1(u^0, \pi),$$

(1.9a)

for the closed string

$$x^\mu(u^0, u^1 + 2\pi) = x^\mu(u^0, u^1).$$

(1.9b)

Because of (1.7) the boundary conditions (1.9a) imply that the end points of the open string move with the velocity of light. The string as a whole in general does not move with this velocity; the trajectories of the endpoints are then light-like helices in M_4.

In view of the aim to carry out a canonical quantization one wants to cast the classical string into a Hamiltonian formalism. This task is complicated by the identities (1.7). The following alternatives are available.

(a)   One eliminated the redundant variables from the identities and one then constructs the usual canonical formalism in terms of the independent variables. This procedure necessarily spoils manifest covariance and the simple, symmetric appearance of the theory. It results in more complicated equations.

(b)   One uses the theory of Hamiltonian systems with constraints[12,13] where the identities are treated as constraints.

In this latter alternative, which shall be adopted in the present lectures, one treats the x^μ_α and p^α_μ as independent canonical variables and one treats the identities as constraints on these variables. One then has a Hamiltonian which is the sum of the canonical one, H_o, and of a linear combination of the constraints (1.7),

$$H = H_o + H_c$$

(1.10)

$$H_c = \int \left[c^\alpha_\beta(u^1)\, p^\alpha \cdot x_\beta + \mu^\alpha_\beta(u^1)\left((p^\alpha)^2 + x^2_\beta\right)\right] du^1.$$

H_c will be found to be independent of u^o.   The constraints are added for a given u^o.   Since there are constraints for each point u^1 on the string, the two c^α_β and the two μ^α_β are functions of u^1.

Because of gauge invariance, the canonical Hamiltonian H_o vanishes identically and one is left with H_c.   Each choice of the c^α_β and μ^α_β yields a particular choice of "gauge", i.e. a particular parametization u^o, u^1. These choices do not necessarily restrict one to a <u>unique</u> choice of gauge: classes of gauge may result.   In that case additional constraints are necessary in order to obtain a unique gauge.   An analogy in electrodynamics is well known: the covariant gauge characterized by the restriction of A^μ to fourvectors still permits various choices such as the Feynman gauge, the Landau gauge, etc.

In the string system a choice of the c^α_β and μ^α_β can be made so that the metric tensor $g_{\alpha\beta}$ becomes diagonal.   This is called the "Orthogonal gauge".   It does not specify the u^o, u^1 uniquely but restricts this parametrization only up to conformal transformations.   Additional constraints are then introduced known as "ghost elimation conditions" (for reasons to be seen later) which restrict the orthogonal gauge further and makes it essentially unique, i.e. these constraints fix u^o and u^1 up to at most an additive constant.

The specification of gauge thus proceeds according to the following diagram (Fig. 2)

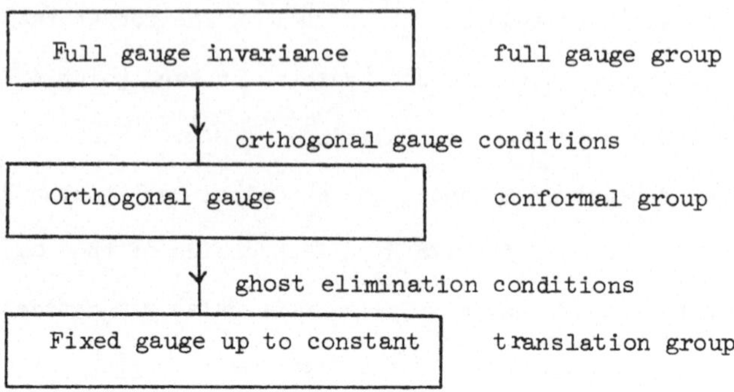

Fig. 2 Gauge fixing

The orthogonal gauge conditions can be expressed in various equivalent forms: as conditions on $g_{\alpha\beta}$ (equation (1.11) below), as a relation between x_0^μ and p_μ^0 , (1.15) below; as constraints, (1.16) or (1.49) below. The ghost elimination conditions are the constraints (1.37) or equivalently (1.51) below.

(1c) Gauge fixing

It is customary to denote the Σ_2 coordinatization u^0 and u^1 in the orthogonal gauge by τ and σ respectively. The orthogonal gauge is then characterized by

$$g_{\alpha\beta} = x'^2 \begin{pmatrix} -1 & 0 \\ 0 & 1 \end{pmatrix} \tag{1.11}$$

In order to achieve this form one needs the choice

$$c_\beta^\alpha = 0, \qquad \mu_1^0 = \tfrac{1}{2} , \qquad \mu_i^1 = 0 \tag{1.12}$$

The Hamiltonian is then (restricting ourselves for definiteness to the open string from here on):

$$H = \frac{1}{2} \int_0^\pi \left(p^2 + x'^2 \right) d\sigma \tag{1.13}$$

with the notation $p_\mu = p_\mu^{\,0}$. Using the canonical equal time PB

$$\left\{ x^\mu(\tau,\sigma), x^\nu(\tau,\sigma') \right\} = 0 \qquad \left\{ p^\mu(\tau,\sigma), p^\nu(\tau,\sigma') \right\} = 0$$

$$\left\{ x^\mu(\tau,\sigma), p^\nu(\tau,\sigma') \right\} = \eta^{\mu\nu} \delta(\sigma,\sigma') \tag{1.14}$$

one verifies that the choice (1.12) leads to

$$\dot{x}^\mu = \left\{ x^\mu, H \right\} = p^\mu \tag{1.15}$$

so that the constraints (1.7) become

$$C_0(\sigma) \equiv p \cdot x' = \dot{x} \cdot x' = 0 \qquad C_1(\sigma) \equiv p^2 + x'^2 = \dot{x}^2 + x'^2 = 0 \tag{1.16}$$

which indeed results in (1.11). This derivation requires three comments.

Firstly, it is convenient to consider the variable $\sigma \in [0, \pi]$ as the restriction to the indicated domain of the variable $\sigma \in (-\infty, \infty)$ with x^μ periodic in σ with period 2π. The boundary condition (1.8) can then be assured by choosing x^μ symmetric in σ. This permits one to regard the open string as a limiting case of the closed string, obtained by the identification

$$x^\mu(\pi + \sigma) = x^\mu(\pi - \sigma)$$

for all τ. For the above derivation this is relevant only insofar as the δ-function in (1.14) must be a periodic δ-function,

$$\delta(\sigma,\sigma') = \sum_{n=-\infty}^{\infty} \left[\delta(\sigma - \sigma' - 2\pi n) + \delta(\sigma + \sigma' - 2\pi n) \right]. \tag{1.17}$$

Secondly, (1.16) are only two of the four identities (1.17), i.e. those containing p_μ^o. But using these one finds $p_\mu^1 = x'^\mu$ so that the other two become identical with the first two.

Thirdly, when the identities (1.17) are now treated as constraints, it means that they become "weak equations". This was indicated by $\overset{w}{=}$ in (1.16) and has the following meaning. Expressions that are weakly zero can be put equal to zero only <u>after</u> all PB involving them have been evaluated. With this in mind one finds that

$$\left\{ C_o(\sigma), C_o(\sigma') \right\} \overset{w}{=} \left\{ C_o(\sigma), C_1(\sigma') \right\} \overset{w}{=} \left\{ C_1(\sigma), C_1(\sigma') \right\} \overset{w}{=} 0 \tag{1.18}$$

and that they commute, (i.e. have vanishing PB) with H. They are therefore independent of τ. Such constraints are called "first class constraints". Note that H itself is an integral over the constraints $C_1(\sigma)$ and thus also vanishes weakly. C_o and C_1 can thus be imposed as initial conditions and will be propagated in an invariant way.

Parametrizations that leave $g_{\alpha\beta}$ invariant up to a multiplicative factor are related by the conformal group of transformations,

$$(\tau, \sigma) \longrightarrow (\tau', \sigma') : \qquad \frac{\partial \tau'}{\partial \tau} = \frac{\partial \sigma'}{\partial \sigma} , \qquad \frac{\partial \tau'}{\partial \sigma} = \frac{\partial \sigma'}{\partial \tau} . \tag{1.19}$$

The name "orthogonal" gauge comes from the diagonal nature of $g_{\alpha\beta}$.

We now have a physical system characterized by the PB (1.14), the Hamiltonian (1.13), and the first class constraints (1.16). Its equations of motion are easily written down. $\dot{x}^\mu$ is given by (1.15) and

$$\dot{p}^\mu = \left\{ p^\mu, H \right\} = x''^\mu . \tag{1.20}$$

Elimination of $\dot{p}^{\mu}$ between these two equation leads to the alternative form

$$\ddot{x}^{\mu} = x''^{\mu} \qquad\qquad (1.20')$$

in agreement with (1.8). The great simplicity of this equation compared to the Euler-Lagrange equations that result from the action (1.4) in an arbitrary gauge, is an example of the usefulness of a suitable gauge choice.

The Euler-Lagrange equations which follow directly from the action (1.4) are quite complicated and highly nonlinear. The fact that the orthogonal gauge choice made them linear is no accident. By a general theorem, minimal surfaces lead to linear equations when such a gauge is chosen[14].

(1d) Realizations of $\mathcal{P}$.

At this point it is desirable to take into account that one is dealing with a free system which must satisfy all the requirements of Poincare invariance. This invariance is conveniently expressed in terms of the generators of infinitesimal translations P^{μ} and of infinitesimal rotations and Lorentz boosts $M^{\mu\nu}$. These generators satisfy the Poincare algebra when expressed in suitable variables x^{μ}, p^{μ} which satisfy a Poisson bracket algebra. It is convenient to introduce the center position of the system, Q^{μ}, whose uniform motion characterizes the fact that the system is free. It is related to the Lorentz generators $M^{\mu\nu}$ by

$$M^{\mu\nu} = Q^{\mu}P^{\nu} - Q^{\nu}P^{\mu} + S^{\mu\nu} \qquad\qquad (1.21)$$

but it is not uniquely defined by it. For various choices of Q^{μ} a suitable choice of the "spin generators" $S^{\mu\nu}$ is possible so that the algebra which

Q^μ, P^μ, and $S^{\mu\nu}$ satisfy ensures the Poincaré algebra. Such a "QPS algebra" is called a __realization__ of the Poincaré algebra $\mathcal{P}$. These realizations have been classified[6] according to the following criteria:

(i)   whether or not Q^μ belongs to the enveloping algebra of $\mathcal{P}$

(ii) whether or not $S^{\mu\nu}$ is translation invariant

(iii) whether or not Q^μ is local, i.e. $[Q^\mu, Q^\nu] = 0$ for all μ, ν.

(iv) whether or not Q^μ is a Lorentz fourvector.

Of these 16 classes of realizations the following three are of special interest:

CCR, the covariant canonical realization in which Q^μ is covariant under Poincare transformations and a manifestly covariant canonical Poisson bracket (PB) algebra of the Q^μ and P^μ exists.

CNR, the covariant __non__canonical realization in which Q^μ is covariant under Lorentz transformations and Q^μ, P^μ satisfy a noncanonical PB algebra.

NWR, the Newton-Wigner realization in which Q^μ is not a fourvector but the three-vectors $\vec{Q}$ and $\vec{P}$ satisfy the canonical PB.

The QPS algebras of these three realization were given in reference 2.

In terms of the above four alternatives they can be characterized by answering all four questions with "yes", except (i), (iii), and (iv) for CCR, CNR, and NWR, respectively. The theory of the free relativistic string was treated according to CCR in ref. 2, according to CNR in reference 4, and according to NWR in reference 1. The resulting physics is the same for all. Since CCR is the simplest and physically most intuitive realization we shall adopt it for the present lectures. This choice is consistent with (1.14).

The QPS algebra of the CCR is

$$\{Q^\mu, Q^\nu\} = 0, \qquad \{Q^\mu, P^\nu\} = \eta^{\mu\nu}, \qquad \{P^\mu, P^\nu\} = 0 \quad (1.22a)$$

$$\{Q^\mu, S^{\rho\sigma}\} = 0, \qquad \{P^\mu, S^{\rho\sigma}\} = 0 \quad (1.22b)$$

$$\{S^{\mu\nu}, S^{\rho\sigma}\} = \eta^{\mu\rho} S^{\nu\sigma} + \eta^{\nu\sigma} S^{\mu\rho} - \eta^{\mu\sigma} S^{\nu\rho} - \eta^{\nu\rho} S^{\mu\sigma}. \quad (1.22c)$$

Because of (1.11b) this realization makes the variables that describe the system as a whole, Q^μ, P^μ, independent of the internal variables (later called ξ^μ, π^μ) that describe internal orbital angular momenta $\xi^\mu \pi^\nu - \xi^\nu \pi^\mu$ whose sum is the total internal angular momentum, the spin of the system, $S^{\mu\nu}$.

The various alternative treatments of the string are shown in the scheme of Fig. 3

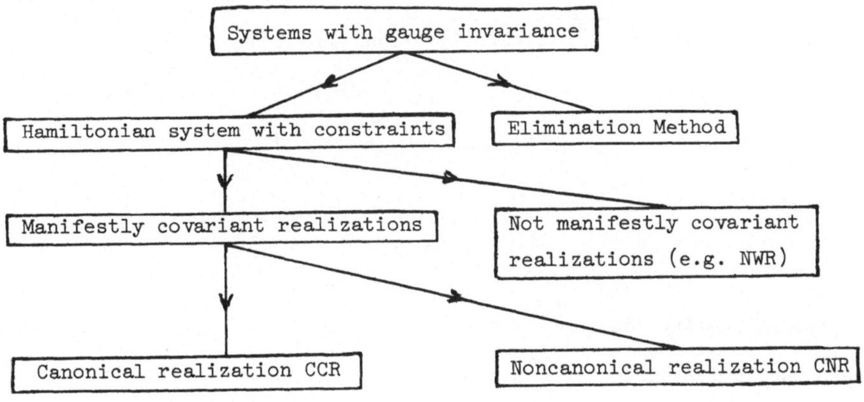

Fig. 3 Alternative ways of treating a gauge invariant system.

(1e) Internal variables

In free nonrelativistic systems it is well known that the internal dynamics of the system is treated best when the trivial motion of the center of mass is separated first. The _relative_ motion of the individual particles can then be studied in the center-of-mass frame. A similar situation holds also in relativistic systems.

Let $x^\mu(\tau,\sigma)$ be the position of a particular point on the string, $Q^\mu(\tau)$ its "center of mass" at that time, then the relative position of that point is defined by

$$\xi^\mu(\tau,\sigma) = x^\mu(\tau,\sigma) - Q^\mu(\tau) \tag{1.23}$$

If the system as a whole behaves like a free particle of mass M then its four-momentum P^μ must satisfy

$$P^2 + M^2 \overset{w}{=} 0. \tag{1.24}$$

This can only be a constraint (weak equation) since otherwise it contradicts the adopted PB (1.22a). P^μ is clearly a constant since the system is free. For $Q^\mu(\tau)$ we want uniform motion,

$$Q^\mu(\tau) = Q^\mu + P^\mu\tau/\pi \tag{1.25}$$

i.e. we want

$$\dot{Q}^\mu = P^\mu/\pi. \tag{1.26}$$

(The proportionality constant $1/\pi$ is chosen for later convenience and simply gives a particular scale to Q^μ and $Q^\mu(\tau)$). This is accomplished by a

Hamiltonian of the form

$$H = \frac{P^2}{2\pi} - H_{int} \tag{1.27}$$

where H_{int} depends only on the internal variables which are assumed to commute with Q and P. Of course H of (1.27) must be just the same as H of (1.13) but written in terms of Q, P, and the internal variables.

The τ-derivative of (1.23) is thus, using (1.15) and (1.26),

$$\pi^\mu(\tau,\sigma) \equiv \dot{\xi}^\mu(\tau,\sigma) = \{\xi^\mu(\tau,\sigma), H\} = p^\mu(\tau,\sigma) - P^\mu/\pi. \tag{1.28}$$

In this way we have defined an internal momentum $\pi^\mu(\tau,\sigma)$ in terms of p^μ and P^μ. On physical grounds we want

$$P^\mu = \int_0^\pi p^\mu(\tau,\sigma)\, d\sigma \tag{1.28}$$

which justifies the choice of the coefficient $1/\pi$ in (1.26). But this implies

$$\int_0^\pi \pi^\mu(\tau,\sigma)\, d\sigma = 0 \tag{1.29}$$

and suggests that we define Q^μ the center of $Q^\mu(\tau)$ at $\tau = 0$, as the average of $x^\mu(\tau,\sigma)$, i.e. as if the string had a uniform mass density 1,

$$Q^\mu = \frac{\int_0^\pi x^\mu(\tau,\sigma)\, d\sigma}{\int_0^\pi d\sigma} = \frac{1}{\pi} \int_0^\pi x^\mu(\tau,\sigma)\, d\sigma. \tag{1.30}$$

From this follows

$$\int_0^\pi \xi^\mu(\tau,\sigma)\, d\sigma = 0 \tag{1.31}$$

which in turn implies (1.29) via (1.28).

Now the Lorentz covariance of x^μ and p^μ requires $M^{\mu\nu}$ to be of the form

$$M^{\mu\nu} = \int_0^\pi \left[x^\mu(\tau,\sigma) p^\nu(\tau,\sigma) - x^\nu(\tau,\sigma) p^\mu(\tau,\sigma) \right] d\sigma. \tag{1.32}$$

In fact, the two equations (1.28) and (1.41) for the Poincaré generators ensure Poincaré covariance of x^μ,

$$\left\{ x^\mu, P^\nu \right\} = \eta^{\mu\nu} \tag{1.33a}$$

$$\left\{ x^\mu, M^{\rho\sigma} \right\} = \eta^{\mu\sigma} x^\rho - \eta^{\mu\rho} x^\sigma. \tag{1.33b}$$

Similarly, p^μ is translation invariant and Lorentz <u>covariant</u>.

The substitution into $M^{\mu\nu}$ of the transformations (1.23) and (1.28) to the internal variables leads now exactly to (1.21) with

$$S^{\mu\nu} = \int_0^\pi \left[\xi^\mu(\tau,\sigma)\, \pi^\nu(\tau,\sigma) - \xi^\nu(\tau,\sigma)\, \pi^\mu(\tau,\sigma) \right] d\sigma \tag{1.34}$$

so that

$$\left\{ Q^\rho, \xi^\nu \right\} = \left\{ Q^\rho, \pi^\nu \right\} = 0 \qquad \text{and} \qquad \left\{ P^\mu, \xi^\nu \right\} = \left\{ P^\mu, \pi^\nu \right\} = 0, \tag{1.35}$$

in order to satisfy the QPS algebra (1.22b). The last two equations are of course the statement of translation invariance for relative variables.

The P.B. of ξ^μ and π^μ follow from those of x^μ and p^μ by substitution of the same transformation. They are consistent with the restrictions on ξ and π, equations (1.31) and (1.29), which are therefore <u>strong</u> equations.

$$\left\{ \xi^\mu(\tau,\sigma), \xi^\nu(\tau,\sigma') \right\} = 0 = \left\{ \pi^\mu(\tau,\sigma), \pi^\nu(\tau,\sigma') \right\} ; \quad \left\{ \xi^\mu(\tau,\sigma), \pi^\nu(\tau,\sigma') \right\} = \eta^{\mu\nu}\left[\delta(\sigma,\sigma') - \tfrac{1}{\pi} \right]. \tag{1.36}$$

One verifies that ξ^μ and π^μ transform covariantly under Lorentz transformations.

The vectors ξ^μ and π^μ are expected to be spacelike vectors since they specify the string on a given hyperplane $\tau = $ const. In fact, in the "center of momentum" frame where $\vec{P} = o$ on expects these variables to have no fourth components, i.e.

$$\xi^\mu = (o, \vec{\xi}), \qquad \pi^\mu = (o, \vec{\pi})$$

in the CM frame. The covariant statement of such a condition is

$$\xi \cdot P \stackrel{w}{=} 0, \qquad \pi \cdot P \stackrel{w}{=} 0. \tag{1.37}$$

It is of course not proven by the above heuristic argument that the conditions (1.37) can be imposed consistently on the system. They certainly can hold only weakly since they would otherwise contradict the PB (1.36). Furthermore, they do not commute with each other. They are thus second class constraints. However, they are independent of time, i.e. they do commute with H.

To see this one only needs to write (1.13) in terms of the internal variables. Using (1.23), (1.28) and the strong equation (1.29) one finds

$$H = \frac{P^2}{2\pi} + \frac{1}{2} \int_0^\pi \left(\pi^2 + \xi'^2\right) d\sigma \tag{1.38}$$

which fixes H_{int} of (1.27). One now verifies that (1.37) commutes with H weakly. These constraints can thus also be imposed as initial conditions.

The Hamiltonian (1.38) provides the separation of the CM and yields the following internal dynamics,

$$\dot{\xi}^\mu = \left\{ \xi^\mu, H \right\} = \pi^\mu \tag{1.39a}$$

$$\dot{\pi}^\mu = \left\{ \pi^\mu, H \right\} = \xi''^\mu. \tag{1.39b}$$

Therefore the internal position satisfies the same equation as the $x^\mu(\tau, \sigma)$,

$$\ddot{\xi}^\mu = \xi''^\mu \tag{1.39'}$$

These equations are to be solved subject to periodicity, to the boundary condition (1.9a),

$$x'^\mu(\tau, 0) = 0 = x'^\mu(\tau, \pi) \qquad or \qquad \xi'^\mu(\tau, 0) = 0 = \xi'^\mu(\tau, \pi) \tag{1.40}$$

- 214 -

and the constraints (1.31) (which are strong), and (1.16) and (1.37) (which are weak).

The general solution of (1.39) is

$$\xi^\mu(\tau,\sigma) = F^\mu(\tau+\sigma) + G^\mu(\tau-\sigma).$$

The strong conditions require $F'^\mu(u) = G'^\mu(u)$ and periodicity of $F(u)$ with period 2π , i.e.

$$\xi^\mu(\tau,\sigma) = F^\mu(\tau+\sigma) + F^\mu(\tau-\sigma), \qquad F^\mu(\tau+2\pi) = F^\mu(\tau). \tag{1.41}$$

(1f) Mode variables

The solution (1.41) permits the Fourier expansion

$$\xi^\mu(\tau,\sigma) = \frac{i}{\sqrt{\pi}} \sum_{n\neq 0} \frac{\alpha_n^\mu}{n} \cos n\sigma\, e^{-in\tau} \tag{1.42}$$

and (1.39a) then gives

$$\pi^\mu(\tau,\sigma) = \frac{1}{\sqrt{\pi}} \sum_{n\neq 0} \alpha_n^\mu \cos n\sigma\, e^{-in\tau} \tag{1.43}$$

Reality of ξ^μ requires

$$\alpha_{-n}^\mu = \alpha_n^{\mu *} \tag{1.44}$$

The α_n^μ are Lorentz four-vectors and are called mode coefficients. The PB of ξ and π induce the PB

$$\{\alpha_m^\mu, \alpha_n^\nu\} = -im\,\eta^{\mu\nu}\delta_{m+n,0} \tag{1.45}$$

One can clearly cast the theory into a form in which the α_n^μ take over the role of ξ^μ and π^μ as fundamental variables. Substitution yields

$$H = \frac{P^2}{2\pi} + \Lambda_0 \tag{1.46}$$

where

$$\Lambda_0 = \frac{1}{2} \sum_{m \neq 0} \alpha_m \cdot \alpha_{-m} \, . \tag{1.47}$$

One then has

$$\dot{\alpha}_n^\mu(\tau) = \left\{ \alpha_n^\mu(\tau), H \right\} = -in\, \alpha_n^\mu(\tau) \tag{1.48}$$

or

$$\alpha_n^\mu(\tau) = \alpha_n^\mu \, e^{-in\tau} \tag{1.48}$$

with $\alpha_n^\mu = \alpha_n^\mu(0)$, in agreement with (1.42).

In terms of these variables the orthogonal gauge conditions (1.16)

become

$$H \overset{w}{=} 0 \tag{1.49a}$$

$$\frac{1}{\sqrt{\pi}} \alpha_n \cdot P + \Lambda_n \overset{w}{=} 0 \qquad (n \neq 0) \tag{1.49b}$$

with

$$\Lambda_n = \frac{1}{2} \sum_{\substack{m \neq 0 \\ m \neq n}} \alpha_m \cdot \alpha_{n-m}, \tag{1.50}$$

and the conditions (1.37) is

$$\alpha_n \cdot P \overset{w}{=} 0 \, . \tag{1.51}$$

The orthogonal gauge conditions in the form (1.49) are known as "Virasoro

conditions". The constraints (1.49) are first class while (1.51) are second

class and are to be added to H and to the PB to specify the system completely.

One observes that the mass M of (1.24) is expressed in terms of the mode

variables only,

$$M^2 = 2\pi \Lambda_o = 2\pi \sum_{1}^{\infty} |\alpha_m|^2 > 0 \tag{1.52}$$

and so is the spin tensor,

$$S^{\mu\nu} = \frac{i}{2} \sum_{n \neq 0} \frac{1}{n} \left(\alpha_n^\mu \alpha_{-n}^\nu - \alpha_n^\nu \alpha_{-n}^\mu \right). \tag{1.53}$$

Finally, one notes that the system is manifestly Poincaré invariant.

A commonly used procedure is to construct PB which are consistent with the weak constraints so that these constraints can then be used as strong equations. This "Dirac bracket" construction is by no means necessary. Nor is it always possible in cases like the present one when one is dealing with a mixture of first class and second class constraints.

II. Quantum Dynamics of the String

The classical theory of the relativistic string has been presented above on three different but equivalent levels

(i)   the x,p level of the individual point variables;

(ii)   the internal $\left(\xi,\pi\right)$ and system variables (Q,P) which afford a separation of the trivial CM motion;

(iii)   the mode variables $\left(\alpha_n\right)$ and system variables (Q,P) which afford not only a separation of the CM motion but also a decomposition into fundamental modes.

It is the last of these that is most closely related to the common quantum mechanical description. The canonical quantization will therefore be carried out with the variables (iii) although it could equally well be done with any of the other sets of variables.

The canonical quantization consists in two heuristic steps. They are necessarily heuristic because a derivation in the logical or mathematical sense of a quantum dynamics from a classical dynamics is clearly impossible.

These two steps are

(α) a replacement of all PB's by $\frac{1}{i}$ times commutator brackets

(β) a choice of ordering of products of noncommuting quantities.

The second step may contain a large amount of arbitrariness. The most common method used is to use Wick ordering of operators, but that method is not always applicable. Such an ordering always leaves an arbitrary additive constant available for determination on physical grounds. That constant is the vacuum expectation value of the product in question.

In the present case the only quantity that involves noncommuting variables is Λ_o ; one can therefore define new Λ_n,

$$\Lambda_n \equiv \frac{1}{2} \sum_m : \alpha_m \cdot \alpha_{n-m} : \tag{2.1}$$

and replace the classical Λ_o by the new Λ_o and a constant,

$$H = \Lambda_o + \frac{1}{2\pi} \left(P^2 + m_o^2 \right) \tag{2.2}$$

This Hamiltonian together with the commutation relations (CR)

$$\left[\alpha_m^\mu, \alpha_n^\nu \right] = m \, \eta^{\mu\nu} \delta_{m,n} \tag{2.3}$$

and the "reality" conditions

$$\alpha_n^{\mu \dagger} = \alpha_{-n}^\mu \tag{2.4}$$

yield the dynamics. This dynamics is represented in an indefinite metric Hilbert space $\mathcal{H}$ on which the α_n^μ act as linear operators. The indefinite metric is a direct consequent of the indefinite metric of M_4 which occurs on the right side of (2.3). This can best be seen by defining the new operators

$$a_n^\mu \equiv \frac{1}{\sqrt{n}} \alpha_n^\mu \quad (n > o), \qquad a_{-n}^\mu \equiv a_n^{\mu \dagger} \tag{2.5}$$

Because of (2.3) these operators are just the fourvector generalizations of creation and annihilation operators, since they satisfy

$$\left[a_m^\mu, a_n^\nu \right] = o, \qquad \left[a_m^\mu, a_n^{\nu \dagger} \right] = \eta^{\mu\nu} \delta_{m,n} \quad (m, n > o) \tag{2.6}$$

In this Hilbert space representation the weak equations of classical physics become conditions on the states, restricting $\mathcal{H}$ to a smaller, allowed space.

The first constraints to be taken into account are the ghost elimination equations (1.51). In quantum mechanics they become (using a_n^μ instead of α_n^μ),

$$a_n \cdot P \, |f\rangle = 0 \qquad (n > 0) \tag{2.7}$$

The restriction to $n > 0$ is sufficient and necessary here to avoid contradiction. It is also sufficient to ensure the weaker condition

$$\langle f | \, a_n \cdot P \, | f \rangle = 0. \tag{2.8}$$

The states $|f\rangle \in \mathcal{H}$ which satisfy (2.7) span a space $\mathcal{F}_+ \subset \mathcal{H}$ which is positive definite. The condition (2.8) would not ensure this properly. To see positive definiteness one refers to the Lorentz invariance of this condition: if it produces a positive definite $\mathcal{F}_+$ for the rest frame, it will do so for any frame. But in the rest frame characterized by

$$\vec{P} \, |f\rangle_R = 0 \tag{2.9}$$

one has from (2.7)

$$a_n^0 \, |f\rangle_R = 0 \tag{2.10}$$

so that only the space parts of the a_n^μ contribute; (2.6) becomes

$$\left[a_m^k, a_n^\ell \right] = 0, \qquad\qquad \left[a_m, a_n^\dagger \right] = \delta^{k\ell} \delta_{m,n} \tag{2.11}$$

and the polynomials of $\vec{a}_n^{\dagger}$ generate a positive definite space when acting on the vacuum state $|0\rangle$.

We are thus justified in calling (2.7) the "ghost elimination conditions", the states of negative norm being referred to as "ghosts". Their classical equivalents are (1.37).

One can now proceed to the orthogonal gauge conditions, the classical conditions (1.49). They become

$$-P^2 |\phi\rangle = \left(2\pi \Lambda_0 + m_0^2\right) |\phi\rangle \tag{2.12a}$$

and

$$\Lambda_n |\phi\rangle = 0 \qquad (n > 0). \tag{2.12b}$$

The first term of (1.49b) need not be included since it vanishes separately. The conditions (2.12) restrict $\mathcal{F}_+$ further. The new space $\Phi \subset \mathcal{F}_+$ is spanned by all $|\phi\rangle \in \mathcal{F}_+$ which satisfy (2.12). Φ is the space of physical states since all constraints are properly observed only in Φ . We have the hierarchy

$$\Phi \subset \mathcal{F}_+ \subset \mathcal{H}. \tag{2.13}$$

The Poincaré invariance of the theory as a whole is manifest and can easily be checked explicitly by means of the generators P^μ and $M^{\mu\nu}$ expressed in terms of Q, P and the a_n^μ . Since all constraints that define Φ are Poincaré invariant, so is the space Φ .

The energy spectrum is given by (2.12a). With M^2 introduced in (1.24),

$$M^2 |\phi\rangle = \left(2\pi \sum_1^\infty n\, a_n^+ a_n + m_0^2\right) |\phi\rangle \tag{2.14}$$

The spin spectrum is obtained by constructing simultaneous eigenfunctions[4,5,7] of $\vec{S}^2, S$ and M^2, where

$$S_k \equiv \tfrac{1}{2} \varepsilon_{klm} S^{lm} \tag{2.15}$$

and $S^{\ell m}$ is the only non zero part in the rest frame in Φ of

$$S^{\mu\nu} = \frac{i}{2} \sum_{1}^{\infty} \left(a_n^{\mu\dagger} a_n^{\nu} - a_n^{\nu\dagger} a_n^{\mu} \right) \tag{2.16}$$

The arbitrary constant m_o^2 is seen to determine the energy of the unexcited string which can be identified with the vacuum. The requirement that we are dealing with a physical system (P^μ is timelike) which was made at the beginning, gives the restriction $M^2 \geqslant 0$ or $m_o^2 \geqslant 0$ the zero value being applicable only to the vacuum.

On the other hand, a plot of Regge trajectories resulting from (2.14) and (2.16) is

$$\alpha(s) = \frac{1}{n} \alpha' \left(s - m_o^2 \right)$$

which shows that these are linear and rising but that a positive intercept cannot be achieved because it would require $m_o^2 < 0$.

Furthermore, the construction of a Veneziano amplitude by a multiperipheral approximation from this string model fails because of this same restriction.

III. Comparison with other Work

In comparing our work with that of others[15] a distinction can be made between the classical and the quantum dynamics of the string.

In the <u>classical case</u> the methods employed by us lead to the usual physical results for any number of space-time dimensions $D > 2$. For $D = 2$ it was first shown by Patrascioiu[16] that the use of null plane coordinates $x^0 \pm x^1$ instead of x^0 and x^1 arbitrarily exclude a non-trivial solution. The proof lies in the observation that the transformation to null coordinates involves a Jacobian that can vanish under certain conditions. These are just the conditions for the existence of longitudinal modes. In fact the conventional string solutions, including the one presented above, gives only transverse modes, $\dot{\vec{x}} \perp \vec{x}\,'$ in the CM frame. In the classical case such solutions can be expressed by a finite number of modes; this is not the case for the longitudinal solutions[9].

The <u>quantum case</u> is more interesting because difficulties occur here in the work of other authors. These difficulties are:

(1) Without ad hoc artifices the theory is Lorentz invariant only in $D=26$ space-time dimensions and only for a leading Regge trajectory intercept of $\alpha_0 = 1$ (corresponding to a particular <u>negative</u> value of m_0^2 in our notation).

(2) The lowest state of the string (first state above the vacuum) is tachyonic with $M^2 < 0$.

(3) The first excited state has zero energy ($M^2 = 0$) and being a spin 1 state has therefore only 2 substates instead of three.

The Syracuse string has none of these difficulties: it is Lorentz invariant for any number of dimensions but is limited to $\alpha_0 \leqslant 0$ corresponding to $m_0^2 \geqslant 0$; the lowest string state has real mass $M^2 > 0$; the first excited state thus has three substates rather than only 2.

The difference of these two theories also appears in applications other than the Regge intercept; for example, the Syracuse results do not permit the construction of a Veneziano amplitude.

However, there are also important similarities:

(a) The spectra differ only by an additive constant which in our case is $m_0^2 \geqslant 0$ and which removes difficulties (2) and (3). The number of physical states are thus also equal in number except for the one state missing in those treatments which exhibit difficulty (3).

(b) the Regge trajectories are linear and are rising with the same slope in all treatments.

The origin of the difficulty encountered in the papers summarized in references[15] lies in the use of null plane coordinates. Their use is called "transverse gauge" when introduced explicitly. However, they are also introduced implicitly in the so-called "covariant quantization." The latter is based on the DDF states[17] which are associated with vibrations in D-2 dimensions. It uses a null vector that plays a key role in the proof of the positivity of the corresponding physical space.

Even when certain authors claim to study the dynamics of the string in the "timelike gauge" (as we do), they do not carry it through but fall back on the DDF construction and thus claim that the difficulties (1) to (3) are still there[18].

While the restriction to D-2 assures a positive Hilbert space, this restriction is not necessary and is in fact too strong. Our ghost elimination condition accomplishes the same goal but restricts only one dimension (the zero components α_n^0 in the CM frame).

The number 26 seems to have no special significance but D=26 and $\alpha_0=1$ just happen to be the fortunate values of these parameters for which the term vanishes that prevents the closure of the Lorentz algebra.

We conclude by a few references to string models of hadrons. The simplest model involves two point masses at the ends of a string in two space-time dimensions[19], and with a different Lagrangian in four space-time dimensions[3,5].

Somewhat more sophisticated are the models which involve quark fields confined to the world lines of the string end points. Such confinement was first suggested by Giles[20]. These models were treated by Bars and Hanson[21] for D=2, and by Bars[22]; see also Andreo[5].

The next level of complexity is one or more quark fields (Dirac fields) confined to a closed string (loop) for D=4. Geometric confinement to a bubble in D=4 reduces to just this case for D=3. The work of Giles and Tye[23] and of Tye[24] is relevant here. The latter work was repeated by me using the Syracuse string[8]. As a consequence no restriction on the number of quark fields emerges. Tye's model requires 22 quark fields.

A successful comparison with charmonium was recently carried out for their model by Giles and Tye[25].

Future applications will have to go beyond these models which correspond to strong coupling limits. The next term in a $\frac{1}{g}$ expansion will have to be taken into account.

References

1. F. Rohrlich, Phys. Rev. Lett. $\underline{34}$, 842 (1975).

2. F. Rohrlich, Nucl. Phys. $\underline{B112}$, 177 (1976).

3. R. Andreo and F. Rohrlich, Nucl. Phys. $\underline{B115}$, 521 (1976).

4. F. Rohrlich, Nuovo Cim. $\underline{37}$ A, 242 (1977).

5. R. Andreo, Ph.D. thesis, Syracuse University, (1977).

6. G. Lanyi, Ph.D. thesis, Syracuse University, (1977).

7. D.C. Salisbury, Ph.D. thesis, Syracuse University, (1977).

8. F. Rohrlich, Phys. Rev. D $\underline{16}$, 354 (1977).

9. R. Andreo and F. Rohrlich, "Longitudinal Vibrations of the Relativistic String," (to be published in Phys. Rev. D $\underline{18}$, (1978)).

10. Y. Nambu, Proc. Conf. Symmetry and Quark Models, Wayne State U. 1969; L. Sussking, Nuovo Cim. $\underline{69}$, 457 (1970); T. Goto, Prog. Theor. Phys. $\underline{46}$, 1560 (1971).

11. For a rather general discussion of Noether's theorems see for example A. Troutman in "Lectures on General Relativity," Brandeis Summer Institute in Theoretical Physics, 1964, Volume 1, p. 172, Prentice-Hall, N.J.

12. P.A.M. Dirac, Can. J. Math. $\underline{2}$, 129 (1950); Proc. Roy. Soc., A $\underline{246}$, 326 (1958).

13. J.L. Anderson and P.G. Bergmann, Phys. Rev. $\underline{83}$, 1018 (1951); P.G. Bergmann and I. Goldberg, Phys. Rev. $\underline{98}$, 531 (1955).

14. J. Douglas, Ann. Math. $\underline{40}$, 205 (1939).

15. The large literature on the subject cannot possibly be cited here. The following two reviews however are fairly complete up to two years ago: C. Rebbi, Phys. Rep. $\underline{12C}$, No. 1 (1974). J. Scherk, Rev. Mod. Phys. $\underline{47}$, 123 (1975).

16. A. Patrascioiu, Nucl. Phys. B $\underline{81}$, 525 (1974).

17. P. DiVecchia, E. Del Guidice and S. Fubini, Ann. Physics $\underline{70}$, 378 (1972).

18. See for example the paper by A.J. Hanson, P. Goddard and G. Ponzano, Nucl. Phys. B $\underline{89}$, 76 (1975).

19. W.A. Bardeen, I. Bars, A.J. Hanson and R.D. Peccei, Phys. Rev. D $\underline{13}$, 2364 (1976).

20. R.C. Giles, Phys. Rev. D $\underline{13}$, 1670 (1976).

21. I. Bars and A.J. Hanson, Phys. Rev. D $\underline{13}$, 1744 (1976).

22. I. Bars, Phys. Rev. Lett. $\underline{36}$, 1521 (1976).

23. R.C. Giles and S-H.H. Tye, Phys. Rev. D $\underline{13}$, 1690 (1976).

24. S-H.H. Tye, Phys. Rev. D $\underline{13}$, 3416 (1976).

25. R.C. Giles and S-H.H. Tye, Phys. Rev. Lett. $\underline{37}$, 1175 (1976); Phys. Rev. D $\underline{16}$, 1079 (1977).

Introduction and Selected Topics in Source Theory

J. Schwinger

University of California, Los Angeles, California

Lecture I

Since Shelly Glashow has cornered the market in
charm and beauty, I shall not intrude on his domain.
If there is to be a catchword for my contribution,
it is - <u>reality</u>.

That introduces the subject of Source Theory:
what is it, and who needs it? It began as a personal
reaction to the ambiguities and obscurities of
operator field theory, but to persuade you that it
is not totally idiosyncratic, I remark that, seated
in this audience, are at least a few people who have
made routine use of it for a fair number of years and
regard it as the only physically sensible language
to use for high energy physics. To them I say - the
blessings of Pele upon you - and I take for granted
that the repetition of a familiar story will be
pleasurable rather than boring. I must also assume
that the vast majority of you are totally untutored
in this art. Accordingly, I intend to answer my
question, not by a Madison Avenue pitch, but by a
systematic, if rapid and somewhat sketchy, development
of this viewpoint, with a few applications designed
to display the simplicity, mathematical and physical,
and the power, of the method.

In an overly simple phrase, Source Theory is a predictive, phenomenological theory, which is to say that it is concerned with the properties of the observed particles, without the explicit intervention of any speculative hypotheses about inner structure, based, if you like, upon a cynical disbelief in the ultimate validity of any of the simple "dynamical" models that are now common currency. But skepticism aside, it is surely legitimate to inquire how far one can go on the basis of external quantum numbers alone, without invoking speculative internal dynamics, for if that does suffice, the evidence for belief in any particular dynamical model has been removed.

How does one go about reconstructing a theory of particles in this phenomenological sense? By paying strict attention to the operational definition of a particle that is provided by the experimenter's manipulations, rather than through some a priori definition of a particle. The outstanding fact about the particles of high energy physics is that they must be created to be studied, and that they are destroyed in the process of detection. In fact, one can take this as the definition of a particle. Even high energy protons are not found in this condition but have to be prepared, be created, in that situation of high energy in order to be studied. And, any act

of detection is ultimately the destruction of the
particle as its properties are handed on to other
forms which are more conveniently studied for detection purposes. I shall use the charged pi meson as
an elementary (and historically first) example of
particles that are artifically created.

<u>Production</u>

We now begin the development of a theory in
which one takes the procedure of the experimenter
as the starting point. Of course, a theory is not
just the entries in a laboratory log book but is an
abstraction and an idealization in which one focuses
on what is important about the particular acts that
are involved. For the example of a pi meson, we
begin with the collision of a proton beam on a proton
target, whereby, in a nuclear reaction that converts
a proton to a neutron, a positively charged pi meson
is produced, see Fig. 1:

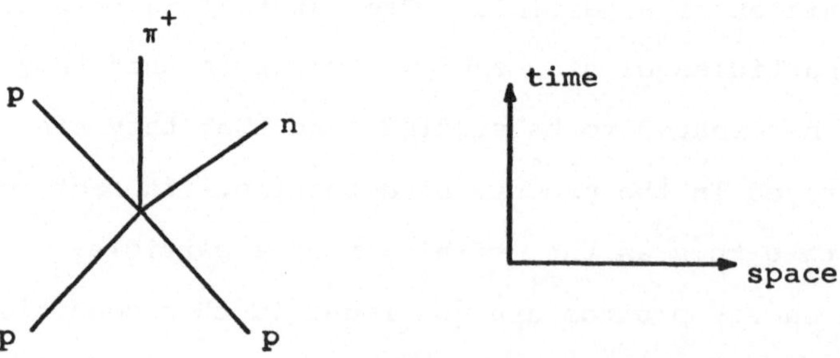

Fig. 1

If one is just interested in creating a π^+ meson
beam, the details of the collision process, or even
the particular reaction that is used, are not relevant. The other particles are there as the means for
transferring the necessary physical properties of
energy, momentum, charge, that characterize this
particular particle. Therefore, the first thing to
do in developing a theory is to abstract from the
details of the realistic collision, details which
have no significant bearing on the creation of a pion
beam. Here is a graphical representation of that
first step in which all that one sees is the thing
of interest, the creation of the pi meson, with
everything else replaced by an idealization, which
I call a source, since the other particles of a
realistic collision are there to be the source of

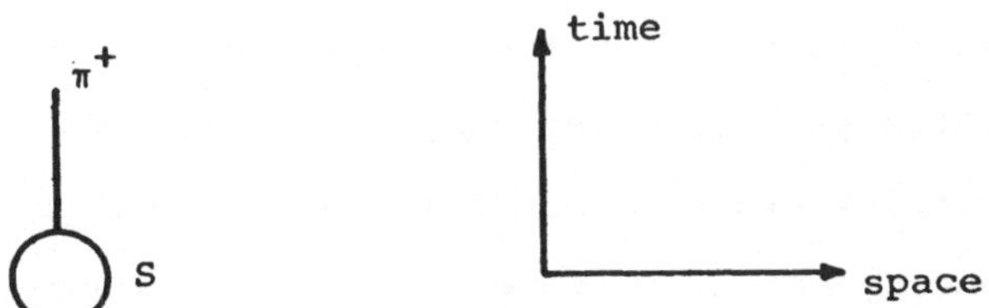

Fig. 2

the properties that characterize the pi meson, through
the net balance between what enters and what leaves
the collision. The circle is intended to suggest

that one has some degree of space-time control over
the creation act. It is a limited control, but we
must be aware, from the beginning, of the causal
aspects of this creation process. The experimenter
aims the proton beam at a target, or he has two
colliding beams with a certain area of intersection,
and he turns the machine on and off at will. Certainly there is some degree of control in space and
time. Correspondingly, the source function, $S(x)$,
is intended to be at least a qualitative measure of
where in space-time the creation act is more probable
(the source function is larger) or less probable (the
source function is smaller).

There is also need for another source function,
not in space and time, but in energy and momentum,
$S(p)$. The experimenter also has the possibility,
through the selection of the collision, and by various
filters, of controlling the energy and momentum of
the emitted particle. He can, in particular, reject
all particles except those falling within a particular
small energy-momentum range. Therefore, there is also
a complementary description of the source - the
idealization of a realistic collision mechanism -
representing the fact that a particular four-dimensional range of momentum may be more significantly
produced than some other range. Accordingly, we

require a source function in coordinate space and a
source function in momentum space, describing, in a
qualitative way, the relative effectiveness of
collisions in creating various states.

Now we must convert this qualitative conception
of source into a quantitative definition. Any con-
tingent microscopic event, such as the creation of
a single particle with specified properties, is
characterized by the value of a wave function or
probability amplitude. In the idealization where the
laboratory recedes totally into the background and
our attention is focused on the presence or absence
of a certain type of particle, the probability
amplitude of immediate interest connects the state
in which no particles are present - the vacuum state -
with a single particle state characterized by the
appropriate quantum numbers. The source, initially,
is a qualitative characterization of the effectiveness
of collisions in creating a particle. What could be
better than to define the source quantitatively as a
measure of the very process the source is designed to
represent? We will therefore define the source
function, to within a factor, as the probability
amplitude that relates the vacuum state [0] to the
single particle state specified by a value of the
momentum $[1_p]$:

$$\langle 1_p | 0 \rangle^S \sim S(p). \tag{1}$$

This must be qualified, however, to refer only to a weak source, which is such that the probability of creating two particles is negligible compared to that of creating a single one. This is not a restriction on the theory - it is merely our starting point.

The use of coordinate and momentum descriptions must be subjected to the requirement of quantum mechanics that these be complementary descriptions; the more you sharpen the coordinate specification, the more diffuse the momentum specification must be, and conversely. Therefore, as a generalization of familiar three-dimensional relations on wave functions, we take, as part of the definition, that the four-dimensional Fourier transform of the space-time source function, $S(x)$, which is the fundamental one, gives the momentum space source function [our units are such that $\hbar = c = 1$],

$$S(p) = \int (dx) e^{-ipx} S(x), \tag{2}$$

$$px = \vec{p} \cdot \vec{x} - p^o x^o,$$

$$(dx) = dx^o dx^1 dx^2 dx^3 = dt (d\vec{x}). \tag{3}$$

Notice particularly the choice of metric, with the minus sign being associated with the time component.

Detection

To speak of a particle being created implies
the existence of some physical means whereby you can
detect the particle in question. One way to detect
a pi meson is through its decay, but that is not
subject to control. We do achieve some measure of
space-time and energy-momentum control by again
utilizing a nuclear reaction;

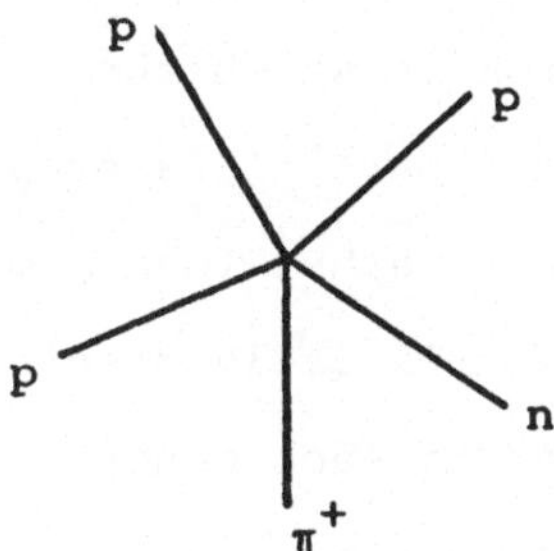

Fig. 3

which has the signature of a pi meson. For every
kind of particle, there will be a characteristic
signature, the physical processes in which it takes
part and are unique to it. Again, in the detection
of a pi meson, the experimenter is not interested in
the details of the process that signals the presence
of the particle. Accordingly, we draw below a picture designed to suggest this abstraction, in which
we have recognized that the pi meson has, within a
region of space and time somewhat subject to our

control and with varying degrees of effectiveness
in selecting this momentum or that momentum,

Fig. 4

been detected. Again, the idealization is in terms
of a source function in which the source now acts as
a sink, that is, the particle's properties are trans-
ferred to the source rather than produced by it. One
does not want to artificially distinguish things that
are clearly related to each other. Therefore, we
adopt a unified concept of source, in which the two
kinds of actions are characterized by the sense of
momentum flow. Thus, we now write the following for
the detection probability amplitude that connects the
single particle and vacuum states:

$$<0|1_p>^S \sim S(-p). \qquad (4)$$

Weak Sources

Next we turn to the factors that must be present
in these definitions of the source. We first note
that a particle is not placed into a precise state
of momentum but rather into a small range of momentum

since momentum is a continuous variable. At best,
a probability must refer to a small momentum cell,
for which we use the invariant measure

$$d\omega_p = \frac{(d\vec{p})}{(2\pi)^3} \frac{1}{2p^o}, \qquad p^o = (\vec{p}^2 + m^2)^{\frac{1}{2}}. \qquad (5)$$

The probability is proportional to $d\omega_p$ and thus the
probability amplitude for a small but finite cell
contains $d\omega_p^{\frac{1}{2}}$ as a factor. Further, the probability
amplitudes for creation and detection are related to
each other. This follows from the fact that the
physically distinct states - in particular, the
vacuum state and any single particle state - are
orthogonal to each other at a common time, which
turns out to require for <u>weak</u> sources, that

$$<0|1_p>^{S*} + <1_p|0>^{S} = 0. \qquad (6)$$

[The reader is reminded that the states appearing
here refer to <u>different</u> times, one of which is
before, the other after, the temporal region where
the source S acts.] That follows generally from
orthonormality properties but you will recognize it
in a more special way from elementary quantum perturbation theory. The probability amplitude for a
process is proportional to the matrix element of
iH' (H' = perturbation part of the Hamiltonian) if
the perturbation is weak. Since H' is Hermitian,

the negative complex conjugate relationship follows
from the factor of i. I now write down the complete
definitions that incorporate the invariant momentum
space measure and, with factors of i, satisfy the
relation between emission and absorption sources in
Eq. (6),

$$\langle 1_p | 0 \rangle^S = i\sqrt{d\omega_p}\; S(p),$$

$$\langle 0 | 1_p \rangle^S = i\sqrt{d\omega_p}\; S(p)^* = i\sqrt{d\omega_p}\; S(-p).$$

$$(7)$$

We have also recalled here that absorption corresponds to a reversed flow of momentum. The consequence
of the equality of the two versions,

$$S(p)^* = \int (dx)\, e^{ipx} S(x)^*,$$

$$S(-p) = \int (dx)\, e^{ipx} S(x),$$

$$(8)$$

is that the space-time source function is real.
Therefore, except for one brief period, we will deal
entirely with real sources. In general, it will not
be a single real source but a collection of real
sources, by means of which we shall realize the
other, discrete, quantum numbers.

Vacuum Amplitude

Let us now focus on the simplest example, which
is a particle without spin, without charge or similar
quantum number, and thus is a carrier of only energy
and momentum. This is not a restriction of the theory

but just the most elementary illustration. We use
$K(x)$ as the standard symbol for the source of a spin
zero particle. Now consider the complete process of
creation and detection, in both aspects of which we
have some measure of control, as in the experimenter's beam of particles. This also requires that we
describe the intervening stage, the undisturbed
flight of the particle from its creation domain to
its detection domain. It is characteristic of this
viewpoint that we do not say - consider this and that
particle - but, rather, we back up one stage and
include the representation of the act whereby the
particles are created and also a representation of
how the particles are detected. That is how we
acquire labels for the particles, through the representation of the creation and detection stages by
means of sources. As we shall see, sources also
serve another function, as the means whereby we move
into explicit dynamics, since the source itself is
an idealization of realistic dynamical processes. I
warn you, however, not to focus too much on the
source, as it is not the object of physical interest,
but a means to cut the Gordian knot, of getting the
theory started at some simple and insensitive point.

A complete process can be described as follows:

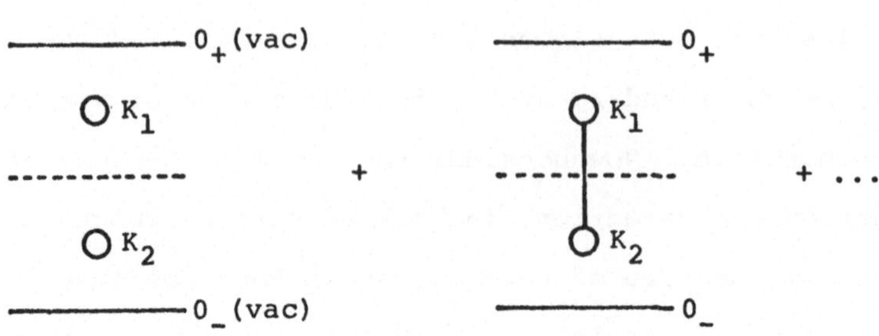

Fig. 5

We begin with the vacuum state. Then we manipulate
a source to potentially emit a particle. The first
possibility, if the source is weak, is that it does
not succeed. We then consider the situation where
the source operates and does create a particle that
propagates without disturbance and is subsequently
detected, thereby returning us to the vacuum state.
Indeed, every complete process begins with the vacuum
state and ends with the vacuum state. All the physics
is in the correlations among the sources. To repeat,
either nothing happens, or one particle is exchanged,
and so on; omitting the exchange of more than one
particle is the restriction to a weak source. I now
write down the quantum mechanical probability amplitude for this whole arrangement,

$$\langle 0_+|0_-\rangle^K = \Sigma \langle 0_+|\rangle^{K_1} \langle |0_-\rangle^{K_2} \tag{9}$$

Notice that subscripts have now appeared, - and +.
These are causal labels, - means before, + means

after, relative to the domain of the specified
source. To speak only of the vacuum state is not
enough. Vacuum states at different times are not
generally the same, if some physical action has
intervened. As we have indicated, the vacuum persistence amplitude should refer only to the total
source, $K = K_1 + K_2$. This is an essential hypothesis
admitting no intrinsic distinctions between different
regions of space and time. Here is the concept of
the uniformity of nature, that the laws of physics
are the same everywhere. The diagram, Fig. 5,
describes a causal arrangement, for you will agree
that setting up a beam of particles represents three
causally consecutive acts - production, propagation,
detection. No change of coordinate system can produce a different sequence; these are invariant
distinctions. What is indicated by the dots, in
Fig. 5, is the existence of a time subsequent to the
act of emission and preceding the act of detection
when we can, conceptually at least, classify the
states that have been produced. The overall vacuum
amplitude is then causally analyzed, first from the
initial vacuum state up to that intermediate time,
during which interval the production source has acted
but the detection of the particles has not yet
occurred, and therefore no reference to detection

can appear. This ingredient is $<|0_->^{K_2}$, which is
the probability amplitude that the source K_2 produces
some multiparticle state. The next step is the
evolution in time from the particular state at the
intermediate stage to the final vacuum state, which
occurs only because of the presence of the detection
source: $<0_+|>^{K_1}$. We then sum over all such inter-
mediate states. The simplification in the situation
of a weak source is that either nothing happens or
just one particle is exchanged:

$$<0_+|0_->^K \simeq \; <0_+|0_->^{K_1}<0_+|0_->^{K_2} + \sum_p <0_+|1_p>^{K_1}<1_p|0_->^{K_2}.$$

$$(10)$$

Here again I have injected the causality concept.
For example, I write $<0_+|0_->^{K_2}$ since the intermediate
time follows the operation of the source K_2 and the
plus subscript on the vacuum state, 0_+, describes
its causal relationship to the source K_2. But that
same state becomes 0_- when it is related to the
subsequently acting source K_1. We do not know what
this vacuum persistence amplitude is; indeed, we are
in the process of constructing it. The only thing
we are sure about is that, if there are no sources
present, the vacuum state persists with certainty
and with zero phase (zero energy-momentum) so that
the vacuum persistence amplitude is unity. In the

presence of a weak source, it deviates from unity
by some small amount,

$$\langle 0_+|0_-\rangle^K = 1 + f(K), \qquad |f(K)| \ll 1. \tag{11}$$

But, where we do know something is the single particle exchange term, involving the probability
amplitudes for emitting and absorbing a particle,
for here are our definitions of the source function:
We put it all together, for a weak source, to get

$$\langle 0_+|0_-\rangle^K \cong 1 + f(K_1) + f(K_2) + i\int (dx)(dx')K_1(x)$$
$$\times [i\int d\omega_p e^{ip(x-x')}]K_2(x'), \tag{12}$$

where, in the single particle exchange term, we have
exhibited in space-time the Fourier transforms $K_2(p)$
and $K_1(-p)$ and replaced the sum over momentum cells
by a momentum space integral. We are now turning
from a momentum specification, since we were concentrating on particles, to the complementary space-time
viewpoint where we see explicitly the causal situation, with the source $K_2(x')$ emitting any possible
particle to be subsequently detected by $K_1(x)$. Notice
how the two plane waves have combined to express the
necessary invariance under rigid displacements.

Now we recall the demand that the vacuum persistence amplitude depends only on the total source
K, to express the space-time uniformity of nature.

Having in mind that K_1 and K_2 are disjoint pieces
of K (= K_1 + K_2), we recognize that $<0_+|0_->^K$ must
differ from unity, for a weak source, by a quadratic
function of K, for that is the structure of the K_1K_2
term:

$$<0_+|0_->^K = 1 + \frac{i}{2} \int (dx)(dx') K(x) \Delta_+(x-x') K(x'). \quad (13)$$

Conversely, by inserting K = K_1 + K_2, we can identify
the various pieces of (12). In doing this, we must
note that the product $K(x)K(x')$ is symmetrical in x
and x', so that the multiplying function $\Delta_+(x-x')$
must also have this symmetry,

$$\Delta_+(x'-x) = \Delta_+(x-x'). \quad (14)$$

Our previous result, (12), made use of a causal
space-time arrangement. The comparison of the special
and the general forms identifies the function $\Delta_+(x-x')$
for the time sequence $x^0 > x^{0'}$, and the symmetry of
$\Delta_+(x-x')$ then provides its form for the situation
$x^{0'} > x^0$:

$$\begin{matrix} x^0 > x^{0'}: \\ x^{0'} > x^0 : \end{matrix} \quad \Delta_+(x-x') = i\int d\omega_p \begin{cases} e^{ip(x-x')} \\ e^{ip(x'-x)} \end{cases}, \quad (p^2+m^2 = 0)$$

$$(15)$$

in which the invariant integral extends over all
possible momenta of a real particle. It is easy to
verify that these statements define an everywhere

invariant function. I think you will recognize that
we have rediscovered the standard propagation function of conventional theories, but here it enters at
a very elementary level prior to the consideration
of differential equations. Rather, conversely, we
can derive the fact that the invariant function Δ_+
satisfies an invariant differential equation,

$$(-\partial^2 + m^2)\Delta_+(x-x') = \delta(x-x') = \int \frac{(dp)}{(2\pi)^4}\, e^{ip(x-x')} \;, \quad (16)$$

since the definition of the continuous function Δ_+
leads to a discontinuity in its time derivative, at
$x^o = x^{o'}$, which produces the four-dimensional delta
function. In giving a Fourier integral representation
for the four-dimensional delta function, we use the
symbol p quite differently, to mean a variable that
ranges over the full four-dimensional space. The
solution to this differential equation is

$$\Delta_+(x-x') = \int \frac{(dp)}{(2\pi)^4}\, \frac{e^{ip(x-x')}}{p^2 + m^2 - i\epsilon}\Bigg|_{\epsilon \to +0} \;, \qquad (17)$$

where the $i\epsilon$ conveys the boundary conditions that
distinguish the particular combination of plane waves
given in (15). It is worth emphasizing that the
four-dimensional integral construction of Δ_+ is not
a primary result, but rather merely expresses in
explicit invariant form a function that is already
completely specified.

Strong Source

Our next task is the removal of the restriction
to weak sources. What we have constructed is a
description of a weak beam of particles. We would
like to describe the beam more realistically under
circumstances in which, at a given time, a large
number of particles are present, but under such
conditions of spatial separation that they do not
interact significantly with each other. I emphasize
that the absence of interactions is achieved physically, not by mentally turning off the interactions.
(The long range Coulomb interaction of charged particles, as always, demands special treatment.) I
want to indicate how to construct such a situation
of many non-interacting particles from what we already
know. Consider the following arrangement,

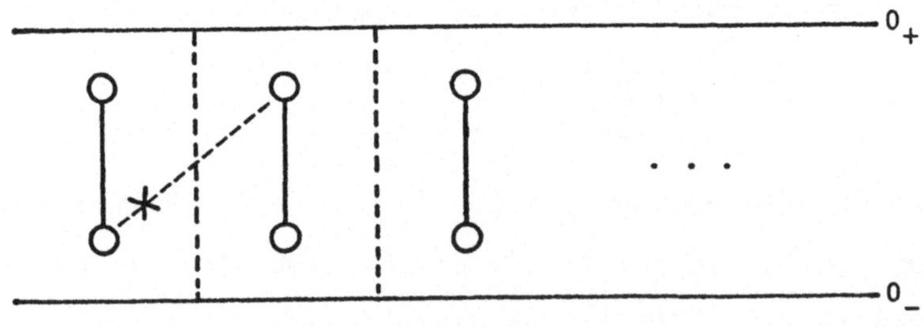

Fig. 6

involving an unlimited number of laboratories, side
by side, in each of which the emission and absorption

of at most one particle is going on independently
because of our ability to control the individual
emission and absorption acts in such a way that no
particle strays from one laboratory to another. Here
are all possible multiparticle states under conditions
of non-interaction. The total source is the superposition of all of the disjoint sources in the
various laboratories, labeled by the index α,

$$K = \Sigma K_\alpha . \tag{18}$$

The absence of physical interference between different laboratories, which characterizes the particular situation being considered, is expressed by

$$\int K_\alpha \Delta_+ K_\beta = 0, \qquad \alpha \neq \beta. \tag{19}$$

We now consider the whole system, for which we are
interested in a probability amplitude, a wave function, that describes a number of physically independent processes. We know that such wave functions
are multiplicative, so that

$$\langle 0_+|0_-\rangle^K = \prod_\alpha [1 + \frac{i}{2}\int K_\alpha \Delta_+ K_\alpha] = e^{i/2 \, \sum_\alpha \int K_\alpha \Delta_+ K_\alpha} \tag{20}$$

where we have rewritten this by taking advantage of
our restriction to individually weak sources. The
challenge to write this entirely in terms of the
total source is answered by

$$\langle 0_+|0_-\rangle = e^{i/2 \int (dx)(dx')K(x)\Delta_+(x-x')K(x')} \tag{21}$$

because of the absence of coupling between different
laboratories, expressed by (19).

It is the spirit of the source theory approach
to regard as generally applicable any space-time
formulation of source couplings, provided the basic
physical constraints are not violated. Here, the
structure (21) is considered to apply to any situation
involving an arbitrary number of particles, provided
the underlying condition of non-interaction is maintained. We emphasize the exponential structure of
(21) by writing

$$<0_+|0_->^K = e^{iW(K)},\qquad(22)$$

where

$$W(K) = \frac{1}{2}\int (dx)\,(dx')\,K(x)\,\Delta_+(x-x')\,K(x'),\qquad(23)$$

reflects the primary process of emission, propagation
and detection, while the exponential represents the
possibility of its indefinite repetition.

Fields

We began with <u>particles</u>, the physical particles,
as their properties are listed in the tables. In
order to construct a theory, we introduced a theoretical construct, the concept of <u>sources</u> which lies
quite close to reality as an abstraction of realistic
collisions. Now I point out that another concept, of
<u>fields</u>, is unavoidable. It is unavoidable because we

are operating in space and time and therefore
are necessarily concerned with what is going on
in various small space-time regions, not inside
the sources but between the sources where various
collisions can be made to occur. If we have an
emitting source and a detecting source defining
a particle beam, we are interested in a measure

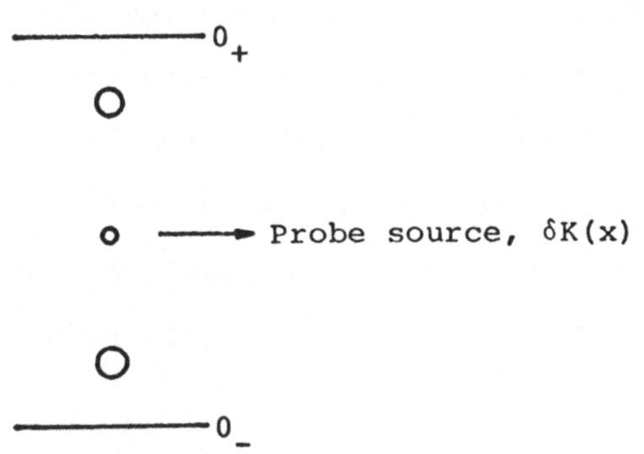

Fig. 7

of the strength of the excitation produced at intermediate points in the beam where a target might
be placed. Such questions in electrostatics are
answered by the effect on a test charge. Therefore,
analogously, we insert a test or probe source, $\delta K(x)$.
Through the action on it, we have some measure of
what is happening in that region. Since all physics
is in the overall vacuum persistence amplitude,
the important physical quantity is W and we ask how

W changes when the probe is introduced. For an
infinitesimal probe source, distributed in space
and time, the effect will be additive in various
space-time regions, as expressed by

$$\delta W = \int (dx) \delta K(x) \phi(x). \tag{24}$$

Here, $\phi(x)$ is a function that characterizes the
pre-existing situation at the point x, just as in
electrostatics, where δW becomes the change in the
electrostatic energy, δK is the test charge density,
and $\phi(x)$ is the electrostatic potential of the given
charge distribution. This is our definition of the
numerical field $\phi(x)$, as a derived concept.

The explicit form of the field thus constructed
from W is

$$\phi(x) = \int (dx') \Delta_+(x-x') K(x'), \tag{25}$$

having in mind the symmetrical quadratic form of W.
This is analogous to the electrostatic potential as
it is related to the charge density and the basic
Coulomb potential. The differential equation obeyed
by Δ_+ implies a differential equation for the field,

$$(-\partial^2 + m^2)\phi(x) = \int (dx')\delta(x-x')K(x') = K(x). \tag{26}$$

This is an inhomogeneous differential equation. The
source, K(x), originally introduced as a particle
source, has now appeared in the role of the source of
the field. I beg you again to appreciate that the

field, $\phi(x)$, is a numerical field, not an operator field.

Action

Being the fundamental quantity that it is, W must have a simple physical interpretation. We can discover it by rewriting W in various forms:

$$W = \tfrac{1}{2}\int K\Delta_+ K = \tfrac{1}{2}\int K\phi = \tfrac{1}{2}\int [(\partial\phi)^2 + m^2\phi^2]. \qquad (27)$$

The first version is an action at a distance form, having in mind the analogy of electrostatic energy. Next is a mixed form involving both field and source with a well-known electrostatic counterpart. The third version, arrived at by use of the differential equation relating source and field, (26), is the analogue of the description of energy as distributed in the field. It has also involved a partial integration that discards contributions from infinitely remote points. Here are three equivalent forms for W. Now consider yet a fourth equivalent combination:

$$W = \int (dx) [K\phi + \mathcal{L}], \qquad (28)$$

$$\mathcal{L} = -\tfrac{1}{2}[(\partial\phi)^2 + m^2\phi^2]. \qquad (29)$$

If K is altered infinitesimally, by δK, the induced change in W can arise either from the explicit appearance of K or from the implicit dependence through the field,

$$\delta W = \int (dx) [\delta K\phi + \delta_\phi (K\phi + \mathcal{L})], \qquad (30)$$

where in the second term, the source is held fixed
while the field changes. Since the first term on
the right equals δW, according to the definition of
the field, (24), the second term must vanish,

$$\int (dx)\, \delta_\phi (K\phi + \mathcal{L}) = \delta_\phi W = 0. \tag{31}$$

In other words, W is stationary with respect to
variations of the field, as a consequence of the
correct field equations. This is Hamilton's Principle
of Stationary Action. What we have thus <u>derived</u>
identifies W as the action of the system. You will
recall the asymptotic connection that exists between
a quantum wave function and classical action. Our
result resembles this, but applies only to the vacuum
persistence amplitude, and is an exact statement,
making no reference to a classical limit.

Lecture II

Probability Test

Throughout, as the theory develops by a process of evolution, we must test every extension of the theory, from a certain limited domain to a wider domain, for consistency with all the physical principles that are available. Here is an example. If $<0_+|0_->^K$ is a vacuum probability amplitude, its absolute square is a probability and, as such, must be a number that never exceeds unity. This probability is conveniently written as

$$|<0_+|0_->^K|^2 = e^{-\operatorname{Re}\int K\frac{1}{i}\Delta_+ K}. \tag{32}$$

We recall the construction given for the propagation function, Δ_+ (Eq. (15)); it shows that $(\frac{1}{i})\Delta_+$ is an invariant momentum space integral of plane waves, in which there is a correspondence in the sign of i with the sign of the time differences. But, the fact that we require the real part of $\frac{1}{i}\Delta_+$ means the algebraic sign of i does not matter, and the formula can be written without exception as

$$\operatorname{Re} \frac{1}{i}\Delta_+(x-x') = \operatorname{Re}\int d\omega_p e^{ip(x-x')}. \tag{33}$$

We now use this in (32) and recognize that we are called upon to do the space-time integrations. These are Fourier transforms yielding $K(p)$ and $K(-p) = K(p)^*$, so that

$$|<0_+|0_->^K|^2 = e^{-\int d\omega_p |K(p)|^2} = e^{-\sum_p |K_p|^2} \leq 1, \qquad (34)$$

where the injunction Re is seen to be superfluous
and, in the latter form, we are summing over momentum
cells and have defined

$$K_p = \sqrt{d\omega_p} \, K(p). \qquad (35)$$

The test is successful - the probability never
exceeds unity, and it can only equal one if the
source is incapable of emitting particles, meaning
that it varies too slowly to contain Fourier components with at least the minimum energy necessary to
create a particle.

Multi-Particle States

There is another consistency test that displays
the vacuum persistence amplitude in two different
roles. We begin with a reversal of the analysis that
led to $<0_+|0_->^K$. Recall that we started with the
restriction to a weak source, but have now arrived
at a formula that is supposed to be valid for a
strong source, subject only to the condition that
the particles do not interact. Now let us turn
around and use the vacuum amplitude to deduce all
possible multi-particle states. The source we
will use consists of two disjoint pieces,

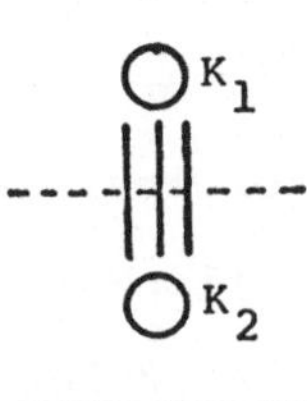

Fig. 8

having a time-like relation to each other, which define a causal situation, Fig. 8. Here is the causal analysis of this arrangement, as it is expressed by the composition property of quantum transformation functions,

$$\langle 0_+|0_-\rangle^K = \Sigma \langle 0_+|\rangle^{K_1}\langle|0_-\rangle^{K_2}. \tag{36}$$

This describes how K_2 acts to produce any multi-particle state which is subsequently detected by K_1 leading to the final vacuum state. Causality implies the physical independence of the emission and absorption processes. It must be possible, therefore, to write the probability amplitude as a sum of terms each of which has such a factored form.

We are presented with e^{iW}, where W is a quadratic function of $K = K_1 + K_2$. The three additive terms in the exponent imply that

$$\langle 0_+|0_-\rangle^K = \langle 0_+|0_-\rangle^{K_1} e^{\sum_p iK^*_{1p} iK_{2p}} \langle 0_+|0_-\rangle^{K_2}, \tag{37}$$

where the first factor comes entirely from the K_1 piece of K, the third factor similarly involves only K_2, and the middle factor evaluates the two equal contributions that are bilinear in K_1 and K_2 in conformity with the causal relation between the two. What we see in the exponent of the middle term is our weak source starting point, the summation of all ways in which a single particle can be exchanged. Its appearance in the exponential is the generalization to an arbitrary number of such independent processes. What we are told to do in (36) is write (37) as a sum of factored terms, thereby separating the emission act from the detection act. Accordingly, we take the exponential and expand it, as follows,

$$e^{\Sigma iK_{1p}^* iK_{2p}} = \prod_p \sum_{n_p=0}^{\infty} \frac{(iK_{1p}^*)^{n_p}}{\sqrt{n_p!}} \frac{(iK_{2p})^{n_p}}{\sqrt{n_p!}} , \qquad (38)$$

where the $n_p!$ denominator has been symmetrically divided. The individual terms of each summation are separated into an emission factor and an absorption factor and that will persist on multiplying together the various summations associated with the distinct momentum cells, leading to products over all momentum cells, which are labeled by the collection of integers, $\{n_p\}$, indicating the particular terms in the various summations that have been selected. You will immediately appreciate what the integer n_p means; the emission source, for example, now acts n_p times to produce a state for which there are n_p particles in the momentum cell labeled p.

Here, then, is the identification of the probability amplitude (wave function) for the creation of a state with specified numbers of particles occupying the various momentum cells (subscript 2 is omitted)

$$\langle\{n\}|0_-\rangle^K = \langle 0_+|0_-\rangle^K \prod_p \frac{(iK_p)^{n_p}}{\sqrt{n_p!}}, \qquad (39)$$

and similarly, with absorption replacing emission,

$$\langle 0_+|\{n\}\rangle^K = \langle 0_+|0_-\rangle^K \prod_p \frac{(iK_p^*)^{n_p}}{\sqrt{n_p!}}. \qquad (40)$$

These are probability amplitudes representing arbitrary multi-particle states. They are generalizations of our starting point which referred to only one particle, being restricted to a weak source. Now, for a one-particle state we have

$$\langle 1_p|0_-\rangle^K / \langle 0_+|0_-\rangle^K = iK_p,$$

$$\langle 0_+|1_p\rangle^K / \langle 0_+|0_-\rangle^K = iK_p^*, \qquad (41)$$

which identify our weak source definitions more precisely as the probability amplitudes for emitting or absorbing one particle, relative to the probability amplitude of emitting no particles. That is consistent because the vacuum persistence amplitude is practically unity for a weak source.

Completeness

Now for the promised consistency check. We have indeed indicated two roles for the vacuum persistence amplitude. First, it gives the vacuum persistence probability. Second, it's the generating function that supplies all multi-particle states in terms of their creation and annihilation amplitudes. The consistency test is whether the total probability of creating any multi-particle state, including the vacuum state, is precisely unity, which is to ask - is the totality of multi-particle states complete? The response is immediate:

$$\sum_{\{n\}} |<\{n\}|0_->^K|^2 = |<0_+|0_->^K|^2 \; \prod_p \; \sum_{n_p=0}^{\infty} \frac{(|K_p|^2)^{n_p}}{n_p!}$$

$$= |<0_+|0_->^K|^2 \; e^{\sum_p |K_p|^2} = 1;$$

$$(42)$$

completeness is verified.

Another remark. The multi-particle states are characterized just by the numbers of particles in the various modes - the possible single particle states - and these numbers run from zero to infinity. We are

obviously dealing with a system of identical particles
obeying Bose-Einstein statistics.

Particle Field

We introduced the concept of field in order to
give a measure of the strength of excitation at some
point causally intermediate between the act of emission and the act of detection, see Fig. 7. We now
want to see the explicit form of the field for such
a situation. The field is related to the total
source by the propagation function. The total source
consists of two disjoint pieces, K_1 and K_2, and the
field correspondingly decomposes,

$$\phi(x) = \phi_1(x) + \phi_2(x). \tag{43}$$

Consider ϕ_2 first,

$$\phi_2(x) = \int (dx') i \int d\omega_p e^{ip(x-x')} K(x'). \tag{44}$$

The propagation function has been written in the
appropriate causal form to represent the fact that
we are evaluating it at a point x that is later in
time than the point x'. The x' integration just
analyzes the space-time structure of the source to
yield the effectiveness of the source in producing
a particle of momentum p, and

$$\phi_2(x) = \sum_p \sqrt{d\omega_p}\, e^{ipx} iK_{2p}, \tag{45}$$

where, again, the momentum integral is broken up into

a sum over cells. Here, identified by the symbol of one particle emission, iK_{2p}, is the field of that particle,

$$\phi_p(x) = \sqrt{d\omega_p}\, e^{ipx}, \qquad (46)$$

an appropriately normalized plane wave. This field describes the flow of momentum p to the point x, from the emission source. The other part of the field, $\phi_1(x)$, is calculated in a similar manner, having in mind the different causal situation in which x is the earlier point,

$$\begin{aligned}
\phi_1(x) &= \int (dx')K_1(x')\, i\int d\omega_p\, e^{ip(x'-x)} \\
&= \sum_p iK_{1p}^*\, \phi_p(x)^*.
\end{aligned} \qquad (47)$$

This appearance of e^{-ipx} in $\phi_1(x)$ represents the flow of momentum out of the point x to the detection source. The total field is the combination of the two,

$$\phi(x) = \sum_p [iK_{1p}^*\, \phi_p(x)^* + \phi_p(x)\, iK_{2p}]. \qquad (48)$$

Notice that the definition of the field of a particle is implicit in K_p, the effective source for emitting a particle into the mode with momentum p,

$$K_p = \int (dx)\, \phi_p(x)^* K(x). \qquad (49)$$

Weak Source Target

Having derived this field structure, we make a simple application of it. Suppose we place a weak

probe source, δK, in the middle, between emission
and absorption sources, as we did to define the
field, Fig. 7. This is intended to be an idealization
of an experiment on the beam of particles, in which
a weak probe source, playing the role of a target,
is placed in the beam. The effect of the disturbance
is contained in

$$i\delta W = i\int \delta K\phi = \sum_p [iK^*_{1p} i\delta K_p + i\delta K^*_p iK_{2p}] \qquad (50)$$

which also exhibits the result of performing the
space-time integrations. We see that the probe
source is coupled to the emission and detection
sources by all possible single-particle exchanges,
in precisely the same manner as in the initial discussion of K_1 and K_2. All the physics is in the
vacuum probability amplitude, which changes by

$$\delta \, \langle 0_+|0_-\rangle^K = i\delta W \langle 0_+|0_-\rangle^K$$
$$= \sum_p [iK^*_{1p} i\delta K_p + i\delta K^*_p iK_{2p}] \qquad (51)$$
$$\times \sum_{\{n\}} \langle 0_+|\{n\}\rangle^{K_1} \langle\{n\}|0_-\rangle^{K_2} .$$

What we have to do, as indicated in Fig. 9, is carry
out a three-stage causal analysis. The physical
question being asked is this: given an initial
multi-particle state, identified by its production
mechanism, what is the probability amplitude - element

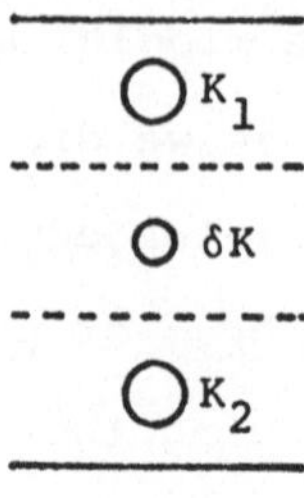

Fig. 9

of the scattering matrix - for a transition, induced
by the presence of δK, to some multi-particle final
state, as we recognize it through a detection process?
This causal analysis is expressed by

$$<0_+|0_->^{K_1 + \delta K + K_2} \qquad (52)$$

$$= \sum_{\{n\}\{n'\}} <0_+|\{n\}>^{K_1}<\{n\}|\{n'\}>^{\delta K}<\{n'\}|0_->^{K_2},$$

where $<\{n\}|\{n'\}>^{\delta K}$, the scattering matrix, is the
quantity of physical interest. The rest constitutes
the machinery for identifying the initial and final
states. The reference to the emission source K_2
in (51) appears in $<\{n\}|0_->^{K_2}$ and in an additional
iK_{2p} factor. We now note that

$$iK_{2p}<\{n\}|0_->^{K_2} = \sqrt{n_p+1} \; <\{n + 1_p\}|0_->^{K_2}, \qquad (53)$$

where the $\sqrt{n_p+1}$ factor is needed to produce the correct normalization of the $\{n + 1_p\}$ particle state.
Similarly, for the detection source, we have

$$<0_+|\{n\}>^{K_1}iK_{1p}^* = \sqrt{n_p+1} \; <0_+|\{n + 1_p\}>^{K_1}. \qquad (54)$$

We now combine (53) and (54) to rewrite the change in the vacuum amplitude, (51), as

$$\delta \langle 0_+|0_-\rangle^K = \sum_{\{n\}p} \sum \left[\langle 0_+|\{n + 1_p\}\rangle^{K_1} \sqrt{n_p+1}\, i\delta K_p \langle\{n\}|0_-\rangle^{K_2} \right.$$

$$\left. + \langle 0_+|\{n\}\rangle^{K_1} i\delta K_p^* \sqrt{n_p+1}\, \langle\{n + 1_p\}|0_-\rangle^{K_2} \right]. \tag{55}$$

This appears in exactly the form of (52), and we learn the probability amplitude for emitting a particle in the presence of $\{n\}$ others,

$$\langle\{n + 1_p\}|\{n\}\rangle^{\delta K} = \sqrt{n_p+1}\, i\delta K_p, \tag{56}$$

as well as the probability amplitude for absorbing a particle from the beam,

$$\langle\{n\}|\{n + 1_p\}\rangle^{\delta K} = \sqrt{n_p+1}\, i\delta K_p^*, \tag{57}$$

or better,

$$\langle\{n - 1_p\}|\{n\}\rangle^{\delta K} = \sqrt{n_p}\, i\delta K_p^*. \tag{58}$$

It is the weak probe simplification that limits the changes to a single particle; these are also generalizations of the initial definitions in which the vacuum state is replaced by an arbitrary multi-particle state. The absolute square of (56), gives the probability that a weak probe emits a particle in the presence of others,

$$\left| \langle\{n + 1_p\}|\{n\}\rangle^{\delta K} \right|^2 = (n_p + 1)|\delta K_p|^2, \tag{59}$$

exhibiting the existence of spontaneous and stimulated emission, a characteristic feature of Bose statistics.

We also have the absorption probability

$$|<\{n - 1_p\}|\{n\}>^{\delta K}|^2 = n_p|\delta K_p|^2. \qquad (60)$$

The appearance of these characteristic quantum factors, $n_p + 1$ and n_p, shows that a correct quantum field theory does not require the use of operator-valued fields; the numerical fields of the source-theoretic formalism suffice.

Photon

It is an essential part of our task, if we are to reach more interesting physical applications, to remove the restrictions of the initial discussion. We begin with spin, and consider a particularly important particle, the photon, which is a kind of degenerate spin one particle. By that I mean that the massless photon is really a unit helicity particle, not all the states of unit spin being realized. Our discussion began with a simple, scalar source that represents a spinless particle. One systematic procedure is to progress from a scalar source to a vector source to a tensor source and so on, with the anticipation that this is also the path from spin 0 to spin 1, to spin 2, ... The quadratic source structure of W describes emission, propagation and detection in space time. The consideration of spin will not change this general pattern but will certainly add detail. Let us begin by merely replacing the scalar source by a vector source,

$$W \overset{?}{=} \tfrac{1}{2} \int (dx)\,(dx')\,J^{\mu}(x)\,D_{+}(x-x')\,J_{\mu}(x'). \qquad (61)$$

Here, I have introduced a different code letter for the propagation function to remind you that we are dealing with a massless particle,

$$D_{+} = \Delta_{+}\ (m = 0). \qquad (62)$$

The question mark suggests that something might be wrong with this. What is wrong refers to the simple probability test that the vacuum persistence probability can never exceed unity. This will be violated as things stand, for, with the replacement $K \to J^{\mu}$, the vacuum probability, (34), becomes

$$|<0_{+}|0_{-}>^{J}|^{2} = e^{-\int d\omega_{p} J^{\mu}(p)^{*} J_{\mu}(p)}$$

$$= e^{-\int d\omega_{p}[\,|\vec{J}(p)|^{2} - |J^{o}(p)|^{2}\,]}. \qquad (63)$$

With no further qualification, this probability can exceed unity. We must restrict the magnitude of the time component so that it never exceeds the magnitude of the space component of the four-vector source,

$$|J^{o}(p)| \leq |\vec{J}(p)|, \qquad (64)$$

and this must be done in a relativistic invariant manner. In doing this, we must keep in mind that the particle under consideration is massless, which means that

$$p^O = |\vec{p}|. \tag{65}$$

Accordingly, we can rewrite (64) as

$$|p^O J^O| \leq |\vec{p}||\vec{J}|. \tag{66}$$

Now I immediately recognize that (66) is certainly satisfied if

$$p^O J^O = \vec{p} \cdot \vec{J}, \tag{67}$$

and this restriction is a relativistically invariant statement,

$$p_\mu J^\mu(p) = 0. \tag{68}$$

Here is one solution to our problem. That there is no other is a little more elaborate to show, but it can be done by considering the massless limit of a massive spin one particle. To see what else this restriction does for us, consider a particular coordinate system in which the photon moves along the third axis, where (67) says that

$$J^O = J_3. \tag{69}$$

Then, the combination that is so critical becomes

$$|\vec{J}|^2 - |J^O|^2 = |J_1|^2 + |J_2|^2$$

$$= \left|\frac{J_1 + iJ_2}{\sqrt{2}}\right|^2 + \left|\frac{J_1 - iJ_2}{\sqrt{2}}\right|^2 \geq 0. \tag{70}$$

Therefore, we find that our source has only <u>two</u> independent components and the particle (truly the photon) has only transverse polarization vectors, or helicity ± 1. All of this has occurred as a consequence of the source restriction, (68) which, in coordinate space, reads

$$\partial_\mu J^\mu(x) = 0. \tag{71}$$

Therefore, the existence of the photon demands an absolute conservation law. We know very well what physical property is absolutely conserved, it is electric charge. From this point on, all further physical tests succeed, as with spin 0.

Energy of Interaction

We now have a theory of the photon with W given by (61) where the source is restricted by (71). I described Source Theory as a phenomenological predictive theory. Let me give you an elementary example of a prediction. An essential point about W is that it refers to the source, $J_\mu(x)$, and not any particular space-time arrangement of that source. To discover new physics, consider another choice than the initial emission and absorption arrangement. Let the source J^μ consist of two pieces, labeled a and b,

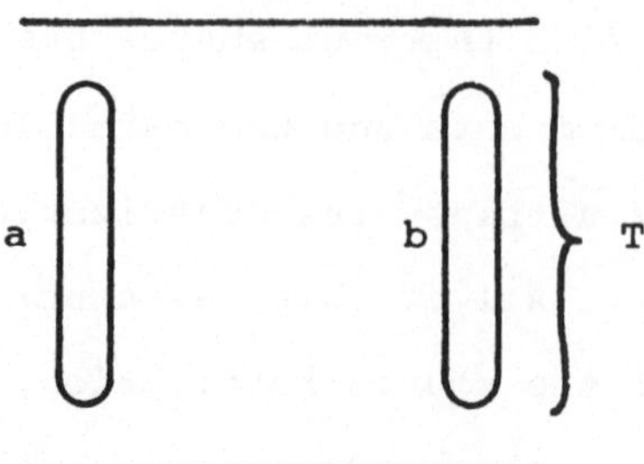

Fig. 10

which sit side by side in space-like configuration.
As indicated in Fig. 10, I want to apply the theory
to a quasi-static situation. In other words, we
begin with the vacuum, we turn on these sources
slowly, and then keep them constant for a long time,
T. We want to see what physical meaning W has under
these conditions, where the fact that the sources do
not change appreciably in time means that their
Fourier transforms are essentially restricted to
zero frequency, so that they are incapable of emitting photons. The quantity W depends on the total
source, $J^\mu = J^\mu_a + J^\mu_b$, and has a piece that refers to
J_a and J_b jointly. The corresponding factor in the
vacuum amplitude is

$$e^{i\int (dx)(dx') J^\mu_a(x) D_+(x-x') J_{\mu b}(x')}, \qquad (72)$$

where, over the long time interval T, we have,
effectively,

$$J^\mu_{a,b}(x) \cong J^\mu_{a,b}(\vec{x}). \tag{73}$$

In performing the two time integrations, we first integrate over the relative time variable that appears in D_+,

$$\int_{-\infty}^{\infty} d(x^0 - x^{0\prime}) D_+(x-x') = 2\int \frac{d\omega_p}{p^0}\, e^{i\vec{p}(\vec{x}-\vec{x}')} \tag{74}$$

$$= \int \frac{(d\vec{p})}{(2\pi)^3}\, \frac{e^{i\vec{p}(\vec{x}-\vec{x}')}}{|\vec{p}|^2} = \frac{1}{4\pi|\vec{x}-\vec{x}'|},$$

which is just the Coulomb potential in rationalized units. Having carried out the integral over the relative time, we are left with an integration over the average time coordinate of the two points. The latter is dominated by T, the arbitrarily long time interval during which this static situation exists. Therefore, the vacuum amplitude contains a factor of

$$e^{-iTE}, \tag{75}$$

which is the phase factor, in a wave function, that announces that the system has energy E, here given by

$$E = \int (d\vec{x})(d\vec{x}')\frac{J^0_a(\vec{x})J^0_b(\vec{x}') - \vec{J}_a(\vec{x})\cdot\vec{J}_b(\vec{x}')}{4\pi|\vec{x}-\vec{x}'|}, \tag{76}$$

$$\vec{\nabla}\cdot\vec{J}_{a,b}(\vec{x}) = 0. \tag{77}$$

The energy thus arrived at is the union of the Coulomb and Ampèrian interactions of charges and closed currents. Evidently, the procedure of extrapolation

from one space-time arrangement to another is
correct.

The Vector Potential

We now return to our general program. For
spin zero, we obtained W in terms of sources, then
introduced the concept of field and derived field
equations, an action and a Lagrange function. We
now repeat this pattern. To define the field, we
consider the effect of a probe source, δJ^μ, with the
essential difference that the restrictive condition
(71) cannot be violated, so any additional
source, by itself, must satisfy the condition,

$$\partial_\mu \delta J^\mu = 0. \tag{78}$$

This changes things somewhat. Recall the spin 0
formula for δW, (24). There, δK is quite arbitrary and that enables us to identify $\phi(x)$ uniquely.
Now, δJ^μ is not arbitrary. To put (78) in a more
useful form, multiply it by an arbitrary scalar
function, $\lambda(x)$, and integrate over all space-time.
After a partial integration, we have

$$\int (dx)\, \delta J^\mu \partial_\mu \lambda = 0. \tag{79}$$

The advantage of this way of writing it is that when
we consider δW in order to define a field, the vector
potential A^μ,

$$\delta W = \int (dx) \, \delta J^\mu(x) A_\mu(x) , \qquad (80)$$

we see that the vector field is arbitrary to the extent of an additional gradient of a scalar function, which is the known gauge variance of the vector potential. Having recognized this degree of arbitrariness, the field is

$$A_\mu(x) = \partial_\mu \lambda(x) + \int (dx') D_+(x-x') J_\mu(x') . \qquad (81)$$

We now do the obvious things. Given a vector relationship, you generate a scalar one by taking the divergence ($\partial J = 0$),

$$\partial^\mu A_\mu = \partial^2 \lambda . \qquad (82)$$

Then, we want to derive a differential equation making use of the differential equation satisfied by D_+, (16) with $m = 0$. With the aid of (82), this gives

$$-\partial^2 A_\mu = -\partial_\mu \partial_\nu A^\nu + J_\mu . \qquad (83)$$

Now I introduce the curl of the vector potential, the field strength,

$$F^{\mu\nu} = \partial^\mu A^\nu - \partial^\nu A^\mu , \qquad (84)$$

so that you will recognize that these are Maxwell's equations,

$$\partial_\nu F^{\mu\nu}(x) = J^\mu(x) . \qquad (85)$$

In arriving at the scalar relation (82), we
have made use of the conservation law (71).
You might therefore think that you have lost some
information, that it would be necessary to add
(71) as an additional requirement. But, as you
all know, it is a consequence of Maxwell's equations
that the differential law of conservation of charge
is an identity,

$$\partial_\mu J^\mu = \partial_\mu \partial_\nu F^{\mu\nu} \equiv 0. \qquad (86)$$

Having arrived at the field equations, the next
point is to re-express W to exhibit its physical
interpretation as an action. We repeat what we did
for spin 0:

$$W = \tfrac{1}{2}\int JDJ = \tfrac{1}{2}\int JA = \tfrac{1}{2}\int A_\mu \partial_\nu F^{\mu\nu} = \tfrac{1}{4}\int F^{\mu\nu}F_{\mu\nu}. \qquad (87)$$

The expression that is an action, in the sense that
it is stationary under field variations, is

$$W = \int (dx)\,[J^\mu A_\mu + \mathcal{L}]. \qquad (88)$$

$$\mathcal{L} = -\tfrac{1}{4}F^{\mu\nu}F_{\mu\nu} = \tfrac{1}{2}(\vec{E}^2 - \vec{H}^2), \qquad (89)$$

which is the standard Lagrange function of Maxwell's
theory.

The further development of the photon theory
is in the spin 0 pattern, but with the additional
quantum number of helicity or a polarization label,
as indicated by

$$K_p \rightarrow J_{p\lambda} = \int (dx) A^{\mu}_{p\lambda}(x)^* J_{\mu}(x), \qquad (90)$$

where $A^{\mu}_{p\lambda}$ is the field of the particular photon labeled by $p\lambda$,

$$A^{\mu}_{p\lambda}(x) = \sqrt{d\omega_p}\; e^{ipx} e^{\mu}_{p\lambda}, \qquad (91)$$

$e^{\mu}_{p\lambda}$ being an orthonormal set of two polarization vectors. This is characteristic of how things develop in general. The field of a particle will be composed of eigenvectors of the various quantum numbers. We also remark that the field at an intermediate point in a causal arrangement, Fig. 7, is constructed analogously to (48), for spin 0,

$$A^{\mu}(x) = \sum_{p\lambda} [iJ^*_{1\,p\lambda} A^{\mu}_{p\lambda}(x)^* + A^{\mu}_{p\lambda}(x) iJ_{2\,p\lambda}]. \qquad (92)$$

First Order Lagrange Function

It is very useful, for some purposes, not to deal with second order differential equations as in a Lagrangian viewpoint, but the first order equations of a Hamiltonian viewpoint. This requires the introduction of more variables. Thus, for spin zero, instead of the single scalar field ϕ, we now introduce an additional vector field, ϕ^{μ}. We take the Lagrange function, (29), and add a particular structure,

$$\mathcal{L} = -\tfrac{1}{2}(\partial\phi)^2 - \tfrac{1}{2}m^2\phi^2 + \tfrac{1}{2}(\phi^{\mu} - \partial^{\mu}\phi)(\phi_{\mu} - \partial_{\mu}\phi). \qquad (93)$$

You might think that we have thereby changed the system, but it is an equivalent description. If I extend the action principle to variations of the vector field, the field equation I deduce is

$$\delta\phi_\mu: \quad \phi_\mu = \partial_\mu\phi \tag{94}$$

which is to say that the vector field is just the gradient of the scalar field. If I exercise the option to restrict the vector field to be that gradient the original Lagrange function is recovered. But if I use (93) as it stands, the terms with two gradients cancel to yield a Lagrange function with a single gradient,

$$\mathcal{L} = -\phi^\mu\partial_\mu\phi + \frac{1}{2}\phi^\mu\phi_\mu - \frac{1}{2}m^2\phi^2. \tag{95}$$

This is the Hamiltonian form, the analog of $p\dot{q}$ - H. The action principle, applied to this $\mathcal{L}$, (95), now yields a pair of first order field equations,

$$\partial_\mu\phi = \phi_\mu, \tag{96}$$

$$-\partial_\mu\phi^\mu + m^2\phi = K, \tag{97}$$

from which the second order differential equation can be recovered. We can do the same thing for the photon by using both A^μ and $F^{\mu\nu}$ as independent variables. Then the first order form of the Lagrange function is

$$\mathcal{L} = -\tfrac{1}{2}F^{\mu\nu}(\partial_\mu A_\nu - \partial_\nu A_\mu) + \tfrac{1}{4}F^{\mu\nu}F_{\mu\nu}, \tag{98}$$

which implies the field equations,

$$\partial_\mu A_\nu - \partial_\nu A_\mu = F_{\mu\nu}, \tag{99}$$

$$\partial_\nu F^{\mu\nu} = J^\mu. \tag{100}$$

Lecture III

Complex Sources

In the discussion of spin-zero particles, we used a real space-time source function. Suppose we had considered a complex source? Because of time limitations, I will discuss this question only briefly. Certainly the general idea of emission, propagation and absorption continues to hold. We therefore ask only this - given a complex source, does it combine with itself or with its complex conjugate in the structure of W?:

$$W \stackrel{?}{=} \int K\Delta_+ K, \tag{101a}$$

$$W \stackrel{?}{=} \int K^*\Delta_+ K. \tag{101b}$$

To answer the question, I point to a simple physical requirement, namely, if no particle creation is possible, the vacuum persistence probability amplitude must be such that the derived probability is one,

$$|e^{iW}|^2 = 1. \tag{102}$$

In other words, W must be real under these circumstances. Then, if we look at the propagation function, Δ_+, in its four-dimensional invariant form, (17),

$$\Delta_+ = \int \frac{(dp)}{(2\pi)^4} \frac{e^{ip(x-x')}}{p^2 + m^2 - i\varepsilon}, \tag{103}$$

the condition that no particle creation occur is
that no Fourier components exist with $p^2 + m^2 = 0$.
In that circumstance, where the denominator is
effectively different from zero, and therefore real,
and, where the integration variables run over all
values, positive and negative, so that i and -i occur
with equal weight, the function Δ_+ is an effectively
real function; it becomes complex only when propagation can occur. We now see that to make a real W
using a real propagation function, it is necessary
to combine the complex source with its complex
conjugate, (101b),

$$W = \int (dx)\,(dx')\,K^*(x)\,\Delta_+(x-x')\,K(x'). \qquad (104)$$

The use of multi-component real sources is much
more convenient than that of complex sources. A
complex source can be written in terms of two real
sources:

$$K_{complex} = \frac{1}{\sqrt{2}}(K_{(1)} - iK_{(2)}) = \phi_+^*\,K_{real}. \qquad (105)$$

In the last form, I regard this combination of real
sources as a projection of a two-component real
source, K_{real}, onto the two-component vector

$$\phi_+ = \frac{1}{\sqrt{2}}\binom{1}{i}. \qquad (106)$$

In addition, there is the complex conjugate structure,

$$K_{complex}^* = \frac{1}{\sqrt{2}}(K_{(1)} + iK_{(2)}) = \phi_-^*\,K_{real}, \qquad (107a)$$

where the associated two-component vector has the
opposite sign of i,

$$\phi_- = \frac{1}{\sqrt{2}}\begin{pmatrix} 1 \\ -i \end{pmatrix}. \tag{107b}$$

The $\phi_\pm$ are the unit orthogonal vectors that are the
two eigenvectors of the antisymmetrical, imaginary,
Hermitian matrix q,

$$q = \begin{pmatrix} 0 & -i \\ i & 0 \end{pmatrix}, \tag{108a}$$

$$q\phi_\pm = \pm\ \phi_\pm, \tag{108b}$$

which identifies the subscripts of $\phi_\pm$ as the eigenvalues of q.

The structure of W in terms of a complex source
is

$$W = \int K_c^* \Delta_+ K_c. \tag{109}$$

You will notice that it is invariant under a constant
phase transformation (α real)

$$K_c \rightarrow e^{i\alpha} K_c, \quad K_c^* \rightarrow e^{-i\alpha} K_c^*. \tag{110}$$

When real sources are introduced, the expression for
W is

$$W = \frac{1}{2}\int [K_{(1)} \Delta_+ K_{(1)} + K_{(2)} \Delta_+ K_{(2)}]$$
$$= \frac{1}{2}\int K\Delta_+ K, \tag{111}$$

where, in the last form, we have adopted a matrix

notation to convey the underlying geometry, which
is that of a two-dimensional Euclidean space called
charge space, anticipating that the additional degree
of freedom is that of charge, as realized by, but
not necessarily restricted to, electric charge. The
invariance expressed by the complex transformation
(110) now appears in real terms as the two-dimensional
rotation conveyed by

$$K \rightarrow e^{iq\alpha}K = K\,e^{-iq\alpha}. \tag{112}$$

Since the final form of (111) generalizes the single
real source structure, the remainder of the discussion
concerning states, completeness, probability requirements, is also analogous. The only difference is
that the particle has acquired a new quantum number,
the eigenvalue of q,

$$K_p \rightarrow K_{pq}, \qquad q = \pm 1. \tag{113}$$

It is evident that this is the situation of particles
and anti-particles with a common mass. The source
for the emission of a particle labeled by quantum
numbers for energy and momentum, p, and charge, q,
will again be the projection of the space-time (two-
component) source onto the field that represents that
particular particle,

$$K_{pq} = \int (dx)\,\phi_{pq}(x)^{*}K(x), \tag{114}$$

$$\phi_{pq}(x) = \sqrt{d\omega_p}\,\phi_q e^{ipx}, \tag{115}$$

and, as before, the field of the particle is an
eigenvector specified by all the appropriate quantum
numbers.

Spin 1/2

The time has come for a rapid run-through of
spin 1/2. To appreciate the implication of the
restriction to real sources, consider an infinitesimal three-dimensional rotation, as described by the
usual unitary matrix

$$1 + i \frac{1}{2} \vec{\sigma} \cdot \delta\vec{\omega}, \qquad (116)$$

where the $\frac{1}{2} \vec{\sigma}$ are the spin angular momentum matrices.
If the real sources are to stay real, these matrices
must all be imaginary,[*] and therefore cannot be the
2 x 2 Pauli matrices. To achieve our goal an additional 2 x 2 real representation of i is needed, thus
requiring 4 x 4 matrices. You know that the complete
set of 16 such matrices can be generated by the basic
set γ^μ, $\mu = 0,1,2,3$, which obey anti-commutation
relations,

$$\frac{1}{2}\{\gamma_\mu, \gamma_\nu\} = -g_{\mu\nu}, \qquad (117)$$

$$g_{00} = -1, \quad g_{11} = g_{22} = g_{33} = +1, \qquad (118)$$

where, in particular,

$$(\gamma_o)^2 = +1. \qquad (119)$$

---

[*]This is illustrated by the transformation (112) with
its use of the imaginary matrix q.

A standard set of 16 Hermitian matrices is given by

$$\gamma^o[1, \ \gamma^\mu, \ \sigma^{\mu\nu}, \ i\gamma^\mu\gamma_5, \ \gamma_5] \qquad (120)$$

with

$$\sigma^{\mu\nu} = \tfrac{i}{2}[\gamma^\mu, \ \gamma^\nu], \qquad (121)$$

$$\gamma_5 = \gamma^o\gamma^1\gamma^2\gamma^3. \qquad (122)$$

These 16 matrices can also be separated into 10 real, symmetrical and 6 imaginary, antisymmetrical matrices:

Real symmetrical: $\qquad\qquad \gamma^o[\gamma^\mu, \ \sigma^{\mu\nu}], \qquad$ (123a)

Imaginary antisymmetrical: $\ \gamma^o[1, \ i\gamma^\mu\gamma_5, \ \gamma_5].$ (123b)

The matrix γ^o is the matrix analogue of the metric tensor. It anticommutes with its three spatial partners $\vec{\gamma}$. As such, it is the space reflection matrix, and its eigenvalue is the space parity.

We want to describe a spin-1/2 particle, with a definite parity, in its rest frame. This is a requirement that has to be imposed, since our framework admits two values of the parity: $\gamma^{o'} = \pm 1$. In order to pinpoint the essential difference, compared with our earlier spin-0 discussion, we consider the weak source coupling between emission and absorption, for a particular particle in its rest frame. This is represented by

$$m \ \eta_1(-p) \ (1 + \gamma^o)\eta_2(p), \qquad (124)$$

where I use the letter η to indicate the source of
spin-1/2 particles. The latter is a four-component
object, which four components can be classified by
a spin eigenvalue ($\sigma_3' = \pm 1$), and, independently,
since $[\gamma^o, \sigma_3] = 0$, the parity eigenvalue ($\gamma^{o\prime} = \pm 1$).
To achieve a definite parity in the rest frame, we
have inserted a projection matrix that selects
$\gamma^{o\prime} = +1$ and rejects $\gamma^{o\prime} = -1$. The factor of m,
the particle mass, appears in (124) for convenience,
and I remind you that $m = p^o$ in the rest frame.
Therefore, (124) can be rewritten as

$$\eta_1(-p)\gamma^o[m + \gamma^o p^o]\eta_2(p), \tag{125}$$

where we recognize that $\gamma^o p^o$ is the reduced form in
the rest frame of the invariant structure, $-\gamma^\mu p_\mu$.
When presented in space-time language, the resulting
expression for W, appropriate to spin-1/2 particles,
is

$$W = \frac{1}{2}\int (dx)(dx')\eta(x)\gamma^o(m - \gamma^\mu \frac{1}{i}\partial_\mu)\Delta_+(x-x')\eta(x'). \tag{126}$$

This quadratic form is, for our immediate purpose, usefully   presented as

$$W = \frac{1}{2}\int \eta_\zeta(x)K_{\zeta\zeta'}(x, x')\eta_{\zeta'}(x'), \tag{127}$$

where ζ is the four-valued spinor index, and summations are understood. I now observe that the kernel
K is antisymmetrical with respect to the full set of

labels carried by the source:

$$K_{\zeta'\zeta}(x',x) = -K_{\zeta\zeta'}(x,x'). \tag{128}$$

This follows from the antisymmetry of γ^0 and the symmetry of $\gamma^0\gamma^\mu$ combined with $\partial'_\mu = -\partial_\mu$. Any attempt to alter this property founders on the probability restrictions; it is inescapable that the kernel is totally antisymmetric. What does this mean? If the η's were the same kind of numbers we have been dealing with for spins 0 and 1, the W function would vanish identically, somewhat contrary to the experimental existence of spin-1/2 particles. The only escape is to adapt the algebraic properties of the source to the symmetry properties of the kernel, which is to say that they must be <u>totally anticommutating numbers</u>,

$$\eta_{\zeta'}(x')\eta_\zeta(x) = -\eta_\zeta(x)\eta_{\zeta'}(x'), \tag{129}$$

familiar mathematically as the elements of a Grassmann algebra.

What are the physical consequences of this total anticommutativity? The sources for the emission of particles with definite quantum numbers, here momentum and spin, are again projections onto the fields characterizing the particles, by means of eigenfunctions associated with the quantum numbers,

$$\psi_{p\sigma}(x) = \sqrt{d\omega_p}\; u_{p\sigma}\, e^{ipx}, \tag{130}$$

where $u_{p\sigma}$ is a spin eigenvector. These particle
emission sources,

$$\eta_{p\sigma} = \int (dx) \psi_{p\sigma}^*(x) \gamma^o \eta(x), \qquad (131)$$

as linear combinations of the space-time source
function, are also totally anticommutative,

$$\{\eta_{p\sigma}, \eta_{p'\sigma'}\} = 0. \qquad (132)$$

Included here is the statement referring to $p\sigma = p'\sigma'$,
namely,

$$(\eta_{p\sigma})^2 = 0. \qquad (133)$$

You will recognize the Pauli Verbot - the exclusion
principle - since the attempt to emit two particles
into the same mode fails! We have been forced to
invent Fermi-Dirac (F.D.) statistics. Here is an
example of the spin-statistics connection, which
can be extended to all values of the spin.

As in our previous discussions, we define fields
by considering the response of W to the introduction
of a weak probe source, $\delta\eta$,

$$\delta W = \int (dx) \delta\eta(x) \gamma^o \psi(x) = \int (dx) \psi(x) \gamma^o \delta\eta(x), \quad (134)$$

where we include γ^o, acting in its role as metric
matrix. The field is thus given by

$$\psi(x) = \int (dx') G_+(x,x') \eta(x'), \qquad (135)$$

where G_+ is the propagation function,

$$G_+(x,x') = (m - \gamma^\mu \tfrac{1}{i} \partial_\mu) \Delta_+(x-x'). \qquad (136)$$

Since these fields are a linear combination of the
sources, they have the same kind of algebraic property
as the sources; they are <u>totally anticommuting num-
bers</u>. For Bose statistics, we use commuting numbers,
while Fermi statistics requires anticommuting numbers.
Indeed these are the algebraic counterparts of the
essential physical characteristics of the two types
of particles, when successive multiplication symbolizes successive emission. Notice that the two forms
of (134) are equivalent in consequence of the spin-
statistics connection, the antisymmetry of γ^o (spin)
compensating the anticommutativity of the two factors
(statistics).

The propagation function G_+ obeys a first order
differential equation that follows from the second
order equation obeyed by Δ_+,

$$(\gamma^\mu \tfrac{1}{i}\partial_\mu + m)G_+ = (m^2 - \partial^2)\Delta_+ = \delta(x-x'), \quad (137)$$

and correspondingly, the field obeys an inhomogeneous
equation in which the particle source has become the
source of the field,

$$(\gamma^\mu \tfrac{1}{i}\partial_\mu + m)\psi(x) = \eta(x); \quad (138)$$

it is the inhomogeneous Dirac equation.

The procedure followed before leads to an action,
and the associated Lagrange function. Here are the
various forms of W:

$$W = \frac{1}{2}\int \eta\gamma^{o}G_{+}\eta = \frac{1}{2}\int \eta\gamma^{o}\psi = \frac{1}{2}\int \psi\gamma^{o}(\gamma^{\mu}\frac{1}{i}\partial_{\mu} + m)\psi(x). \qquad (139)$$

The particular combination that has the stationary property is

$$W = \int (dx)\,[\eta\gamma^{o}\psi + \mathcal{L}], \qquad (140a)$$

$$\mathcal{L} = -\frac{1}{2}\psi\gamma^{o}(\gamma^{\mu}\frac{1}{i}\partial_{\mu} + m)\psi, \qquad (140b)$$

from which the inhomogeneous Dirac equation can be recovered.

Currents

Let us return to the charge space discussed in the context of spin zero and the invariance under phase transformations, (112). We recall the structure of the action that yields first order field equations,

$$W = \int [K\phi - \phi^{\mu}\partial_{\mu}\phi + \frac{1}{2}\phi^{\mu}\phi_{\mu} - \frac{1}{2}m^{2}\phi\phi]. \qquad (141)$$

It is invariant under the constant phase source transformation

$$K \to e^{iq\alpha}K \qquad (142)$$

together with that of the fields,

$$\left.\begin{array}{c}\phi\\ \phi^{\mu}\end{array}\right\} \to e^{iq\alpha}\left\{\begin{array}{c}\phi\\ \phi^{\mu}\end{array}\right. . \qquad (143)$$

Suppose we deliberately destroy that invariance by letting the rotation angle α be a function of space and time, $\alpha \to \alpha(x)$. For simplicity, consider an infinitesimal transformation, with the parameter

$\delta\alpha(x)$. Then the infinitesimal changes in the source and the fields are

$$\delta K(x) = iq\delta\alpha(x)K(x), \qquad (144a)$$

$$\left.\begin{array}{c}\delta\phi \\ \delta\phi^{\mu}\end{array}\right\} = iq\delta\alpha(x)\left\{\begin{array}{c}\phi \\ \phi^{\mu}.\end{array}\right. \qquad (144b)$$

The induced change in W can be computed in two ways. According to the action principle, only the explicit change in K matters,

$$\delta W = \int\phi\delta K = \int\phi iqK\delta\alpha. \qquad (145a)$$

But, alternatively, δW differs from zero only because ∂_{μ} now acts on $\delta\alpha(x)$,

$$\delta W = -\int\phi^{\mu}iq\phi\partial_{\mu}\delta\alpha, \qquad (145b)$$

thereby creating, as the coefficient of $-\partial_{\mu}\delta\alpha$, the vector,

$$j^{\mu} = \phi^{\mu}iq\phi. \qquad (146)$$

What is its physical significance? For simplicity, we will assume that $\delta\alpha(x)$ is zero inside the source so that $\delta W = 0$ according to (145a). Then, everywhere outside the source, the divergence of j^{μ} is zero,

$$\partial_{\mu}j^{\mu} = 0. \qquad (147)$$

Accordingly, this vector describes the flow of a conserved quantity. Recalling the antisymmetry of q and that $\phi_{\mu} = \partial_{\mu}\phi$, we can write j^{μ} as

$$j^{\mu} = \phi q\frac{1}{i}\partial^{\mu}\phi, \qquad (148)$$

which makes explicit the reference to flux: $\frac{1}{i}\partial^{\mu}$,
symbolizing momentum, or four-dimensional velocity,
and the reference to what is flowing: q, a charge.
Thus j^{μ} is a charge flux vector.

A similar discussion applies to spin-1/2, now
referring to particle and antiparticle, so that
sources and fields have 4 x 2 components. The
action,

$$W = \int (dx)\,[\psi\gamma^{o}\eta - \tfrac{1}{2}\psi\gamma^{o}(\gamma^{\mu}\tfrac{1}{i}\partial_{\mu} + m)\psi], \qquad (149)$$

is subjected to infinitesimal phase transformations

$$\delta\eta(x) = iq\delta\alpha(x)\eta(x), \quad \delta\psi(x) = iq\delta\alpha(x)\psi(x), \qquad (150)$$

so that

$$\delta W = \int (dx)\,\psi\gamma^{o}iq\eta\delta\alpha = -\int (dx)\,j^{\mu}\partial_{\mu}\delta\alpha, \qquad (151)$$

where

$$j^{\mu} = \tfrac{1}{2}\psi\gamma^{o}\gamma^{\mu}q\psi \qquad (152)$$

is the conserved charged flux vector.

Currents as Photon Sources

In these two examples we recognize vector objects,
associated with particles, that obey the fundamental
characteristic of photon sources, $\partial_{\mu}J^{\mu} = 0$. Thus
far, the primary role of sources has been as a book-
keeping device, a way of identifying various particle
states. But the source, as an idealization of realistic collisions, is also a model of physical interaction

mechanisms, emphasizing what is common to all such
mechanisms. As such, it provides a starting point
for the development of dynamics. We illustrate this
with the photon and the electron-positron: electrodynamics.

In the initial situation, of non-interacting
particles of two different types, here, photons and
spin-1/2 electrons (positrons), the physical independence of the two types, as expressed by the
multiplicativity of the respective vacuum amplitudes,
implies simple additivity of the corresponding
actions:

$$W = \int (dx) [J^\mu A_\mu - \tfrac{1}{4} F^{\mu\nu} F_{\mu\nu}]$$
$$+ \int (dx) [\psi\gamma^o \eta - \tfrac{1}{2}\psi\gamma^o (\gamma^\mu \tfrac{1}{i}\partial_\mu + m)\psi].$$

(153)

We are going to introduce interactions by asserting
that the vector constructed from the electron field
ψ can also act as a photon source. But this requires
a small change in (152) if we are to maintain the
identification of the photon source vector as the
electric charge vector, since the charge unit is e,
not one. Accordingly, we adopt

$$j^\mu = \tfrac{1}{2}\psi\gamma^o\gamma^\mu eq\psi$$

(154)

as the electric current vector for the electron-
positron system.

The statement that j acts physically in the manner of J, which is conveyed by the substitution $J \rightarrow J + j$, yields the first level of dynamics, as expressed by the action

$$W = \int (dx) [(J + j)^{\mu} A_{\mu} - \tfrac{1}{4} F^{\mu\nu} F_{\mu\nu}]$$
$$+ \int (dx) [\psi\gamma^{o}\eta - \tfrac{1}{2}\psi\gamma^{o}(\gamma^{\mu}\tfrac{1}{i}\partial_{\mu} + m)\psi] \tag{155}$$

or

$$W = \int (dx) [J^{\mu} A_{\mu} + \psi\gamma^{o}\eta + \mathcal{L}], \tag{156}$$

with

$$\mathcal{L}(\psi,A) = -\tfrac{1}{4} F^{\mu\nu} F_{\mu\nu} - \tfrac{1}{2}\psi\gamma^{o}[\gamma^{\mu}(\tfrac{1}{i}\partial_{\mu} - eqA_{\mu}) + m]\psi. \tag{157}$$

Here we see that our physical way of looking at things has yielded a familiar result - the appearance of the gauge covariant derivative that conveys the locking of the electromagnetic gauge transformation,

$$A_{\mu} \rightarrow A_{\mu} + \partial_{\mu}\lambda, \tag{158a}$$

with the charge-bearing electron field phase transformation,

$$\psi \rightarrow e^{ieq\lambda}\psi. \tag{158b}$$

What follows is a very cursory glimpse into how quantum electrodynamics is evolved in source theory. I want to emphasize its characteristic inductive procedure as contrasted with the almost universally

accepted deductive ideal that does such violence
to the reality of research activity.

<u>Extended Sources</u>

The initial non-interactive stage of electrodynamics is suggested by the causal diagrams in
Fig. 11. Here we recognize electron and photon

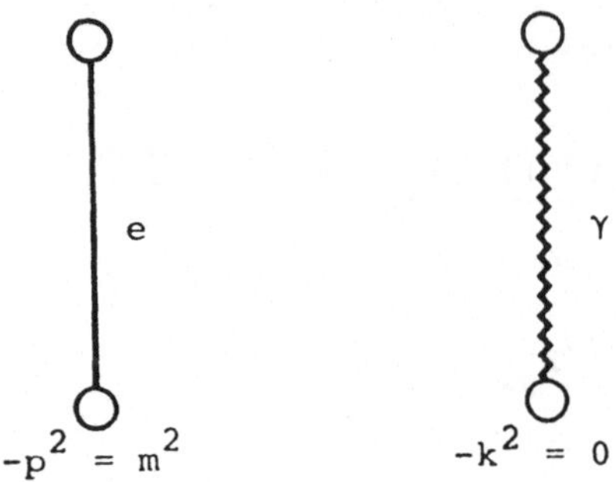

Fig. 11

sources that emit and absorb just the right balance
of energy and momentum to exchange an electron or a
photon over arbitrary temporal intervals. Now, a
source that is capable of emitting exactly the right
energy for a given momentum also has the potential
to put out either too much or too little energy.
What happens when a source emits too much energy
to produce an electron? It is customary to make a
distinction between real particles, for which $-p^2 =
m^2$ is satisfied, and virtual particles, where the
relation is not satisfied, but there is still an

analogous physical excitation. This excitation

cannot propagate very far as such because it does

not have the correct balance of energy and momentum.

In anthropomorphic language, it then tries to get

rid of the excess energy and does so by any of the

dynamical means available to it. For electrons, the

dynamical mechanism is the emission of a photon, as

illustrated* in Fig. 12a. This is the quantum

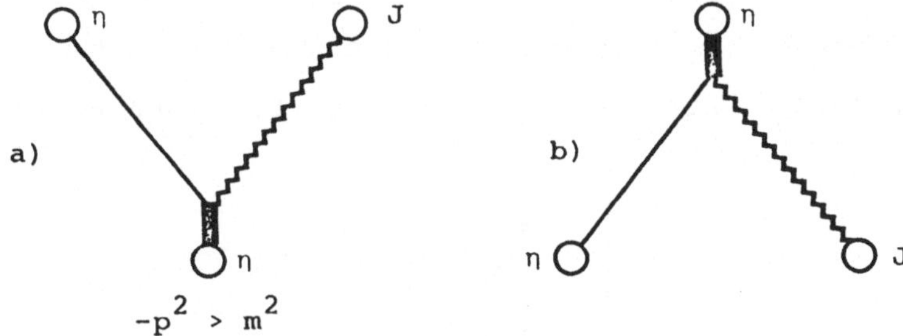

Fig. 12

transcription of the classical fact that an accel-

erated charge radiates. Through this extension of

the concept of source, freeing it from the initial

restriction on energy and momentum, we have recognized

the existence of processes involving three sources.

This is a physical transcription of the mathematical

$\psi\psi A$ coupling that appears in (157). The three-source

-----

*In source diagrams, a light thin line corresponds
to real particle propagation while a dark heavy line
represents a virtual particle.

coupling will apply to any space-time arrangement, for example, that shown in Fig. 12b. And, once we have introduced a three-source coupling, we can generate couplings involving any larger number of sources.

Scattering Processes

We illustrate this by constructing a coupling among four electron sources. The couplings of Fig. 12a and b provide effective photon emission and absorption sources, which are combined in the complete causal arrangement of Fig. 13. When this

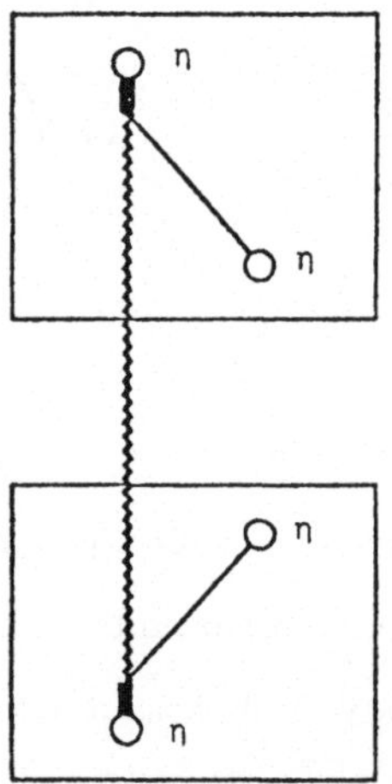

Fig. 13

coupling is exhibited in space-time form, it will no longer refer to the initial causal arrangement and can be applied to any arrangement of four electron

sources. If I pick the sources to emit and absorb
real particles, I arrive at the physical processes
shown in Fig. 14. You will recognize that these

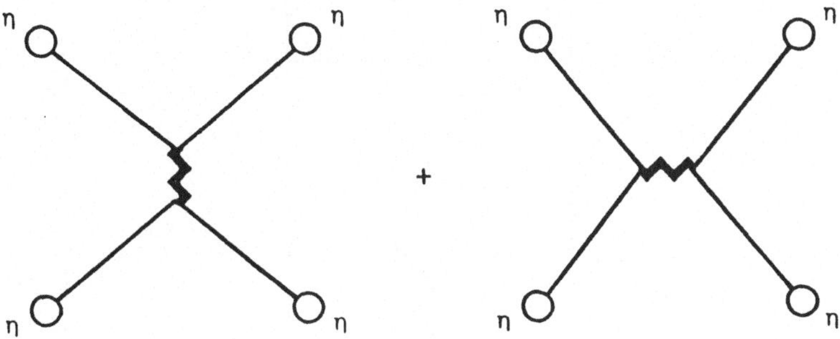

Fig. 14

pictures have the same connectivity as that in Fig.
13, expressing their common origin in a single space-
time coupling. In the second process of Fig. 14, two
electron sources have emitted two electrons, of either
charge, which interact by exchanging a space-like
virtual photon and are subsequently detected. This
is relativistic Coulomb scattering, proceeding through
a single photon exchange. In the first arrangement
of Fig. 14, the photon is time-like. Obviously, this
is only possible for particles of opposite charge,
unlike Coulomb scattering which involves no restriction on the individual charges.

Another way to generate a four-source coupling
is by exchanging an electron, as shown in Fig. 15.

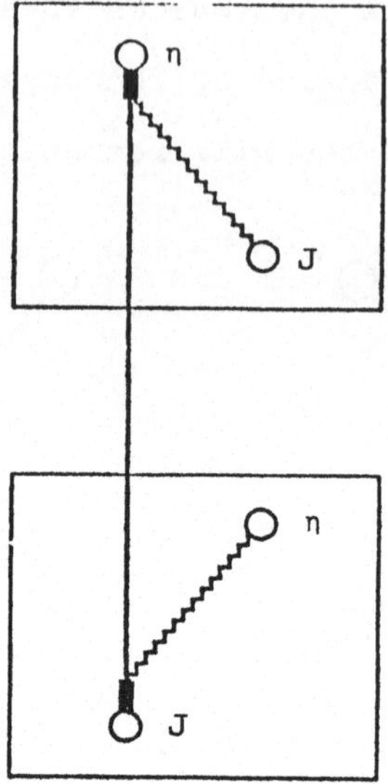

Fig. 15

This leads to a mechanism for electron-photon
scattering, see Fig. 16, which includes both time-
like and space-like exchanged electrons.   The same

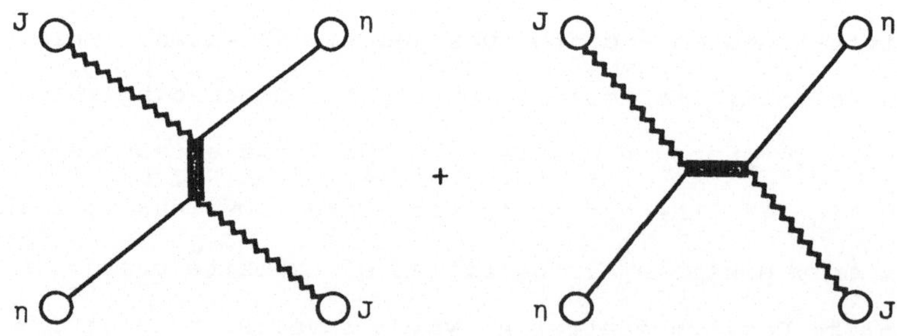

Fig. 16

four-source coupling also leads to electron-positron
annihilation into two photons, Fig. 17 (as well as
the inverse process).

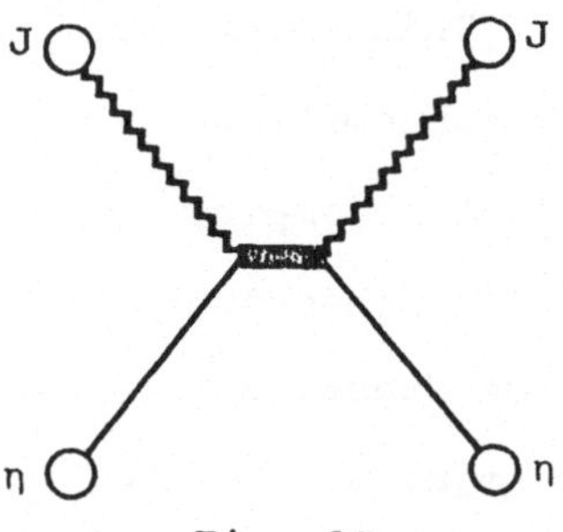

Fig. 17

Modified Propagation Functions

This illustrates the first stage of dynamics, wherein we recognize the existence of an unlimited number of physical processes in an initial, skeletal description. The second stage introduces yet other processes and improves the description of the first stage. It involves the recognition of two-particle exchange. In consequence of the extension of the source concept, electron sources can also exchange an electron and a photon. This process is shown in Fig. 18, along with the starting point, single

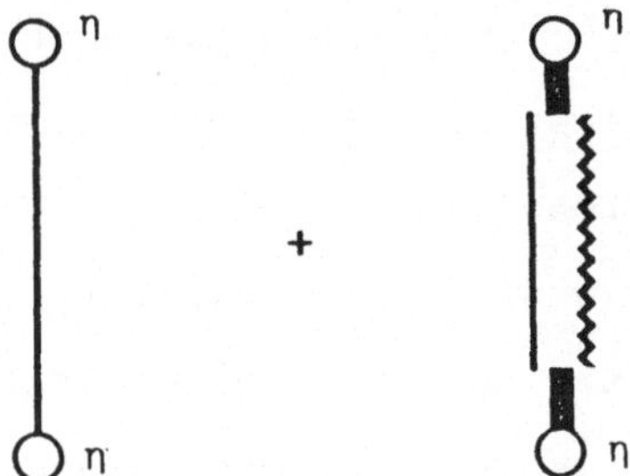

Fig. 18

particle exchange. This generalization has its
counterpart in the mathematical scheme since the
propagation functions we employ describe how an
excitation travels from one source to another. The
original propagation function referred only to a
single particle of mass m. However, we must supplement it to represent that, propagating between the
two sources, there can be a composite of an electron
and a photon, which, as a two-particle system, has
a mass that ranges from the electron mass up to
arbitrarily large values. This forces a modification
of the propagation function,

$$G_+ \rightarrow \overline{G}_+.$$

A direct experimental test of this modified propagation function as it stands is difficult, but it
suffices to consider an external weak homogeneous
magnetic field. Then the modified propagation
function describes a particle with an additional
magnetic moment, over and above what the Dirac
equation implies, and it is the well-known result
of $\alpha/2\pi$ magnetons.

There is a similar possibility of two-particle
exchange for the photon. Our general three-source
coupling implies that a photon source, through the
intermediary of a virtual photon, can emit, or detect,

an electron-positron pair. Indeed, these mechanisms
are evident in the first of the two ways electron-
positron scattering takes place (Fig. 14). This
process, as well as single-photon exchange, is
suggested in Fig. 19. To represent this additional

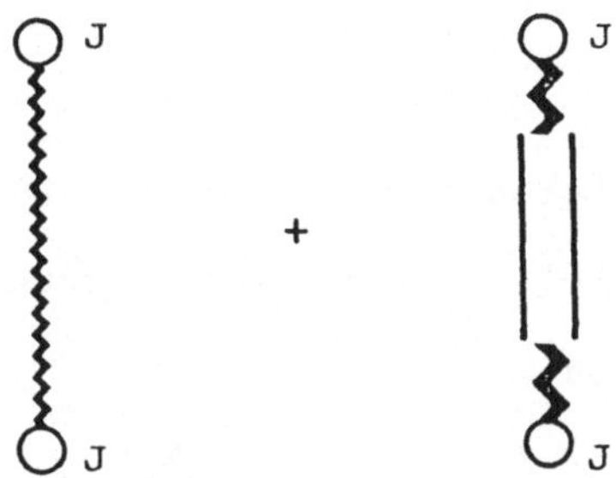

Fig. 19

coupling between the two sources, the propagation
function must be generalized; it must refer not only
to the particle of mass zero but also to excitations
with a mass threshold at twice the electron mass and
ranging upward without limit. Therefore, photon
sources, which are also electric current vectors,
couple together in a modified way, as expressed by

$$D_+ \rightarrow \overline{D}_+,$$

leading to an alteration in the Coulomb potential.
The customary term for this is vacuum polarization.
The X-ray studies on mu-mesic atoms have provided
direct confirmation of the reality of vacuum polarization.

<u>Form Factors</u>

Another aspect of the second dynamical level
is that the initial, primitive, interaction which
is local, now becomes non-local. As we have been
discussing, the primitive interaction describes how
a photon source, through a virtual photon, emits a
non-interacting electron-positron pair. We now make
explicit the interaction between the electron and
positron (Fig. 20). The simplest process is already

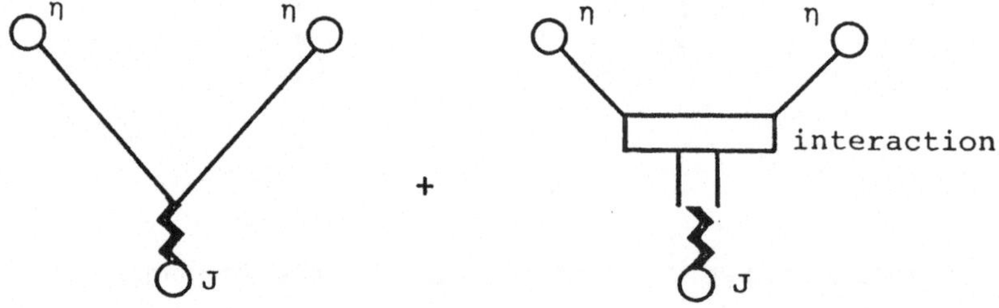

Fig. 20

shown in Fig. 14. Let's look at the two mechanisms

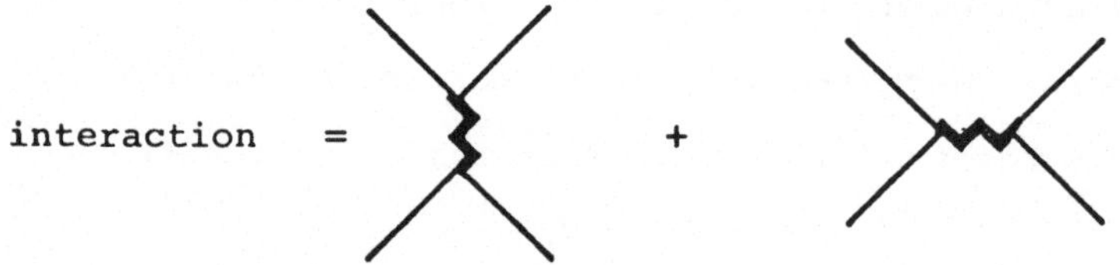

Fig. 21

separately. The first one, involving a time-like
virtual photon, leads to a three-source coupling
shown in Fig. 22. You will recognize here that,

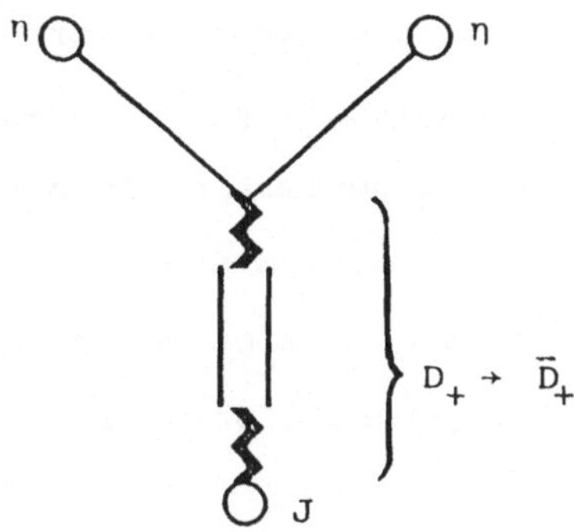

.Fig. 22

connecting the photon source with the final electron-
positron emission point is precisely the mechanism
for the modification of the photon propagation function, Fig. 19. That had to appear since the idealized
source description includes what all realistic sources
have in common. Its effect is to alter the electro-
magnetic field, $A \rightarrow \bar{A}$ in consequence of $D_+ \rightarrow \bar{D}_+$, but
the $\psi\psi\bar{A}$ coupling is still local.

The second mechanism, that of Coulomb scattering,
involves the specific nature of the effective virtual
photon detection source. It implies the three-source
coupling shown in Fig. 23. Now the coupling between

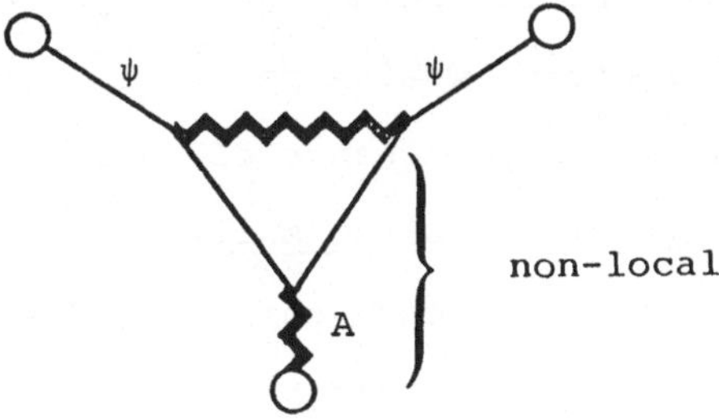

Fig. 23

the photon field and the final electron fields in
non-local. Such non-locality is represented by
form factors that are automatically obtained as a
superposition of propagation functions characterized
by the variable mass of the exchanged particles. The
form factor multiplying the primitive interaction has
this structure:

$$F(k^2) = 1 + k^2 \int dM^2 \; \frac{a(M^2)}{k^2 + M^2} \; , \qquad (159)$$

where k is the virtual photon momentum. Unity here
corresponds to the initial local coupling. The factor
of k^2 multiplying the spectral integral is the representation of the requirement that the primitive interaction, as the starting point, is correct for real
photons and is not to be changed in that circumstance,

$$F(k^2 = 0) = 1. \qquad (160)$$

As illustrated here, it is characteristic of source
theory that an initially correct description for a
certain limited circumstance must be maintained for
that circumstance as the domain of applicability is
widened. Therefore, what appears in this theory is
not renormalization, a change of basis in response
to a wider dynamics, but _normalization_, the main-
tenance of what is already correctly described.

A single spectral integral is not the only way
to represent a form factor. In the example just

considered, we can choose another causal arrangement,
Fig. 24, another space-time situation with correspondingly different momenta, that also produces the
three-source coupling, and leads to a double spectral
form. As the diagram indicates, to realize this
process, the electron sources must be virtual since

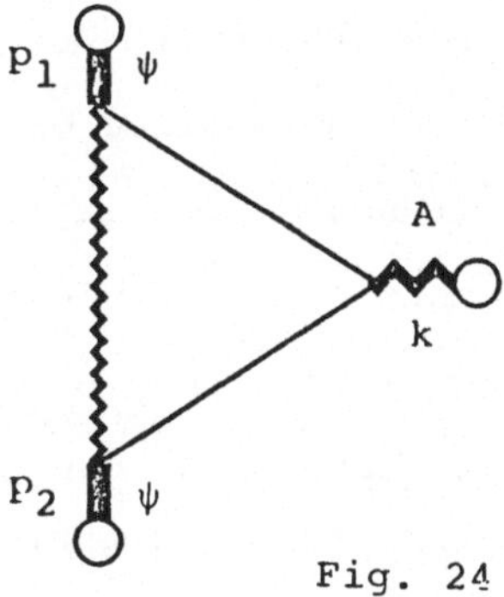

Fig. 24

they are effectively required to emit and absorb a
two-particle system. The emitted electron is scat-
tered by the electromagnetic field of the photon
source. Here we have two excitations propagating
among the sources, one before and one after the
injection of the momentum k, as represented by the
double spectral form

$$\int dM_1^2 dM_2^2 \; \frac{a(M_1^2, M_2^2, k^2)}{(p_1^2 + M_1^2)(p_2^2 + M_2^2)}. \tag{161}$$

There are checks of consistency that validate the
notion of a unique space-time coupling which can

be investigated in a variety of causal arrangements. In particular, if we specialize the above structure to the previous situation, Fig. 23, where $p_{1,2}^2 = -m^2$, we must, and do, recover the single spectral form in k^2, (159).

All electrodynamic results that have been obtained by these source theoretic methods agree with those produced by renormalized operator field theory. That is as it should be; it is not the purpose of this reconstruction of electrodynamics to obtain daringly new - and wrong - predictions. Rather, it is to test whether one can so simplify the physical and mathematical basis of relativistic particle theory that future investigations into the domains of strong and weak interactions will not be needlessly impeded by the dead weight of physically irrelevant hypotheses, and unnecessarily elaborate and abstract mathematical techniques.

<u>Appendix</u>

We have illustrated graphically how the exchange
of an electron-positron pair between photon sources
produces a modification in the photon propagation
function (Fig. 19). Here we will perform the explicit
calculation that the figure is supposed to suggest.
We start by noting that through the intermediary of
a virtual photon,

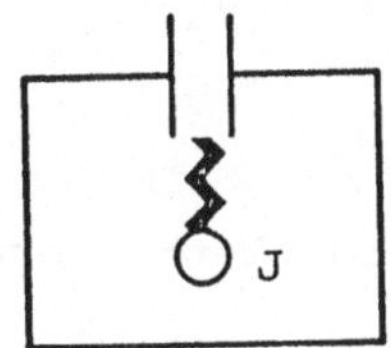

Fig. A-1

a photon source can act as an electron-positron
particle emission source. The relevant vacuum
amplitude is

$$\langle 0_+|0_-\rangle = e^{iW} = e^{i(\ldots + W_{int})} \to iW_{int}, \qquad (A1)$$

where

$$iW_{int} = i\int jA = i\int \tfrac{1}{2}\psi\gamma^0\gamma^\mu\, eq\psi A_\mu. \qquad (A2)$$

The equivalent two-particle source can be obtained
by comparison with the description of non-interacting
particles,

$$V(ac)A(mpl) \rightarrow e^{i\int \eta_1 \gamma^0 G_+ \eta_2} = e^{i\int \psi \gamma^0 \eta}, \qquad (A3)$$

where ψ is the field of the particle and η is the emitting (or detection) source. The two-particle term of (A3) is

$$\frac{1}{2}[i\int \psi \gamma^0 \eta]^2 = i\,\frac{1}{2}\int \psi(x)\gamma^0 i\eta(x)\eta(x')\gamma^0 \psi(x'). \qquad (A4)$$

Comparing this with (A2), we find the equivalent two-particle source,

$$i\eta(x)\eta(x')\gamma^0\Big|_{equiv} = \delta(x-x')eq\gamma^\mu A_\mu(x)\begin{cases} 2 \text{ for emission} \\ 1 \text{ for detection.}\end{cases}$$

$$(A5)$$

The description of the complete process of creation, propagation and detection of the two-particle system is

$$VA = \frac{1}{2}[i\int \eta_1 \gamma^0 G_+ \eta_2]^2 = \frac{1}{2}\,i\int \eta_1\gamma^0 G_+ i\eta_2\eta_2\gamma^0 G_+\eta_1. \qquad (A6)$$

When the last factor, η_1, is brought to the front, it passes three sources producing a sign change (F.D.),

$$VA = -\frac{1}{2}\int i\eta_1\eta_1\gamma^0 G_+ i\eta_2\eta_2\gamma^0 G_+, \qquad (A7)$$

which is now a diagonal sum (integral) of a matrix. The effective two particle sources, (A5), are then inserted, yielding

$$VA = -\frac{1}{2}\int (dx)(dx')\,tr_{4x2}[eq\gamma A_1(x)G_+(x-x')eq\gamma A_2(x')G_+(x'-x)],$$

$$(A8)$$

where the trace refers to the discrete spin and charge

labels. The trace over charge space is simply

$$\text{tr}_{4\times2}[\,(eq)^2...] = 2e^2\text{tr}_4[\qquad\qquad] \qquad (A9)$$

and, subsequently, we omit the subscript referring
to spin indices.

The vacuum amplitude has been derived for the
situation of a causal arrangement, $x^o(A_1) > x^{o'}(A_2)$.
Therefore, the electron propagation functions are

$$G_+(x-x') = (m-\gamma\tfrac{1}{i}\partial)\,i\!\int d\omega_p\,e^{ip(x-x')}$$

$$= i\int d\omega_p\,(m-\gamma p)\,e^{ip(x-x')}, \qquad (A10)$$

$$G_+(x'-x) = (m-\gamma\tfrac{1}{i}\partial')\,i\int d\omega_{p'}\,e^{ip'(x-x')}$$

$$= -i\int d\omega_{p'}\,(-m-\gamma p')\,e^{ip'(x-x')}. \qquad (A11)$$

Putting this information, (A9) - (A11), into (A8),
we find

$$VA = -e^2\!\int (dx)\,(dx')\,d\omega_p d\omega_{p'}\,A_1^\mu(x)\,e^{i(p+p')x}$$

$$\times \text{tr}[\gamma_\mu(m-\gamma p)\gamma_\nu(-m-\gamma p')]e^{-i(p+p')x'}A_2^\nu(x'). \qquad (A12)$$

The two-particle excitation carries energy-
momentum k (= p+p') and is characterized by a variable mass, $-k^2 = M^2 \geq (2m)^2$. These features of the
excitation can be incorporated by inserting the unit
factor

$$1 = \int (2\pi)^3\delta(p+p'-k)\,\frac{(dk)}{(2\pi)^3}, \qquad (A13)$$

$$\frac{(dk)}{(2\pi)^3} = d\omega_k \, (-k^2 = M^2) \, dM^2. \tag{A14}$$

After performing the space-time integrations in (A12) and introducing (A13) and (A14), we can write the vacuum amplitude in a form that separates externally controllable aspects from internal dynamics:

$$VA = -e^2 \int dM^2 d\omega_k \, A_1^\mu(-k) \, I_{\mu\nu}(k) \, A_2^\nu(k), \tag{A15}$$

where the dynamics is contained in

$$I_{\mu\nu}(k) = \int d\omega_p d\omega_{p'} \, (2\pi)^3 \delta(p+p'-k) \, tr[\gamma_\mu(m-\gamma p)\gamma_\nu(-m-\gamma p')]. \tag{A16}$$

The structure $I_{\mu\nu}$ must be gauge invariant, and one can verify algebraically that

$$k^\mu I_{\mu\nu} = 0, \qquad I_{\mu\nu}k^\nu = 0. \tag{A17}$$

Therefore, the general tensor structure of $I_{\mu\nu}$ must be of the form $(k^2 = -M^2)$

$$I_{\mu\nu} = (g_{\mu\nu} + \frac{k_\mu k_\nu}{M^2}) \, I(M^2), \tag{A18}$$

where $I(M^2)$ can be obtained as the diagonal sum

$$I(M^2) = \frac{1}{3} g^{\mu\nu} I_{\mu\nu}. \tag{A19}$$

Therefore, we have

$$I(M^2) = 1/3 \int d\omega_p d\omega_{p'} \, (2\pi)^3 \delta(p+p'-k) \, tr[\gamma^\mu(m-\gamma p)\gamma_\mu(-m-\gamma p')] \tag{A20}$$

with the following evaluation of the trace:

$$\mathrm{tr}[\gamma^{\mu}(m-\gamma p)\gamma_{\mu}(-m-\gamma p')]$$

$$= \mathrm{tr}[(-4m - 2\gamma p)(-m-\gamma p')] \tag{A21}$$

$$= 4(4m^2 - 2pp') = 4(M^2 + 2m^2).$$

Performing the remaining phase space integral, we find, for $M > 2m$,

$$I(M^2) = \frac{4}{3}(M^2 + 2m^2)\frac{1}{(4\pi)^2}\sqrt{1 - 4m^2/M^2}. \tag{A22}$$

A propagation function is identified through the coupling of two sources. Accordingly, we replace the fields in the gauge invariant VA, (A15), by sources, using the Lorentz gauge, $\partial A = 0$, for convenience,

$$-\partial^2 A = J \rightarrow k^2 A(\pm k) = J(\pm k) \tag{A23}$$

or,

$$A(\pm k) = - \frac{1}{M^2} J(\pm k), \tag{A24}$$

where

$$kJ = 0. \tag{A25}$$

The expression for the vacuum amplitude is then $(\alpha = \frac{e^2}{4\pi})$

$$VA = i \frac{\alpha}{3\pi}\int\frac{dM^2}{M^2}(1 + \frac{2m^2}{M^2})\sqrt{1 - 4m^2/M^2}\; J_1^{\mu}(-k)i\; d\omega_k\; J_{2\mu}(k). \tag{A26}$$

At this point we shift the emphasis back to space-time by writing

$$\int J_1^\mu(-k)\, i\, d\omega_k\, J_{2\mu}(k)$$

$$= \int (dx)(dx') J_1^\mu(x) \, [i \int d\omega_k e^{ik(x-x')}] J_{2\mu}(x'), \tag{A27}$$

where we recognize the causal form of $\Delta_+(x-x',M^2)$,

$$x^0 > x^{0'}: \quad \Delta_+(x-x',M^2) = i \int d\omega_k e^{ik(x-x')}. \tag{A28}$$

On introducing the general form of $\Delta_+(x-x',M^2)$, we remove the reference to the particular source arrangement, and, in terms of the total source, $J = J_1 + J_2$,

$$\int J_1 \ldots J_2 \rightarrow \frac{1}{2} \int J \ldots J, \tag{A29}$$

arrive at the generally valid version of (A26),

$$VA = i \int \frac{1}{2}\, J^\mu (\bar{D}_+ - D_+) J_\mu, \tag{A30}$$

where the modified propagation function is

$$\bar{D}_+(x-x') = D_+(x-x') + \frac{\alpha}{3\pi} \int_{(2m)^2}^\infty \frac{dM^2}{M^2}(1 + \frac{2m^2}{M^2})$$

$$\times \sqrt{1 - 4m^2/M^2}\; \Delta_+(x-x', M^2). \tag{A31}$$

The four-dimensional Fourier transform of this yields

$$\bar{D}_+(k) = \frac{1}{k^2 - i\varepsilon} + \frac{\alpha}{3\pi} \int_{(2m)^2}^\infty \frac{dM^2}{M^2}(1 + \frac{2m^2}{M^2})$$

$$\times \sqrt{1 - 4m^2/M^2}\; \frac{1}{k^2 + M^2 - i\varepsilon}. \tag{A32}$$

As mentioned in Lecture 3, this will lead to a modification of the Coulomb potential, now given by

$$\frac{1}{4\pi r} + \frac{2\alpha}{3\pi} \int_{2m}^\infty \frac{dM}{M}(1 + \frac{2m^2}{M^2})\; \sqrt{1 - 4m^2/M^2}\; \frac{e^{-Mr}}{4\pi r}. \tag{A33}$$

Lecture IV

Partial Symmetry

The objective of the previous discussion is to set the stage for considering areas of less certainty than electrodynamics - strong and weak interactions. Strong interactions are characterized by the existence of approximate conservation laws. As you know, isotopic spin invariance is broken by, at least, mass splittings within multiplets. Also, SU_3 symmetry is broken by the mass differences between the different multiplets. The scale involved varies from mass splittings of a few MeV, for isospin, to hundreds of MeV, for SU_3. But, what I want to consider specifically is the situation in which a symmetry is not broken by mass differences but by the mass itself. That is, the symmetry becomes apparent only when the mass of the physical particle is ignored. The particle in question is the pion, where the approximation of ignoring the mass is not entirely outrageous since it is by far the least massive of the hadrons. The pion has unit isotopic spin and its field, for which we use the symbol Π, is a three-component vector in the three-dimensional iso-space. The Lagrange function in the second order form appropriate to a spin zero particle, Eq. (29), is

$$\mathcal{L} = -\tfrac{1}{2}(\partial\Pi)^2 - \tfrac{1}{2}m_\Pi^2\Pi^2, \qquad\qquad (162)$$

where we have already introduced an approximation
by ignoring the mass difference between charged and
neutral pions. We do not actually neglect the pion
mass so much as to set it aside to look at a part
of the Lagrange function,

$$\mathcal{L} \to -\tfrac{1}{2}(\partial\Pi)^2, \qquad\qquad (163)$$

at which point a new invariance comes to light.
With only the derivative of the pion field appearing,
this part is invariant under the displacement of the
pion field by a constant,

$$\Pi(x) \to \Pi(x) + \text{const.} \qquad\qquad (164)$$

By partial symmetry I shall mean a symmetry that is
characteristic of a significant part of the Lagrange
function but not of all of it - a significant part
that may dominate the physics under appropriate
conditions. Inasmuch as the physical particles and
their interactions are all manifestations of the
same dynamics, it is not unreasonable to suppose
that if you recognize a partial symmetry in individual particle properties, it would not have survived
unless it was also characteristic of the interactions
the particle undergoes.

Pion-Nucleon Phenomenology

Let's apply this idea of partial symmetry to
low energy pion-nucleon phenomenology. The first
question is the form of the coupling of single pions
to nucleons. Here we have the alternatives suggested
by (N symbolizes the nucleon field)

$$\sim N\ldots N\Pi, \qquad \sim N\ldots N\partial\Pi, \qquad (165)$$

which choice will seem more natural for spin zero
particles when you recall the first order formulation,
(95), where the fields ϕ and ϕ^μ (= $\partial^\mu\phi$) are treated
on an equal footing. Which of the two possibilities
is preferable from the point of view of invariance
under the partial symmetry? Clearly the first is
not invariant while the second is. Therefore, we
chose the derivative form, as the interaction that
sustains the partial symmetry. The detailed pion-
nucleon coupling is

$$\mathcal{L}_{\pi N} = \frac{f}{m_\pi}\,\frac{1}{2}\,N\gamma^{o}i\gamma^{\mu}\gamma_5\,\vec{\tau}\nu N\cdot\partial_\mu\,\vec{\Pi}. \qquad (166)$$

The nucleon field carries indices that refer to the
Dirac matrices, antisymmetrical isotopic spin matrices
$\vec{\tau}$, and the antisymmetrical nucleonic charge matrix ν.
This interaction term describes low energy pion
emission and absorption in p-states, for which the
following value of the constant f is appropriate,

$$f = 1.0. \qquad (167)$$

But it does not describe low energy πN scattering
completely as it gives only the p-wave part; the
s-wave scattering must be added in separately. The
additional term in the Lagrange function that represents the observed isotopic dependence in low energy
s-wave scattering is

$$\mathcal{L}_{\pi N} = \ldots + \left(\frac{f_o}{m_\pi}\right)^2 \tfrac{1}{2} \bar{N}\gamma^o\gamma^\mu\ \vec{\tau}N\cdot\partial_\mu\vec{\Pi}\times\vec{\Pi}, \qquad (168)$$

where the coupling constant f_o is

$$f_o = 0.8. \qquad (169)$$

Is this term invariant under the partial symmetry?
Here we will find it simpler to discuss infinitesimal
constant displacements,

$$\vec{\Pi} \rightarrow \vec{\Pi} + \delta\vec{\varphi}. \qquad (170)$$

The Lagrange function now does change under this
transformation, but in a very special way,

$$\delta\mathcal{L}_{\pi N} = \left(\frac{f_o}{m_\pi}\right)^2 \tfrac{1}{2}\bar{N}\gamma^o\gamma^\mu\ \vec{\tau}N\cdot\partial_\mu(\vec{\Pi}\times\delta\vec{\varphi}). \qquad (171)$$

We recognize here a structure resembling an electromagnetic gauge transformation (158a), and we can
counter this change in the Lagrange function by a
corresponding phase transformation on the nucleon
field,

$$\delta N = i\left(\frac{f_o}{m_\pi}\right)^2 \vec{\tau}\cdot\vec{\Pi}\times\delta\vec{\varphi}\ N, \qquad (172)$$

a transformation that is somewhat unusual because
it is non-linear.

Chiral Group

If we now take this set of three transformations,
(172), and combine them with the three isotopic spin
transformations,

$$\delta N = i \; \frac{1}{2} \vec{\tau} \cdot \delta \vec{\omega} \; N, \tag{173}$$

the total set of six has, in a certain context, the
group property. Indeed, this system has the structure
of 0_4, the six-parameter rotation group in four
Euclidean dimensions. The identification of the 0_4
parameters is

$$\delta\omega_{12} = \delta\omega_3, \text{ etc.},$$
$$\lambda = \frac{2f_o}{m_\pi}. \tag{174}$$
$$\delta\omega_{34} = \lambda\delta\varphi_3, \text{ etc.},$$

It is well-known that this group, the four-dimensional
orthogonal group, can be factored into two such three-
dimensional groups,

$$0_4 = 0_3(+) \times 0_3(-), \tag{175}$$

where the parameters of the two independent three-
dimensional groups $0_3(\pm)$ are

$$\delta\omega_{\pm 3} = \delta\omega_{12} \pm \delta\omega_{34}, \text{ etc.} \tag{176}$$

In doing this, we have combined scalars $(\delta\omega_{12})$ with
pseudoscalars $(\delta\omega_{34})$, which means that a spatial

reflection interchanges the two factor groups. We
are evidently talking about the <u>chiral group</u> which
we have inferred, not from weak interactions or
speculations on inner structure, but from the low
energy strong interaction phenomenology of the pion-
nucleon system. Once we have recognized the chiral
group, we can make the parameters functions of space
and time and thereby generate vectors - currents.
The currents associated with the chiral group that
are inferred from the strong interactions are
obviously relevant to the chiral properties observed
in weak interactions. We take one of these, say the
+ combination, and derive the current:

$$2j_+^\mu(\Pi,N) = \partial^\mu\Pi \times \Pi + \frac{m_\pi}{2f_o} \partial^\mu\Pi$$

$$+ \frac{1}{2}N\gamma^o\gamma^\mu \frac{1}{2}\tau(1 - \frac{f}{f_o} i\gamma_5\nu)N. \tag{177}$$

It is naturally suggested that this is the hadronic
current that multiplies the appropriate current for
leptons. In confirmation of this, we note that the
pion coupling constant in $\pi \rightarrow \mu\nu(e\nu)$ decay, experimentally about 94 MeV, is given by $m_\pi/2f_o \sim 90$ MeV,
and that the relative coefficient in the combination
of vector and axial vector nucleon terms, experimentally 1.24, is here

$$- \frac{G_A}{G_V} = \frac{f}{f_o} \sim 1.2. \tag{178}$$

Pion Chiral Transformations

When I described the chiral group structure
I referred to "a certain context," for there is a
mathematical inconsistency in what we have done. I
have said that the transformations on the nucleon
field, together with those on the pion field, form
a group having the structure of the four-dimensional
rotation group, which is a compact, or closed, group.
On the other hand, we have done this in terms of
pion transformations that include displacements,
which do not form a compact group. In other words,
contrary to the non-Abelian structure of a rotation
group, translations are Abelian. This is mathematically, but not physically, inconsistent because the
discussion was limited to single-pion processes.
Now, if we wish to remove the mathematical incom-
pleteness, we shall make physical predictions about
pion-pion interactions, specifically concerning low
energy scattering properties. A suitable non-linear
generalization of the infinitesimal displacement is

$$\delta \Pi = \delta \varphi + \frac{1}{4}\lambda^2 (2\pi \delta\varphi \cdot \Pi - \delta\varphi \, \Pi^2). \tag{179}$$

To check that this replaces displacements by
rotations, it is helpful to introduce the unit

four-dimensional vector with components

$$\vec{u} = \frac{\lambda\vec{\Pi}}{1+\frac{1}{4}\lambda^2\Pi^2}, \quad u_4 = \frac{1-\frac{1}{4}\lambda^2\Pi^2}{1+\frac{1}{4}\lambda^2\Pi^2}, \quad \sum_1^4 u_\alpha^2 = 1. \qquad (180)$$

The transformations induced by the $\delta\Pi$ of Eq. (179) are the four-dimensional rotations that act on the fourth axis,

$$\delta\vec{u} = \lambda\delta\vec{\varphi}u_4, \quad \delta u_4 = -\lambda\delta\vec{\varphi}\cdot\vec{u} \qquad (181)$$

while the isotopic rotations $\delta\Pi = -\delta\omega\times\Pi$ constitute the rotations that leave the fourth axis invariant,

$$\delta\vec{u} = -\delta\vec{\omega}\times\vec{u}, \quad \delta u_4 = 0. \qquad (182)$$

We are now asked to construct a Lagrange function that is invariant under 0_4, in the sense of a partial symmetry. The obvious 0_4 generalization of $-\frac{1}{2}(\partial\Pi)^2$ is

$$\mathcal{L} = -\frac{1}{2\lambda^2} \sum_1^4 (\partial u_\alpha)^2 = -\frac{1}{2}(\partial\Pi)^2 + \frac{\lambda^2}{4}(\partial\Pi)^2\Pi^2 + \dots \; . \qquad (183)$$

The expansion in powers of the pion field yields a specific quartic term which describes, in part, low energy pion-pion scattering.

Broken Chiral Symmetry

But before discussing this scattering, we must turn our attention to the pion mass term which breaks the chiral symmetry:

$$\delta[-\frac{1}{2}m_\pi^2\Pi^2] = -m_\pi^2\Pi\cdot\delta\varphi\,(1 + \frac{1}{4}\lambda^2\Pi^2). \qquad (184)$$

How shall we generalize the mass term to incorporate it into the partial symmetry scheme? We first note that Π^2 can be conveniently expressed in terms of u_4,

$$u_4 = \frac{1-\frac{1}{4}\lambda^2\Pi^2}{1+\frac{1}{4}\lambda^2\Pi^2} = 1 - \frac{1}{2}\lambda^2\Pi^2 + \dots . \tag{185}$$

It is reasonable that the pion mass term should be written in terms of the fundamental variables, u_α. It destroys the four-dimensional rotational symmetry $(0_4 \to 0_3)$ so that it must depend explicitly on u_4. Two simple possibilities are:

1. (Weinberg) A linear dependence on u_4:

$$- \frac{1}{2}m_\pi^2\Pi^2 \to - \frac{m_\pi^2}{\lambda^2}(1-u_4) = - \frac{1}{2}m_\pi^2 \frac{\Pi^2}{1+\frac{1}{4}\lambda^2\Pi^2}, \tag{186}$$

which recovers the pion mass term along with a quartic structure $\frac{1}{8}m_\pi^2\lambda^2(\Pi^2)^2$, that contributes to pion-pion scattering. However, the linear u_4 dependence not only produces the reduction $0_4 \to 0_3$, it also breaks reflection invariance in the four-dimensional Euclidean space, which need not be correct. Therefore:

2. A quadratic dependence on u_4 is assumed,

$$-\frac{1}{2}m_\pi^2\Pi^2 \to - \frac{m_\pi^2}{2\lambda^2}(1-u_4^2) = -\frac{1}{2}m_\pi^2 \frac{\Pi^2}{(1+\frac{1}{4}\lambda^2\Pi^2)^2}. \tag{187}$$

The quartic term here is $\frac{1}{4}m_\pi^2\lambda^2(\Pi^2)^2$, twice that of the first hypothesis.

Pion-Pion Scattering, Comparison with Experiment

A decision between these two simple alternatives can only be made by experiment. We calculate the implications of the two hypotheses for the low energy pion-pion scattering lengths, a_0 and a_2, where the subscripts indicate the possible total isotopic spin quantum numbers for s-waves. The results are ($f_0^2/4\pi = 0.05$)

$$1. \quad m_\pi a_0 = \frac{7}{2}\frac{f_0^2}{4\pi} = 0.18, \quad m_\pi a_2 = -\frac{f_0^2}{4\pi} = -0.05,$$

$$2. \quad m_\pi a_0 = 6\frac{f_0^2}{4\pi} = 0.30, \quad m_\pi a_2 = 0,$$

(188)

where one verifies that the combination $2a_0 - 5a_2$ is independent of the particular symmetry breaking hypothesis. In principle, the most clear cut experiment is the decay $K^+ \rightarrow e^+ + \bar{\nu} + \pi^+ + \pi^-$ which is influenced by the s-wave $\pi\pi$ phase shift for isotopic spin zero. The most recent experiment (L. Rosselet et al., Phys. Rev. D15, 574 (1977)) gives a value of

$$m_\pi a_0 = 0.3 \pm 0.05,$$

(189)

which is interesting, but hardly definitive. However, one does not measure the zero kinetic energy value directly. Measurements are made at finite kinetic energies and then extrapolated to zero. It is better to compare with the actual experimental values of the phase shifts. The theoretical prediction is expressed in terms of the center of mass energy of the dipion system, M, and the corresponding three-dimensional relative momentum of the pions, $q = \sqrt{\frac{1}{4}M^2 - m_\pi^2}$. The phase shift is

$$\tan\delta_0 \cong \delta_0 = \frac{f_0^2}{4\pi}\,\frac{2q}{M}\begin{cases} 1: & \frac{7}{2} + 4\,\frac{q^2}{m_\pi^2}, \\[2ex] 2: & 6 + 4\,\frac{q^2}{m_\pi^2}. \end{cases} \tag{190}$$

The dependence on momentum is the same for both as that comes from the derivative term which is a chiral invariant. Theory and experiment are compared in the following table.

	M(MeV)	289	303	317	335	367
1.	δ_0	0.05	0.08	0.11	0.14	0.21
2.	δ_0	0.08	0.13	0.17	0.21	0.29
Exp.	δ_0	0.07±0.13	0.21±0.07	0.13±0.05	0.20±0.04	0.27±0.04

The cited masses are bin averages, an indication of the bin size being given by noting that M = 289 means the range from 280 to 296. Also, the approximation $\tan\delta_0 \cong \delta_0$ is still reasonably valid at the highest energy; the use of the tangent would give $\delta_0 = 0.28$

instead of 0.29, at M = 367. An inspection of the
higher mass results, where the relative errors are
smallest, argues in favor of the second hypothesis,
but there is certainly need for much more data.

Pion Field Redefinition

Before continuing, we give another form to the
second chiral invariance-breaking term, which we
henceforth accept, by noting that

$$- \frac{m_\pi^2}{2\lambda^2}(1-u_4^2) = - \tfrac{1}{2}m_\pi^2(\tfrac{1}{\lambda}\vec{u})^2. \tag{191}$$

This suggests the utility of redefining the pion
field so that

$$\tfrac{1}{\lambda}\vec{u} = \vec{\Pi}, \tag{192}$$

thereby reducing the mass term to precisely $-\tfrac{1}{2}m_\pi^2\Pi^2$
With that redefinition, we now have

$$u_4 = \sqrt{1-\lambda^2\Pi^2} \simeq 1 - \tfrac{1}{2}\lambda^2\Pi^2 + \ldots \tag{193}$$

and the infinitesimal transformation law (179) is
replaced by

$$\delta\Pi = \delta\varphi\sqrt{1-\lambda^2\Pi^2} = \delta\varphi(1 - \tfrac{1}{2}\lambda^2\Pi^2 + \ldots). \tag{194}$$

Also, the chiral invariant derivative structure now
reads

$$-\tfrac{1}{2}(\partial\Pi)^2 - \frac{1}{2\lambda^2}(\partial u_4)^2 = -\tfrac{1}{2}(\partial\Pi)^2 - \tfrac{1}{8}\lambda^2(\partial\Pi^2)^2 + \ldots. \tag{195}$$

The terms responsible for low energy $\pi\pi$ scattering have changed their appearance:

$$\frac{1}{4}\lambda^2 (\partial\Pi)^2\Pi^2 + \frac{1}{4}m_\pi^2\lambda^2(\Pi^2)^2 \to - \frac{1}{8}\lambda^2(\partial\Pi^2)^2, \quad (196)$$

but the same phase shifts are predicted.

Rho Meson

Our next example of partial symmetry refers to the ρ meson and isotopic spin transformations. We have derived the Lagrange function for the massless photon, (98). The structure of the Lagrange function for a spin one particle with a mass is

$$\mathcal{L} = -\frac{1}{2}\rho^{\mu\nu}\cdot(\partial_\mu\rho_\nu - \partial_\nu\rho_\mu) + \frac{1}{4}\rho^{\mu\nu}\cdot\rho_{\mu\nu} - \frac{1}{2}m_\rho^2\,\rho^\mu\cdot\rho_\mu, \quad (197)$$

where we have also introduced isotopic scalar products for the unit isotopic spin ρ-field. By ignoring the $\rho^\pm-\rho^0$ mass difference, the structure becomes invariant under three-dimensional isotopic transformations (rotations),

$$\delta\rho = - \delta\omega \times \rho. \quad (198)$$

We now break the invariance by considering the parameters of the rotation in isotopic space to be functions of space and time, $\delta\omega \to \delta\omega(x)$. That yields the conserved iso-current,

$$j_T^\mu = \rho^{\mu\nu} \times \rho_\nu. \quad (199)$$

Gauge Transformations

Now, if the mass term is set aside, (197) becomes

(with a three-fold multiplicity) the photon Lagrange
function. Then certainly we have a new invariance,
generalizing the gauge invariance characteristic of
the photon - the ρ field becomes arbitrary to the
extent of gradients of space-time scalars. We pursue
the analogy between the ρ meson and the photon by
accepting that the ρ plays a special dynamical role
in the sense that the three isotopic spin components
of ρ are coupled to the three components of isotopic
spin, as the photon is coupled to electric charge.
But, to incorporate this hypothesis, that the vector
source of the ρ field is the isospin current, we must
recognize that, inasmuch as the ρ also carries isotopic spin, it contributes to the source of its own
field. Thus, the field equation for ρ, derived from
$\delta\rho$ variations in $\mathcal{L}$, would then read

$$\partial_\nu \rho^{\mu\nu}(+ \, m_\rho^2 \, \rho^\mu) = g \, \rho^{\mu\nu} \times \rho_\nu + \ldots, \qquad (200)$$

where, on the right-hand side, the ρ field source
includes the isotopic current of the ρ field, multiplied by a coupling constant g that is the analogue
of e. In order to derive from a Lagrange function
the field equation (200) with the indicated self-
coupling, we must add the additional term,
$-\frac{1}{2}g \; \rho^{\mu\nu} \cdot \rho_\mu \times \rho_\nu$, to produce the complete Lagrange
function for the ρ meson only:

$$\mathcal{L}_\rho = -\frac{1}{2}\rho^{\mu\nu} \cdot (\partial_\mu \rho_\nu - \partial_\nu \rho_\mu + g\,\rho_\mu \times \rho_\nu)$$

$$+ \frac{1}{4}\rho^{\mu\nu} \cdot \rho_{\mu\nu} - \frac{1}{2}m_\rho^2\,\rho^\mu \cdot \rho_\mu . \tag{201}$$

This has followed from simple physical ideas and not by mathematical considerations of non-Abelian gauge transformations. Indeed, we now turn around and derive the non-Abelian transformations. We have considered coordinate dependent iso-rotations and photon-like gauge transformations. These are combined in the infinitesimal transformation laws

$$\delta\rho_\mu = -\delta\omega \times \rho_\mu + \frac{1}{g}\,\partial_\mu\,\delta\omega, \tag{202a}$$

$$\delta\rho_{\mu\nu} = -\delta\omega \times \rho_{\mu\nu}, \tag{202b}$$

which are closed and form a non-Abelian gauge group. This is a partial symmetry, for the mass term violates the symmetry, as expressed by

$$\delta\mathcal{L}_\rho = -\frac{m_\rho^2}{g}\,\rho^\mu \cdot \partial_\mu\,\delta\omega . \tag{203}$$

The principle of stationary action then implies an exact conservation law,

$$\partial_\mu\left(\frac{m_\rho^2}{g}\,\rho^\mu\right) = 0, \tag{204}$$

exhibiting the structure that more generally plays the role of a conserved current. To pave the way for considering the interaction with ρ of other isospin-bearing particles, we rewrite (202a) as

$$\delta\rho_\mu = \frac{1}{g}(\partial_\mu + g\,\rho_\mu \times)\delta\omega, \qquad (205)$$

thereby exhibiting the gauge covariant derivative that serves to introduce interactions on replacing derivatives by gauge covariant derivatives.

As a simple example, consider the effect of that substitution on the first order form of the Lagrange function of the pion,

$$\mathcal{L}_\pi \to -\,\Pi^\mu_\bullet(\partial_\mu + g\,\rho_\mu \times)\Pi + \frac{1}{2}\,\Pi^\mu\cdot\Pi_\mu - \frac{1}{2}m_\pi^2\Pi^2.$$

$$(206)$$

The coupling term here is $g\,\rho^\mu\cdot\Pi_\mu \times \Pi$, $\Pi_\mu = \partial_\mu\Pi + g\,\rho_\mu \times \Pi$, which exhibits the ρ field coupled to the iso-current of the pion. Here is a mechanism whereby ρ decays into two pions, $\rho \to \pi + \pi$, and a standard calculation supplies a value of the coupling constant implied by the known width of 150 MeV, namely

$$\frac{g^2}{4\pi} = 2.9. \qquad (207)$$

Hadronic Electromagnetic Properties

We next turn to the area of hadronic-electro-magnetic couplings where these ideas are particularly effective. The isotopic spin non-Abelian gauge invariance is broken by the ρ mass term, (203). We can restore the invariance for a special transfor-mation by introducing the electromagnetic field. Suppose we consider an isotopic rotation around the

third axis only, $\delta\omega_3 \neq 0$. The change in the third
component of the ρ field is then produced only by
the derivative term in (205),

$$\delta\rho_3^\mu = \frac{1}{g} \partial^\mu \, \delta\omega_3, \tag{208}$$

analogous to an electromagnetic gauge transformation.
Therefore, if we pick an appropriately related gauge
transformation for the electromagnetic field,

$$\delta A^\mu = \frac{1}{e} \partial^\mu \, \delta\omega_3, \tag{209}$$

the combination of the two, $\rho_3^\mu - \frac{e}{g} A^\mu$, is invariant,
not in general, but for rotations around the third
axis. The ρ mass term now becomes

$$-\frac{1}{2}m_\rho^2 \, \rho^\mu \cdot \rho_\mu \rightarrow -\frac{1}{2}m_\rho^2 (\rho_1^\mu \rho_{1\mu} + \rho_2^\mu \rho_{2\mu})$$

$$-\frac{1}{2}m_\rho^2 (\rho_3^\mu - \frac{e}{g} A^\mu)(\rho_{3\mu} - \frac{e}{g} A_\mu). \tag{210}$$

We have thereby related the electromagnetic field
to hadronic systems, at least insofar as electric
charge is connected with the third component of
isotopic spin.

What is the justification of the factor e/g?
It is so written that e shall be the known electric
charge unit. We can see this by thinking of a simple
system, such as π mesons and ρ mesons. Let us
redefine the third component of the ρ field,

$$\rho_3 \rightarrow \rho_3 + \frac{e}{g} A, \tag{211}$$

so the ρ mass term looks as it did before, thereby
transferring elsewhere the explicit appearance of
the electromagnetic field. Examining the π-ρ coupling,
we see that it acquires an additional term,

$$g \; \rho^\mu \cdot \Pi_\mu \times \Pi \to \ldots + e \; A^\mu (\Pi_\mu \times \Pi)_3 , \qquad (212)$$

and we recognize here the pion electric current vector

$$e (\Pi_\mu \times \Pi)_3 = e \; \Pi_\mu \; iq\Pi = j_\mu \qquad (213)$$

with

$$q = \begin{pmatrix} 0 & -i & 0 \\ i & 0 & 0 \\ 0 & 0 & 0 \end{pmatrix}, \qquad (214)$$

thus confirming the e/g factor. This is not the
only effect of the ρ field redefinition. In particular field strengths are also involved. But such
effects, which introduce form factors into the
coupling between the electromagnetic field and
hadrons, are unimportant at low momenta, where only
the total charge is relevant.

<u>Omega Decay</u>

For an application of this description of the
electromagnetic properties of hadrons, we consider
the particle ω which decays into three pions. It
is customary to assume the ω couples to ρ and π
thereby explaining the ω decay as $\omega \to \pi + (\rho \to \pi + \pi)$.
The coupling in question is

$$\mathcal{L}_{\omega\rho\pi} = - \frac{g}{m_\rho} \, {}^*\omega^{\mu\nu} \, \rho_{\mu\nu} \cdot \Pi, \qquad (215)$$

where the dual operation,

$$^*\omega^{\mu\nu} = \frac{1}{2} \, \varepsilon^{\mu\nu\lambda\kappa} \, \omega_{\lambda\kappa}, \qquad \varepsilon^{0123} = +1, \qquad (216)$$

introduces a pseudo tensor to match the pseudoscalar nature of Π. The particular value of the coupling constant will be justified by its consequences. The response of this structure to the redefinition (211) is the introduction of an electromagnetic coupling (effectively, we let $\rho_3 \rightarrow \frac{e}{g} A$)

$$\mathcal{L}_{\omega\gamma\pi} = - \frac{e}{m_\rho} \, {}^*\omega^{\mu\nu} \, F_{\mu\nu} \, \Pi_3. \qquad (217)$$

It predicts the rate of the decay $\omega \rightarrow \gamma + \pi^0$ to be

$$\Gamma_{\omega\rightarrow\gamma+\pi^0} = \frac{\alpha}{6} \left(\frac{m_\omega}{m_\rho} \right)^2 \left[1 - \frac{m_\pi^2}{m_\omega^2} \right]^3 \qquad m_\omega = 0.89 \text{ MeV} \qquad (218)$$

which is to be compared with the measured value of 0.88 ± 0.05 MeV.

$\underline{U_3}$ Extension and Neutral Pion Decay

To complete the description of electromagnetic properties of hadrons, corresponding to the construction $Q = T_3 + \frac{1}{2} y$, we must supplement the neutral ρ meson by the other neutral members of the multiplet, ω and ϕ. In a simple U_3 theory, the relative coupling strengths of the ω and ϕ are indicated by the following effective substitutions,

$$\rho_3 \rightarrow \frac{e}{g} A; \quad \omega \rightarrow \frac{1}{3} \frac{e}{g} A; \quad \phi \rightarrow - \frac{\sqrt{2}}{3} \frac{e}{g} A, \qquad (219)$$

leaving open the question of mass splitting effects.
If we now return to the $\omega\pi\gamma$ coupling, (217), and
consider the electromagnetic field induced through
the ω meson, we get

$$\mathcal{L}_{\pi\gamma\gamma} = - \frac{1}{3} \frac{e^2}{g} \frac{1}{m_\rho} {}^*F^{\mu\nu} F_{\mu\nu} \Pi_3 \qquad (220)$$

which describes the two photon decay of the neutral
pion. The calculated width is

$$\Gamma_{\pi^0 \rightarrow \gamma+\gamma} = \frac{1}{9} \frac{\alpha^2}{g^2/4\pi} (\frac{m_\pi}{m_\rho})^2 m_\pi = \frac{24.4}{g^2/4\pi} \text{ eV.} \qquad (221)$$

The current experimental value for Γ is 8 ± 0.4 eV,
implying

$$\frac{g^2}{4\pi} = 3.0 \pm 0.15, \qquad (222)$$

which compares well with the previously obtained
value of 2.9, (207).

Vector-Pseudoscalar Couplings

Once we have introduced the wider U_3 framework,
we can discuss a number of other processes involving
the vector and pseudoscalar particles. The 3×3
arrays that represent the fields of the vector (v)
and pseudoscalar (p) multiplets are (only the neutral
members are displayed, and a simplified choice is
used for v)

$$v \cong \begin{pmatrix} \dfrac{\omega+\rho^{O}}{\sqrt{2}} & & \\ & \dfrac{\omega-\rho^{O}}{\sqrt{2}} & \\ & & \phi \end{pmatrix}, \qquad (223)$$

$$p = \begin{pmatrix} \dfrac{1}{\sqrt{2}}(\eta\ \cos\theta + \eta'\ \sin\theta + \pi^{O}) & & \\ & \dfrac{1}{\sqrt{2}}(\eta\ \cos\theta + \eta'\ \sin\theta - \pi^{O}) & \\ & & -\sin\theta\ \eta + \cos\theta\ \eta' \end{pmatrix}. $$

$$(224)$$

The fields η and η' are mixed together in terms of an angle that differs from the conventional one, where the emphasis is on SU_3. Apart from symmetry-breaking factors, the known $\omega\rho\pi$ coupling can be generalized into a quadratic combination of all nine vector fields coupled to the nine pseudoscalar fields,

$$*\omega^{\mu\nu}\ \rho_{\mu\nu}\cdot\Pi \rightarrow \frac{1}{\sqrt{2}}\ *v^{\mu\nu}_{ab}\ v_{\mu\nu bc}\ p_{ca}, \qquad (225)$$

where summations over the three-valued unitary indices are understood. We then infer a number of other couplings on inserting (223) and (224) (the tensor indices are omitted)

$$*\omega\rho\cdot\Pi + \frac{1}{2}\cos\theta(*\rho\rho + *\omega\omega)\eta - \frac{1}{\sqrt{2}}\sin\theta\ *\phi\phi\eta$$

$$(226)$$

$$+ \frac{1}{2}\sin\theta(*\rho\rho + *\omega\omega)\eta' + \frac{1}{\sqrt{2}}\cos\theta\ *\phi\phi\eta'.$$

Eta Decay

There are both electromagnetic and strong interaction statements here. Let me concentrate on an electromagnetic process, specifically, the $\eta \rightarrow \gamma + \gamma$ decay. The current rate for this decay is

$$\Gamma_{\eta \rightarrow \gamma + \gamma} = 324 \pm 46 \text{ eV} \qquad (227)$$

so the ratio to the π rate (221) is

$$\frac{\Gamma_{\eta}}{\Gamma_{\pi}} = 40.5 \pm 6.1. \qquad (228)$$

This is not the significant comparison, however, because the coupling constant [cf Eq. (215)] contains an inverse mass, and therefore the decay constant is proportional to the cube of the decaying particle's mass, as shown in Eq. (221) for π decay. Removing this cubic dependence on the mass, we find

$$\frac{\Gamma_{\eta}/m_{\eta}^3}{\Gamma_{\pi}/m_{\pi}^3} = \frac{40.5 \pm 6.1}{67.24} = 0.602 \pm 0.091. \qquad (229)$$

Using the substitutions (219) in the couplings (226), we find this ratio, without the inclusion of symmetry breaking mass factors, to be

$$\frac{\Gamma_{\eta}/m_{\eta}^3}{\Gamma_{\pi}/m_{\pi}^3} = [3(\tfrac{1}{2} \cos\theta (1 + \tfrac{1}{9}) - \tfrac{1}{\sqrt{2}} \sin\theta \, \tfrac{2}{9})]^2 \qquad (230)$$

$$= (5/3 \cos\theta - \frac{\sqrt{2}}{3} \sin\theta)^2.$$

The theoretical prediction is seen to be sensitive to the value of the mixing angle. The situation of "ideal" mixing, would occur if η', the ninth meson, were so far removed in mass that it became a pure unitary singlet, corresponding to

$$\sin\theta = \sqrt{\tfrac{2}{3}}, \quad \cos\theta = \tfrac{1}{\sqrt{3}}; \quad \theta = 54.7^{\circ}. \qquad (231)$$

For this choice, the predicted ratio is 1/3. But to illustrate the sensitivity of this ratio to the angle, we note that

$$\begin{aligned}
\theta &= 49.1^{\circ}, \qquad \text{ratio} = 0.54, \\
\theta &= 47.9^{\circ}, \qquad \text{ratio} = 0.59, \qquad (232) \\
\theta &= 46.7^{\circ}, \qquad \text{ratio} = 0.64.
\end{aligned}$$

Thus, without symmetry breaking mass factors, a mixing angle about 7 degrees below the "ideal" angle accounts for the observed ratio. But clearly the question of additional mass factors is significant owing to the sensitivity to small changes.

<u>Leptonic Decay and Mass Factors</u>

The experimental handle on symmetry breaking mass factors comes from the observations on the decay rates of the vector mesons into $e^{+}e^{-}$, $\Gamma_{v \to e^{+}e^{-}}$. The observed rates are

$$\begin{aligned}
\Gamma_{\rho \to ee} &= 6.54 \pm 0.77 \text{ keV}, \qquad (233a) \\
\Gamma_{\omega \to ee} &= 0.76 \pm 0.17 \text{ keV}, \qquad (233b) \\
\Gamma_{\phi \to ee} &= 1.31 \pm 0.10 \text{ keV}. \qquad (233c)
\end{aligned}$$

If these particles were mass degenerate, the substitutions, (219), would imply a common value for the three combinations

$$\Gamma_\rho \; : \; 9\Gamma_\omega \; : \; \frac{9}{2}\Gamma_\phi \; = \; 6.5\pm0.8 \; : \; 6.8\pm1.5 \; : \; 5.9\pm0.5, \tag{234}$$

and, within the experimental errors, they are indeed the same; thus

$$\frac{\frac{9}{2}\Gamma_\phi}{\Gamma_\rho} = 0.9 \pm 0.1. \tag{235}$$

In the real world, however, where there are significant mass splittings, $m_\phi/m_\rho = 1.32$, the intrinsic proportionality between the leptonic decay rate and the vector particle mass requires the following comparison,

$$\Gamma_\rho \; : \; 9\,\frac{m_\rho}{m_\omega}\,\Gamma_\omega \; : \; \frac{9}{2}\,\frac{m_\rho}{m_\phi}\,\Gamma_\phi \tag{236}$$

$$= 6.5 \pm 0.8 \; : \; 6.7 \pm 1.5 \; : \; 4.5 \pm 0.4$$

leading to

$$\frac{9}{2}\,\frac{m_\rho}{m_\phi}\,\Gamma_\phi/\Gamma_\rho = 0.7 \pm 0.1 \tag{237}$$

which is not very satisfactory. Accordingly, if (234) is accepted as being essentially constant, the correct conclusion is that the effective substitutions that introduce electromagnetism are

$$\rho_3 \to \frac{e}{g}\, A, \quad \omega \to \frac{1}{3}\sqrt{\frac{m_\rho}{m_\omega}}\, A, \quad \phi \to -\frac{\sqrt{2}}{3}\sqrt{\frac{m_\rho}{m_\phi}}\, A. \qquad (238)$$

On the other hand, there are reasons to believe that the correspondence should involve the mass ratio rather than their square roots. The test of that idea is given by

$$\Gamma_\rho \;:\; 9\,\frac{m_\omega}{m_\rho}\,\Gamma_\omega \;:\; \frac{9}{2}\,\frac{m_\phi}{m_\rho}\,\Gamma_\phi$$

$$= 6.5 \pm 0.8 \;:\; 6.9 \pm 1.5 \;:\; 7.8 \pm 0.6, \qquad (239)$$

or

$$\frac{9}{2}\,\frac{m_\phi}{m_\rho}\,\Gamma_\phi / \Gamma_\rho = 1.2 \pm 0.2, \qquad (240)$$

which shows that this possibility is not ruled out. It is clear* that such mass splitting effects must be included in the discussion of electromagnetic decays.

---

*We may be taking the colliding beam data, or its simple analysis, too seriously. The use of formulas illustrated by

$$\Gamma_{\rho \to ee} = \frac{1}{3}\,\alpha^2\,\frac{m_\rho}{g^2/4\pi}$$

gives

$$\rho : g^2/4\pi = 2.1 \pm 0.3, \quad \phi : g^2/4\pi = 2.3 \pm 0.2,$$

if we adopt the square root mass factor in converting the hadronic and electromagnetic fields. These numbers are significantly lower than are other determinations. On the other hand, the observed rate of $\phi \to \mu\mu$, which should differ negligibly from that of $\phi \to ee$ is currently about 25% lower, leading to

$$g^2/4\pi = 3.0 \pm 0.4.$$

Confinement, Nonlocal Field Theory, and Electromagnetic Interation

W. Becker and D. Großer

Institut für Theoretische Physik der Universität Tübingen,
Auf der Morgenstelle 14, D—7400 Tübingen

I. Introduction

Confinement is usually thought to evolve as a property of the exact solution of a field theory; it does not exist to any finite order of perturbation theory. Recently, an interesting alternative possibility to describe confinement has been suggested by Dubnickova and Efimov [1]. These authors introduce a Lagrangian for the _free_ quark field (called virton field in this context) such that the field operator itself comes out identically zero, the causal propagator, however, is nonzero. A theory like this cannot be formulated in local Lagrangian field theory, but is shown to be possible in the framework of nonlocal Lagrangians. It is now tempting to introduce an interaction for the virtons such that simple perturbation theory in the interaction picture can be applied. Since there are no one virton states, virton-virton scattering amplitudes must vanish identically to any finite order of perturbation theory. Nontrivial results for hadron scattering are only obtained by solving a nonperturbative bound state problem. Hence perturbation theory is not useful if only strong interactions are present. In this note, we introduce an electromagnetic coupling of the virtons to the photon field. As in the case of strong interactions, perturbation theory must not give nontrivial results for amplitudes with external virtons, otherwise confinement would be lost. The hope is, however, that it might yield non-zero results for amplitudes with only external photons.

In § 2 we introduce an electromagnetic coupling which satisfies the requirements of charge conservation. For this purpose, we use a procedure due to Kroll [2]. The introduction of multiphoton vertices is an unavoidable consequence of charge conservation. We make sure that photons cannot create virton pairs, hence confinement is insofar warranted. In § 3 we deal with charged closed loops, particularly with vacuum polarization. It turns out that the special form of the vertices dictated by charge conservation spoils the pleasant property of nonlocal field theory to be finite. The amplitudes are even nonrenormalizable of infinite order unless the virton propagator is suitably restricted. If this is done, the contribution of charged closed loops as well as of diagrams containing them as subdiagrams, turns out to vanish. For all other diagrams a nonlocal photon propagator must be used in order to ensure convergence. We then have a finite theory of virtons interacting with the electromagnetic field which does not yield, however, nontrivial S-matrix elements to any finite order of perturbation theory. It is conjectured that confinement is likely to be destroyed by electromagnetic self interactions.

II. Charge Conservation and Electromagnetic Coupling

The free virton field Lagrangian is given by

$$\mathcal{L}_0(x) = \bar{\Upsilon}(x)\, Z(i\not\partial)\, \Upsilon(x) \tag{1}$$

In order that the field equation

$$Z(i\not\partial)\, \Upsilon(x) = 0 \tag{2}$$

has only the trivial solution $\Upsilon(x) \equiv 0$, $Z(z)$ is chosen as an entire analytic function without zeroes. The Green's function of the virton field is then

$$G(x-y) = Z^{-1}(i\partial) \, \delta(x-y) \tag{3}$$

The simplest choice of Z consistent with all requirements of nonlocal field theory turns out to be

$$Z(z) = - M \exp\left(- \ell z - \frac{L^2}{4} z^2\right) \tag{4}$$

with ℓ and L constants of dimension of length which specify the theory and M with dimension of mass setting the scale for the virton field. For all details we refer to [1].

We now try to couple the virton field to the electromagnetic field. In view of Ward's identity it is immediately clear that there is an intimate connection between the virton propagator and the virton-photon vertex. Charge conservation is more extensively exploited by the Kazes-Chang-Mani identities [3]. In order that these identities are respected, multiphoton vertices must be introduced whenever the charged particle propagator is different from a polynomial of first degree. The general recipe how to construct the multiphoton vertices corresponding to a given propagator has been given by Kroll [2]. It generalizes the minimal electromagnetic interaction and, as the latter, it is not the only possibility consistent with charge conservation, but the simplest.

The multiphoton vertices $V_n(q; k_1, \ldots k_n)$ where q denotes the incoming virton momentum and k_i the incoming photon momenta are obtained as follows. Kroll introduces a linear operation $d_\mu(k)$ acting on functions of a Dirac matrix $\not{p}$ such that

$$d_\mu(k) F(\not{p}) G(\not{p}) = \left(d_\mu(k) F(\not{p})\right) G(\not{p}) + F(\not{p}+\not{k}) d_\mu(k) G(\not{p}) \tag{5}$$

Specifying

$$d_\mu(k)\,\text{const} = 0 \tag{6}$$

$$d_\mu(k)\,\not{p} = \gamma_\mu \tag{7}$$

in view of (5) and its linearity the action of $d_\mu(k)$ is defined on power series of $\not{p}$. Due to (5) and (6)

$$d_\mu(k)\, F^{-1}(\not{p}) = - F^{-1}(\not{p}+\not{k}) \left[d_\mu(k)\, F(\not{p})\right] F^{-1}(\not{p}) \tag{8}$$

hence $d_\mu(k)$ is defined on the inverse of power series, too. It is easily shown that on this class of functions

$$k^\mu d_\mu(k)\, F(\not{p}) = F(\not{p}+\not{k}) - F(\not{p}) \tag{9}$$

and
$$d_{\mu_1}(k_1)\, d_{\mu_2}(k_2)\, F(\not{p}) = d_{\mu_2}(k_2)\, d_{\mu_1}(k_1)\, F(\not{p}) \tag{10}$$

Now the multiphoton vertices $V_n(q; k_1 \ldots k_n)$ are defined recursively by

$$V_n(\not{p}; k_1 \ldots k_n) = - d(k_n)\, V_{n-1}(\not{p}; k_1 \ldots k_{n-1}) \tag{11}$$

where the tensor indices have been suppressed. The recursion
starts with the inverse propagator $Z(\not{q})$

$$\dot{V}_0(\not{k}) = -Z(\not{k}) \qquad\qquad (12)$$

It is understood that $\not{q}$ is replaced by $\not{k}$ everywhere including $2q_\mu = \{\not{q}, \gamma_\mu\}$, then all d-operations are carried
out and $\not{k}$ is replaced by $\not{q}$ afterwards. The first step
leads to ambiguities, if there are other Dirac matrices
besides $\not{k}$, because $d_\mu(k)\not{k} z^2 \neq d_\mu(k) z^2 \not{k}$. This does not occur
in our case. Eq. (9) is crucial for the validity of the
Kazes-Chang-Mani identities.

Satisfactorily, virtons cannot escape by radiating
photons, since

$$\bar{\Psi}(q + \Sigma k_i)\, V_n(q; k_1 \ldots k_n) = V_n(q; k_1 \ldots k_n)\Psi(q) = 0 \qquad (13)$$

For n = 1, we have

$$V_1(q; k)\,\Psi(q) = -Z(q+k)\left[d(k)\, Z^{-1}(q)\right] Z(q)\,\Psi(q) = 0$$

by using the field equation (2) and the fact that the expression
in front of $Z(q)$ is again an entire function. Eq. (13) follows
for all values of n by induction. It is, however, an open
question if virtons can escape due to their self interactions.
This is briefly discussed below.

In the following we shall need the one- and two-photon
vertices which are obtained in a straightforward but tedious way

$$V_{1\mu}(q;k) = \alpha_\mu(k)\, z(\not{p})$$

$$= -M\left\{\left(e^{-\ell(\not{q}+\not{k})-\frac{L^2}{4}(q+k)^2} - e^{-\ell\not{q}-\frac{L^2}{4}q^2}\right) t_\mu\right.$$

$$\left. - 2\,\frac{\gamma_\mu\, qk - q_\mu \not{k}}{2qk+k^2}\,\frac{1}{\sqrt{q^2}}\,\sinh\ell\sqrt{q^2}\; e^{-\frac{L^2}{4}q^2}\right\} \tag{14}$$

where
$$t_\mu = \frac{\not{k}\,\gamma_\mu + 2q_\mu}{2qk+k^2}\quad,\quad \bar{t}_\mu = \gamma^0\, t_\mu^{+}\,\gamma^0 \tag{15}$$

$$V_{2\mu\nu}(q;k,-k) = -\alpha_\mu(k)\,\alpha_\nu(-k)\, z(\not{p})$$

$$= M\left\{\tfrac{1}{2}\left(-\tfrac{L^2}{2}e^{-\ell\not{q}} - \ell\,\tfrac{\not{q}}{q^2}e^{-\ell\not{q}} + \frac{\not{q}}{\sqrt{q^2}^3}\,\sinh\ell\sqrt{q^2}\right)\left(g_{\mu\nu} - (2qk+k^2)\bar{t}_\nu\, t_\mu\right)e^{-\frac{L^2}{4}q^2}\right.$$

$$- \left(e^{-\ell\not{q}-\frac{L^2}{4}q^2} - e^{-\ell(\not{q}+\not{k})-\frac{L^2}{4}(q+k)^2} - \frac{\sinh\ell\sqrt{(q+k)^2}}{\sqrt{(q+k)^2}}\,e^{-\frac{L^2}{4}(q+k)^2}\,\not{k}\right)\bar{t}_\nu\, t_\mu$$

$$- \left(\frac{\sinh\ell\sqrt{(q+k)^2}}{\sqrt{(q+k)^2}}\,e^{-\frac{L^2}{4}(q+k)^2} - \frac{1}{\sqrt{q^2}}\,\sinh\ell\sqrt{q^2}\, e^{-\frac{L^2}{4}q^2}\right)\gamma_\nu\, t_\mu \tag{16}$$

$$\left. + \left(\mu \leftrightarrow \nu,\; k \leftrightarrow -k\right)\right\}$$

V_2 has been given in the only case which is needed below i.e. when the photon momenta sum to zero. It should be noted that the apparent poles of V_1 and V_2 at $2qk \pm k^2 = 0$ actually cancel.

III. Vacuum Polarization and General Closed Loop

As a consequence of eq. (13), all amplitudes with external virtons vanish to any finite order of perturbation theory. In what follows we investigate amplitudes with only external photons. We are exclusively concerned with electromagnetic interactions. The nth order closed loop without radiative corrections is given by [2]

$$\Pi_n^o \ (k_1, \ldots, k_n) = \frac{(-1)^{n+1}}{n} \int d^4q \ \mathrm{Tr} \sum_{j=0}^{n-1} (n-j) \ \bar{P}_n \tag{17}$$

$$\left(d \, (k_{j+1}) \ldots d(k_n) \ Z \, (q_j) \right)\left(d \, (k_n) \ldots d(k_j) \ Z^{-1}(q) \right)$$

where

$$q_j = q + \sum_{i=1}^{j} k_i \quad (j \geqslant 1), \qquad q_0 = q$$

$\bar{P}_n$ denotes the sum over all inequivalent permutations of the photon momenta $k_1, \ldots, k_n$; i.e. permutations of the $k_{j+1}, \ldots, k_n$ and the $k_1, \ldots, k_j$ among themselves are excluded. The notation is such that $dk_1 \ldots dk_j \ Z$ means Z for j = o. As above we have suppressed all tensor indices. Due to Furry's theorem only the case of even n is of interest.

From

$$d(k_1) \ldots d(k_n) \ Z \, (q) \ Z^{-1}(q) \ = \ 0$$

we obtain

$$\sum_{j=0}^{n} \bar{P}_n \left(d \, (k_{j+1}) \ldots d(k_n) \ Z \, (q_j) \right)\left(d(k_1) \ldots d(k_j) \ Z^{-1}(q) \right) = 0 \tag{18}$$

Subtracting $(n/2)$ times eq. (18) from eq. (17) yields

$$\Pi_n^0(k_1 \ldots k_n) =$$

$$= \frac{(-1)^{n+1}}{n} \int d^4q \; \mathrm{Tr} \sum_{j=0}^{n} \left(\frac{n}{2}-j\right) \bar{P}_n \left(\alpha(k_{j+1}) \ldots \alpha(k_n) Z(q_j)\right)\left(\alpha(k_1) \ldots \alpha(k_j) Z^{-1}(q)\right)$$

$$= \frac{(-1)^{n+1}}{n} \int d^4q \; \mathrm{Tr} \sum_{j=0}^{\frac{n}{2}} \left(\frac{n}{2}-j\right) \bar{P}_n \left\{ \left[\alpha(k_{j+1}) \ldots \alpha(k_n) Z(q_j)\right]\left[\alpha(k_1) \ldots \alpha(k_j) Z^{-1}(q)\right] \right.$$

$$\left. - \left[\alpha(k_{j+1}) \ldots \alpha(k_n) \bar{Z}^{-1}(q)\right]\left[\alpha(k_1) \ldots \alpha(k_j) Z(\bar{q}_j)\right] \right\}$$

$$(19)$$

where in view of $\quad \sum k_i = 0$

$$\bar{q}_j = q - \sum_{i=1}^{j} k_i$$

For $n = 2$ only the term with $j = 0$ survives

$$\Pi_{\mu\nu}(k) = \Pi^0_{2\,\mu\nu}(k, -k)$$

$$= -\frac{1}{2} \int d^4q \; \mathrm{Tr} \left\{ \left(\alpha_\mu(k)\, \alpha_\nu(-k) Z(q)\right) Z^{-1}(q) - \left(\alpha_\mu(k)\, \alpha_\nu(-k) Z^{-1}(q)\right) Z(q) \right\}$$

$$= -\frac{1}{2} \int d^4q \; \mathrm{Tr} \left\{ V_{2\,\mu\nu}(q; k, -k) Z^{-1}(q) - (\ell \to -\ell, \; L^2 \to -L^2) \right\}$$

$$(20)$$

In eq. (20) the second order vacuum polarization, which

consists of the usual contribution and a seagull term, has been expressed in terms of the two photon vertex only. We are now going to evaluate eq. (20) explicitly. Inserting eq. (16) straightforward calculation leads to

$$
\Pi_{\mu\nu}(k) = \int d^4 q \, \mathrm{Tr}\left\{ \frac{L^2}{4}\left(g_{\mu\nu} - (2qk + k^2)\, \bar{t}_\nu t_\mu \right) \right.
$$

$$
+ \sinh \frac{L^2}{4}(2qk + k^2)\left[- \frac{\sinh \ell \hat{q}}{\hat{q}} \, \frac{\sinh \ell \widehat{q+k}}{\widehat{q+k}} \, \not{q}\, \gamma_\nu \, t_\mu \right.
$$

$$
\left. + \left(\cosh \ell \hat{q} \, \cosh \ell \widehat{q+k} - \frac{\hat{q}}{\widehat{q+k}} \sinh \ell \hat{q} \, \sinh \ell \widehat{q+k} \right) \bar{t}_\nu t_\mu \right]\right\}
$$

$$
+ \left(\mu \leftrightarrow \nu, \; k \leftrightarrow -k \right)
\tag{21}
$$

where $\hat{q} = \sqrt{q^2}$, $\widehat{q+k} = \sqrt{(q+k)^2}$. In nonlocal field theory all integrals are understood to be evaluated in Euclidean metrics [1,4]. In the present case, this is not sufficient to render the integrals in (21) finite nor even renormalizable. The reason is that within closed loops $\exp\left(-\frac{L^2}{4} q^2\right)$ cancels and the remaining $\sinh\left(\frac{L^2}{4}(k^2 + 2qk) \right)$ causes the integral to be nonrenormalizable of infinite order. This cancellation is a consequence of charge conservation, which enforces by the form of the vertices that within a closed loop then are equal powers of Z and Z^{-1}. Needless to say, the cancellation had not occured, if one had naively taken $e\gamma_\mu$ for the vertex as in usual QED.

Obviously, the only way-out is

$$
L^2 = 0
\tag{22}
$$

with the consequence that

$$\Pi_{\mu\nu}(k) = 0 \qquad (23)$$

With this choice of L^2 the propagator has the remarkable property

$$\widetilde{Z}(-q) = \widetilde{Z}^{-1}(q) \qquad (24)$$

As we shall now show, eq. (24) has the consequence that the sum of all closed loop diagrams as well as the sum of all diagrams containing closed loops vanishes to each order of perturbation theory.

For this purpose, we need some further properties of the d-operation

$$\prod_{i=1}^{n} d(k_i)\, F(q)\Big|_{q \to q+a} = \prod_{i=1}^{n} d(k_i)\, F(q+a) \qquad (25)$$

$$\prod d(k_i)\, F(q)\Big|_{q \to -q} = (-1)^n \prod d(-k_i)\, F(-q) \qquad (26)$$

$$\prod d(k_i)\, F(q+a)\Big|_{q \to -q} = (-1)^n \prod d(-k_i)\, F(-q+a) \qquad (27)$$

The proof of eqs. (25) – (27) proceeds easily by induction as that of eqs. (9) and (10). Eq. (25) has been tacitly used before. If we substitute now $q \to -q$ in the second term of eq. (19), then with regard to eq. (26) and (27), this term comes to read

$$- \Big[d(-k_{j+1})\dots d(-k_n)\widetilde{Z}(q)\Big]\Big[d(-k_n)\dots d(-k_j)\widetilde{Z}^{-1}(q)\Big] \qquad (28)$$

All terms of eq. (19) are separately invariant if the sign of all k_i's is changed. Applying this substitution to eq. (28) and shifting the integration variable according to $q \rightarrow q + \sum_{i=1}^{j} k_i$ we find that the second term in eq. (19) cancels the first one. The shift is allowed if a gauge invariant regularization scheme such as dimensional regularization is used. Hence we have for all n

$$\Pi_n^0 (k_1, \ldots, k_n) = 0 \qquad (29)$$

Next we consider the n-th order closed loop with m $(2m < n)$ internal photon propagators, but without further closed charged loops. It is given by

$$\Pi_n^m (k_1, \ldots k_{n-2m}) = \frac{1}{(2m)!} \int d^4 k_{n-2m+1} \, d^4 k_{n-2m+3} \cdots d^4 k_{n-1} \times$$

$$\times \Pi_n^0 (k_1, \ldots, k_{n-2m}, k_{n-2m+1}, -k_{n-2m+1}, \ldots, k_{n-1}, -k_{n-1}) \times$$

$$\times D(k_{n-2m+1}^2) \cdots D(k_{n-1}^2)$$

$$(30)$$

Since in obtaining eq. (29) we never made use of $k_i^2 = 0$, Π_n^m vanishes as well. The same considerations apply to diagrams which contain closed loops. So, finally, we need not consider any diagrams with any charged closed loop insertions because an appropriate sum of them always vanishes.

We remarked at the end of section 2 that all amplitudes with external virtons vanish to any finite order of perturbation theory. This does not mean, however, that the corresponding Green's functions are zero. The vanishing does not occur before they are multiplied with the virton wave functions. Since we had to put $L^2 = 0$ the factor $\exp(-L^2 q^2/4)$ which guaranteed convergence is now missing. Therefore we are faced with the problem of convergence for the remaining diagrams with only open charged lines. This problem can be solved by choosing a nonlocal generalization of the photon propagator

$$D_{\mu\nu}(x) = i\, g_{\mu\nu} \int \frac{d^4k}{(2\pi)^4} \frac{K^2(\ell^2 k^2)}{k^2}\, e^{-ikx} \tag{31}$$

with K an entire analytic function [4]. The resulting theory is finite without renormalization, and only diagrams without charged closed loops need to be considered.

We briefly point out a problem which might arise with regard to the virton self energy. Due to the latter the virton propagator is modified to

$$Z'(p) = Z(p) - \Sigma(p)$$

Hence confinement is spoiled except that $\Sigma(p)$ just cancels $Z(p)$, the remainder being an entire function without zeroes. This condition will be difficult to meet since in perturbation theory $\Sigma(p)$ starts to order of α.

We remarked above that (11) is not the only choice for the multiphoton vertices which is consistent with charge conservation. To each order, additional "intrinsic vertex modi-

fications" $W_n(q; k_1 \ldots k_n)$ can be introduced subject
to the condition that they are symmetric and transversal
with respect to all photon momenta [2]. Moreover, they must
be real and free of singularities if they are not to violate
unitarity. The W_n then influence the V_m with $m > n$. An
example of an intrinsic vertex modification is a Pauli anomalous magnetic moment term. With allowance for $W_n \neq o$ closed
loops vanish no longer. They are, however, real, because all
ingredients are real and free of singularities. In particular,
the nonlocal photon propagator (31) is real in Euclidean
metrics. Hence e.g. the ratio $R = \sigma(e^+e^- \to$ hadrons) $/$
$\sigma(e^+e^- \to \mu^+ \mu^-) \sim$ Im Π vanishes in this theory to
all orders. Nonzero contributions will not appear before
perturbation theory (with respect to electromagnetic or,
probably, strong interactions) is summed so that bound states
of virtons contribute in the intermediate state.

IV. Conclusions

We have coupled a nonlocal Lagrangian field theory which
is to incorporate confinement from the outset to the electromagnetic field. Charge conservation leads to the necessity
of omitting the Gaussian term in the virton propagator. Then,
however, the combined contribution of diagrams with charged
closed loops vanishes. Therefore the expectation that such a
theory might allow for nontrivial perturbation theory calculations has been shown to fail. Moreover, it seems not to be

sure, that confinement survives the summation of virton
self energies. If not some kind of partial confinement might
result.

Propagators similar to eq. (4) may emerge from an exact
solution of non-abelian gauge theories [5]. The problems
with convergence pointed out above and, hence, the vanishing
of charged closed loops do not arise in that case. The reason
is that the vertices derived by Kroll's procedure may only
be used as elementary vertices if they result from an "ad hoc"-
modification of the charged particle propagator. If the propagator has been obtained by a (partial) summation of self
energies, then deriving Kroll's vertices and using them as
elementary vertices would lead to wrong multiplicities for
perturbation theory diagrams. Hence the cancellation of the
Gaussian term does not occur in that case.

We expect, however, the same difficulty with the self
energy: confinement accomplished with respect to strong
interactions may be destroyed by switching on electromagnetic
interactions.

References

[1] A. Z. Dubnickova, G. V. Efimov, Dubna preprints
E2-10371 (1977) and P2 - 10734 (1977)

[2] N. M. Kroll, Nuovo Cimento 45A (1966), 65

[3] E. Kazes, Nuovo Cimento 13 (1959), 1226
N. P. Chang, H. S. Mani, Phys. Rev. 134B (1964), 896

[4] G. V. Efimov, Ann. Phys. (N. Y.) 71 (1972), 466

[5] J. M. Cornwall, G. Tiktopoulos, preprint UCLA /76/TEP/22

Supergravity and a Problem Raised by ITS S-Matrix

W. Becker and D. Großer

Institut für Theoretische Physik der Universität Tübingen, West Germany

Supergravity has attracted much interest because it offers
a chance for constructing a renormalizable theory which describes
strong, electromagnetic, and weak interaction and gravitation.
Such a construction would have to extend pure supergravity by
including internal symmetries whose gauging would yield the non-
gravitational interactions. In this talk we will describe pure
supergravity, mention its nice features, and point out a difficulty associated with its S-matrix.

Let us begin with some remarks about supersymmetry. A gene-
rator of a symmetry or a supersymmetry is an operator in the
Hilbert space of physical states which commutes with the S-matrix
and which acts additively on states of several incoming particles.
It has been shown by Haag, Łopuszański, and Sohnius [1] that if
the space-time symmetries form the Poincaré group then all one
can have in addition is a finite number of internal symmetry
generators which commute with the Poincaré generators and a
finite number of two-component spinors Q_α^L ($\alpha = 1,2$; $L = 1,\ldots,\nu$)
and their conjugates $(Q_\alpha^L)^+$ which can be chosen such that they
satisfy the following algebraic relations

$$\{ Q_\alpha^L, (Q_\beta^M)^+ \} = \delta^{LM}\, \sigma^\mu_{\alpha\beta}\, P_\mu$$
$$[Q_\alpha^L, P_\mu] = 0 \tag{1}$$
$$[Q_\alpha^L, M_{\mu\nu}] = \tfrac{1}{2}\, (\sigma_{\mu\nu})_\alpha{}^\beta\, Q_\beta^L$$

where $\sigma_{\mu\nu} = \tfrac{i}{2} \left(\sigma^\mu \bar\sigma^\nu - \sigma^\nu \bar\sigma^\mu \right)$, $\sigma^\mu = (1,\sigma_i)$, and $\bar\sigma^\mu = (1,-\sigma_i)$. The
Q_α^L and their conjugates are called generators of supersymmetries.
Their remarkable properties are: they have anticommutation rather
than commutation relations among themselves and nontrivial commu-

tation relations with the Lorentz generators. These commutation relations imply that the corresponding supermultiplets contain bosons and fermions. Wess and Zumino [2] have given an example of a renormalizable Lagrangian field theory exhibiting supersymmetry. They noticed that due to supersymmetry remarkable cancellations of divergent terms occured.

These remarkable cancellations brought about the idea that supersymmetry might help to construct a finite theory of gravitation. The S-matrix of pure gravitation is known to be one-loop finite. However, any attempt to couple another field to gravitation lead to divergences at the one-loop level. In the simplest case the generators of supersymmetries will form one two-component spinor Q_α and its conjugate Q_α^+. Then with massless particles the spin content of the supermultiplets is $(j, j+\frac{1}{2})$. As the spin of the graviton is 2 there are the two possibilities: $(\frac{3}{2}, 2)$ and $(2, \frac{5}{2})$. As has been shown by Grisaru, Pendleton, and van Nieuwenhuizen [3] the case $(2, \frac{5}{2})$ leads to a theory which is consistent only if it is free.

Supergravity was first constructed by Freedman, van Nieuwenhuizen, and Ferrara [4]. These authors coupled a Rarita-Schwinger field ψ_μ which is a Majorana spinor and the vierbein field $V^a{}_\mu$ in the usual minimal way. The resulting theory is not locally supersymmetric but could be made so by adding a quartic contact term. However, the proof of local supersymmetry is very complicated and involves a computer calculation. This proof is much easier in a formulation of supergravity given by Deser and Zumino [5]. These authors used the first-order formulation in which the connection $\omega_{\mu ab}$ is an independent variable. Then the action is

$$I = \int d^4x \left[-\frac{1}{16\pi G} VR - \frac{1}{2} \varepsilon^{\lambda\mu\nu\rho} \bar{\psi}_\lambda \gamma_5 \gamma_\mu D_\nu \psi_\rho \right] \qquad (2)$$

where

$$V = \det V^a{}_\mu$$

$$R = V^{a\mu} V^{b\nu} R_{\mu\nu ab}, \quad R_{\mu\nu ab} = \left(\partial_\mu \omega_{\nu ab} + \omega_{\mu a}{}^c \omega_{\nu cb} \right) - (\mu \leftrightarrow \nu) \quad (3)$$

$$D_\nu \psi_\rho = \partial_\nu \psi_\rho + \frac{1}{2} \omega_{\nu ab} \sigma^{ab} \psi_\rho, \quad \sigma^{ab} = \frac{1}{4} [\gamma^a, \gamma^b]$$

This is not the usual minimal coupling because $D_\nu \psi_\rho$ is defined according to the spin $\frac{1}{2}$ content of ψ_ρ only. Nevertheless the action is a scalar because it involves only the curl.

This theory is generally covariant and locally Lorentz invariant and supersymmetric. It avoids acausal propagation of signals at the classical level which has plagued previous couplings of a spin $\frac{3}{2}$ field. The most remarkable property of supergravity is the fact that it yields an S-matrix which has been shown to be finite up to the two-loop level [6]. There are doubts whether this miracle continues to the three-loop level and explicit calculations seem to be too complicated for the near future. One might hope that additional symmetries in extended versions of supergravity will lead to an S-matrix which is finite to every order. However, the cancellations which render the S-matrix finite take place only on shell. This means that Green's functions are divergent. Therefore the existence of the S-matrix is of vital importance. We will show in the following that the usually adopted naive concept of the S-matrix which rests on scattering states being tensor products of one-particle states leads to a problem.

Consider the scattering process $1 + 2 \to 3 + 4$. Let $F_{\lambda_3 \lambda_4 \lambda_1 \lambda_2}$ (s,t,u) denote the corresponding helicity amplitude which is related to the S-matrix by

$$S = 1 + (2\pi)^4 \, \delta^4(P_i - P_f) \prod_{i=1}^{4} (2E_i)^{-\frac{1}{2}} F \tag{4}$$

where E_i is the center-of-mass energy of the ith particle. Consider in particular elastic graviton-graviton scattering. Since in the Born approximation no fermions may be exchanged the amplitude for this process is that of Einstein theory. This has been calculated in different ways [7]:

$$F_{2222} = 8\pi i \, G \, \frac{s^3}{t u} \tag{5}$$

By supersymmetry this amplitude is related to $F_{2 \frac{3}{2} 2 \frac{3}{2}}$ and this relation leads to difficulties.

Since the generators of supersymmetry commute with the S-matrix one has

- 355 -

$$\langle p_3\, 2,\, p_4\, \tfrac{3}{2}\, |\, [\, Q_\alpha^+,\, S-1\,]\, |\, p_1\, 2,\, p_2\, 2\, \rangle = 0 \qquad (6)$$

The algebraic relations (1) and the structure of the supermulti-
plet $(\tfrac{3}{2}, 2)$ imply that the Q_α and the Q_α^+ annihilate a particle
or change its helicity by $\pm\tfrac{1}{2}$. Therefore if we choose $p_1^\mu = (E,o,o,E)$
and $p_2^\mu = (E,o,o,-E)$ with $E > o$ then we obtain from (6)

$$\langle p_3\, 2,\, p_4\, \tfrac{3}{2}\, |\, S-1\, |\, p_1\, 2,\, p_2\, \tfrac{3}{2}\, \rangle = c\, \frac{\sqrt{p_4^o - p_4^3}}{\sqrt{2E}}\, \langle p_3\, 2,\, p_4\, 2\, |\, S-1\, |\, p_1\, 2,\, p_2\, 2\, \rangle \qquad (7)$$

where c is a phase factor. (4), (5) and (7) imply that in Born
approximation

$$\langle p_3\, 2,\, p_4\, \tfrac{3}{2}\, |\, S-1\, |\, p_1\, 2,\, p_2\, \tfrac{3}{2}\, \rangle = c\, (2\pi)^4 \delta^4 (P_i - P_f) \prod_{i=1}^{4} (2E_i)^{-\frac{1}{2}}\, 8\pi i\, G\, \frac{s^3}{tu}\, \sqrt{\frac{-u}{s}} \qquad (8)$$

Consider backward scattering: $p_3 = p_2$ and $p_1 = p_4$. In this
case the state vectors on the left hand side of (8) are eigenvectors of the angular momentum component J_3 belonging to different eigenvalues. Therefore the left hand side is zero. The
right hand side, however, does not vanish if $G \neq o$.

The phenomenon is clearly related to the mass of the external
as well as the exchanged particles being zero. This circumstance
causes the kinematical zeroes and the dynamical poles to coincide.

A similar difficulty arises with the photon-graviton amplitude [7]

$$F_{2121} = 8\pi i\, G\, \frac{s^2}{t} \qquad (9)$$

which also should vanish at $u = o$ because of angular momentum
conservation, but instead is finite. In a way, this is even more
disturbing, since this is an amplitude for physical mass zero
particles. Hence one is really forced to face the problem.

The same difficulty shows up in Compton scattering off
massless electrons. However, the situation here is different
from that in supergravity. Massless QED is in bad shape as a
physical theory because the infrared problem cannot be solved
in the usual way [8]. This is not a real problem because massless

particles which carry an electric charge do not exist. The interest in massless QED comes from the fact that the Green's functions exist and approximate those of QED for certain momenta. In supergravity the Green's functions do not exist because of ultraviolet divergences. The remarkable cancellations which render the S-matrix finite up to the two-loop level take place only on shell. Therefore the existence of the S-matrix is of vital importance for supergravity. The above difficulty shows that at least it does not exist in the naive sense which is usually adopted.

References

[1]   R. Haag, J. T. Łopuszański, and M. Sohnius, Nucl. Phys. B88 (1975) 257

[2]   J. Wess and B. Zumino, Phys. Lett. 49B (1974) 52

[3]   M. T. Grisaru, H. N. Pendleton, and P. van Nieuwenhuizen, Phys. Rev. D15 (1977) 996

[4]   D. Z. Freedman, P. van Nieuwenhuizen, and S. Ferrara, Phys. Rev. D13 (1976) 3214

[5]   S. Deser and B. Zumino, Phys. Lett. 62B (1976) 335

[6]   S. Deser, J. H. Kay, and K. S. Stelle, Phys. Rev. Lett. 38 (1977) 527

[7]   F. A. Berends and R. Gastmans, Nucl. Phys. B88 (1975) 99; M. T. Grisaru, P. van Nieuwenhuizen, and C. C. Wu, Phys. Rev. D12 (1975) 397

[8]   S. Weinberg, Phys. Rev. 140B (1965) 516, footnote 3 and Sect. III.

Quantum Mechanical Corrections to the Classical Maxwell Lagrangian

W. Dittrich

Institut für Theoretische Physik der Universität Tübingen,
Tübingen, W.-Germany

In the following discussion we want to neglect all but electromagnetic interactions. In particular we want to study quantum mechanical corrections for electromagnetic processes generated by constant fields. Such a study must necessarily employ non-perturbative techniques, as e.g. advocated by Schwinger[1], Symanzik[2] or Fried[3], since we want to treat the external electromagnetic field exactly. From the work of Heisenberg and Euler[4], and Weisskopf[5], we know that classical electromagnetism becomes a non-linear theory due to the probability of pair creation. However, even before the advent of quantum mechanics people were interested in modifying the classical Maxwell theory. Here the non-linear model of Born and Infeld[6] comes to mind. Although forgotten for many years their model enjoys at present a remarkable renaissance for various reasons: some people like to use it as an effective Lagrangian to determine the eigenvalues of electrons bound to superheavy nuclei[7], others are more interested in investigating the existence of a finite classical electron radius and more recently, people have looked at the Born-Infeld Lagrangian as predecessor of string models[8]. Originally Born and Infeld were interested in a non-polynomial modification of the classical Maxwell Lagrange function, constructed from the invariants $(\vec{E}^2 - \vec{H}^2)$, $(\vec{E} \cdot \vec{H})^2$ in order to obtain a finite value for the self-energy (mass) of the electron. In their theory the electron is a stable particle. Furthermore, their model can produce a finite minimum electrostatic electron radius $r_{min} \cong 3.5 \cdot 10^{-13}$ cm and an associated maximum electrostatic field strength $E_{max} = \dfrac{e}{r^2_{min}}$ $\cong 4 \cdot 10^{15}$ ESU . However, the Born-Infeld Lagrangian

was entirely ad hoc and the physical picture that emerged was
never fully understood. But meanwhile quantum mechanics was born
and the problem was again looked at from the viewpoint of quantum electrodynamics (QED). Here the classical expression for the
Lagrangian $\mathcal{L} = \frac{1}{2}(\vec{E}^2 - \vec{H}^2)$ acquires corrections which are cau-
sed by the interaction between electromagnetic fields and thus
violate the superposition principle. The scattering of light by
light or Delbrück scattering are familiar examples for this
quantum mechanical non-linearity. Photon-photon scattering at
low energies, e.g., can be described by the following effective
Lagrangian density

$$\mathcal{L} = \frac{2\,\alpha^2}{45}\,\frac{1}{m^4}\left[(\vec{E}^2 - \vec{H}^2)^2 + 7\,(\vec{E}\cdot\vec{H})^2\right]\;,$$

where $\alpha = \frac{1}{137}$ and m = electron mass. This result for $\mathcal{L}$ can
be obtained by either evaluating the scattering process as indi-
cated in fig. 1

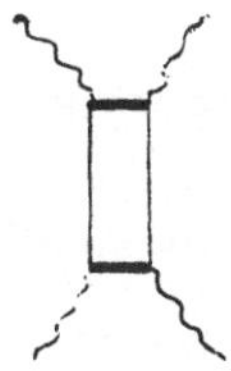

Fig. 1

or by computing the exact Lagrangian first (one-loop process to
all orders in the external field) and then extracting the quartic correction thereof. Among the various computations of this
one-loop effective Lagrangian starting with Heisenberg and Euler[4]
and Weisskopf[5], Schwinger[9] has probably influenced most of the

work that has been performed since. It is in Schwinger's paper where the proper-time methods was used to deal with external field problems. The author[10] and certainly a lot of other people (e.g. Brown and Duff[11]) have invented their own techniques to produce the relevant Green's function of the problem: charged particle in an external constant electromagnetic field. Here we want to use that information to evaluate the vacuum persistence amplitude (loop factor) which represents the effect that an arbitrary number of external photon lines can have on a single charged-particle loop. It is this process (fig. 2) which yields the first non-linear correction to the classical Maxwell Lagrangian.

Fig. 2   Non-linear correction

This single-loop approximation represents the first non-linear correction to any field theory. In particular in QED we know a closed-form solution for the cases of constant electric, constant magnetic, and plane wave field (laser).

The process which summarizes the effect that an external field environment $A_\mu(x)$ can have on a single fermion loop, is given analytically by

$$\langle 0_+|0_-\rangle^A = e^{i W^{(1)}[A]} \tag{1}$$

where $\quad i W^{(1)}[A] \equiv L[A] = - \text{Tr} \ln (1 - e\gamma \cdot A G_+)^{-1}$ (2)

and $\quad G_+[A] = G_+ (1 - e\gamma \cdot A G_+)^{-1}$.

If we employ the proper-time representation[9] for $G_+[A]$,

$$G_+[A] = (m - \gamma\Pi) i \int_0^\infty ds \, e^{-is[m^2 - (\gamma\Pi)^2]} ,$$ (3)

where $\quad \Pi = \frac{1}{i}\partial - e A \quad , \quad -(\gamma\Pi)^2 = \Pi^2 - \frac{e}{2}\sigma F$,

and use the functional derivative of $\quad i W^{(1)}[A] \quad$,

$$i \frac{\delta W^{(1)}}{\delta A^\mu (x)} = - e \, \text{tr} \left[\gamma_\mu \, G_+ (x,x|A) \right]$$ (4)

we find

$$i \int \mathcal{L}^{(1)} (x) \, dx = i \, W^{(1)} = -\frac{1}{2} \int_0^\infty \frac{ds}{s} \, e^{-ism^2} \text{Tr} \left\{ e^{is(\gamma\Pi)^2} \right\}$$ (5)

If there is only an external constant magnetic field present, e.g., $F_{12} = - F_{21} \equiv H$, one finds

$$\mathcal{L}^{(1)}[H] = - \frac{2}{16\pi^2} \int_0^\infty \frac{ds}{s^3} \, e^{-m^2 s} \left[(eHs)\cot h \, (eHs) - 1 - \frac{1}{3}(esH)^2 \right]$$ (6)

One can integrate (6) with the result

$$\mathcal{L}^{(1)}[H] = - \frac{2}{64\pi^2} \left\{ (2m^4 - 4m^2(eH) + \frac{4}{3}(eH)^2) \left[\ln \left(\frac{m^2}{2eH}\right) + 1 \right] \right.$$
$$\left. - 3m^4 + 4m^2(eH) - (4eH)^2 \zeta' (-1;\frac{m^2}{2eH}) \right\}$$ (7)

Incidentally, in the limiting case $\quad \frac{eH}{m^2} \gg 1 \quad$ one finds the ratio

$$\frac{\mathcal{L}^{(1)}}{\mathcal{L}^{(0)}} \simeq -\frac{\alpha}{3\pi}\ln(\frac{eH}{m^2}) \quad , \qquad \mathcal{L}^{(0)} = -\frac{1}{2}H^2 \quad , \tag{8}$$

which is Weisskopf's old result[5]. The other interesting case, in which only a pure constant electric field is applied, yields $\mathcal{L}^{(0,1)}\left[E\right] = \mathcal{L}^{(0,1)}\left[H \to \frac{1}{i}E\right]$. In a recent paper by Rohrlich[12] one finds several limiting applications of the exact expression (6), in particular for strong electric fields.

The results of ref. 12 can be further supplemented by allowing for radiative corrections[13] to the single loop (fig. 3):

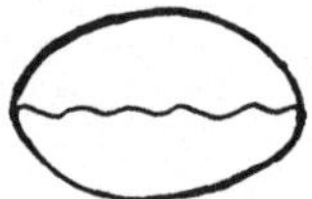

Fig. 3   Second-order correction

This is analytically expressed by

$$\mathcal{L}^{(2)}(x) = \frac{e^2}{2}\int (dx')\,\mathrm{tr}\left[\gamma_\mu\, G_+(x,x'|A)\,\gamma_\nu\, G_+(x',x|A)\right] D_+^{\mu\nu}(x-x') \tag{9}$$

Again, we are interested in a constant external magnetic field which we assume to be in z-direction, $F_{12} = -F_{21} = H$. Then we find for the case of strong fields[13,14]

$$\mathcal{L}^{(2)}\left[H\right] \simeq \frac{\alpha^2 H^2}{8\pi^2}\frac{5}{3}\ln\frac{eH}{m^2} \quad , \qquad \frac{eH}{m^2} \gg 1 \tag{10}$$

or $\quad \mathcal{L}^{(2)}/\mathcal{L}^{(0)} \simeq (\frac{\alpha}{\pi})^2\left[-\frac{5}{12}\ln\frac{eH}{m^2}\right]$ $\tag{11}$

The other limiting case, in which only a pure electric
field is present, yields

$$\mathcal{L}\left[E\right] = \frac{1}{2} E^2 \left\{ 1 - (\frac{\alpha}{\pi})\frac{1}{3}\left[\ln(\frac{eE}{m^2}) - i\frac{\pi}{2}\right] - (\frac{\alpha}{\pi})^2\frac{5}{12}\left[\ln(\frac{eE}{m^2}) - i\frac{\pi}{2}\right] \right\} \qquad (12)$$

The imaginary part is indicative of pair creation. Leaving this
effect aside for the moment, the real part enables us to derive
the displacement

$$D(E) = \frac{\partial \mathcal{L}}{\partial E} = \varepsilon E$$

$$= E\left\{1 - \frac{\alpha}{6\pi} - \frac{5}{24}(\frac{\alpha}{\pi})^2 - \left[\frac{\alpha}{3\pi} + \frac{5}{12}(\frac{\alpha}{\pi})^2\right]\ln\frac{eE}{m^2}\right\} \qquad (13)$$

which agrees with Rohrlich's result to first order in α. However, eq. (13) does not confirm his ansatz (formula (3.10) of
ref. 12) by which he includes radiative corrections. We find in-
stead a term proportional to $(\frac{\alpha}{\pi})^2 \ln\frac{eE}{m^2}$ without squared loga-
rithm, which just means that the following ratio is independent
of the magnitude of the field:

$$\mathcal{L}^{(2)} / \mathcal{L}^{(1)} \sim \frac{\alpha}{\pi} \; .$$

However, further radiative corrections will again be of the order
of $\frac{\alpha}{\pi} \ln\frac{eE}{m^2}$.

References

1. J. Schwinger, Proc. Nat. Acad. of Sciences (USA) $\underline{37}$, 452 (1951); and Harvard Lectures (1954)

2. K. Symanzik, Z. Naturforschung $\underline{9a}$, 10 (1954), 809

3. H. M. Fried, "Functional Methods and Models in Quantum Field Theory", Cambridge, Mass.: MIT Press (1972)

4. W. Heisenberg, H. Euler, Z. Physik $\underline{98}$, 714 (1936)

5. V. Weisskopf, K. Danske Vidensk. Selsk., Mat. - Fys. Meddr. $\underline{14}$, No. 6 (1936)

6. M. Born, Proc. R. Soc. $\underline{A143}$, 410 (1934); M. Born and L. Infeld, ibid. $\underline{A144}$, 425 (1934)

7. J. Rafelski and W. Greiner, Nuovo Cimento $\underline{13B}$, 135 (1973)

8. H. B. Nielsen and P. Olesen, Nucl. Phys. $\underline{B57}$, 367 (1973)

9. J. Schwinger, Phys. Rev. $\underline{82}$, 664 (1951)

10. W. Dittrich, J. Phys. $\underline{A9}$, 1171 (1976)

11. M. R. Brown and M. J. Duff, Phys. Rev. $\underline{D11}$, 2124 (1975)

12. M. Greenman and F. Rohrlich, Phys. Rev. $\underline{D8}$, 1103 (1973)

13. W. Dittrich, J. Phys. $\underline{A10}$, 833 (1977)

14. V. I. Ritus, P. N. Lebedev Physical Institute, Moscow, Preprint No. 125 (1975)

Spoor of a Fixed Point in QED

M. P. Fry

Max-Planck-Institut für Physik und Astrophysik,
München, Fed. Rep. of Germany

I. Introduction

Suppose quantum electrodynamics is in the domain of attraction of an ultraviolet-stable fixed point. Then the renormalized photon-propagator $D_{\mu\nu}$ is asymptotic, within a constant, to the free photon-propagator at large space-like momenta [1-3]

$$\alpha\, D_{\mu\nu}(q) \underset{q^2 \to \infty}{\sim} g_{\mu\nu}\, \frac{\alpha_o}{q^2} + q_\mu q_\nu \text{ terms} . \tag{1}$$

The asymptotic coupling constant α_o is assumed to be finite. By positivity [2]

$$\alpha_o > \alpha,$$

where α is the physical fine structure constant. Thus a clear signal for an UV-stable fixed-point in electrodynamics is the damping at large momentum transfers of the radiative correction logarithms in $D_{\mu\nu}$ that are present to any finite order of perturbation theory.

The effective coupling constant $g(\lambda)$ controlling the transition to the scaling limit (1) of $D_{\mu\nu}$ is given by the equations

$$\frac{d\, g(\lambda)}{d\, \ln\lambda} = \beta(g(\lambda)),$$

$$g(1) = \alpha,$$

where ß is the Callan-Symanzik function for quantum electrodynamics defined for conventional on-shell renormalization. The parameter λ scales the momentum q appearing in (1): $q \to \lambda q$, $\lambda \to \infty$. The assumption that quantum electrodynamics has an UV-stable fixed-point implies that ß has at least one zero on the positive real axis. Let g_∞ denote the first such zero. Then, if $\alpha < g_\infty$, the effects of g_∞ on physical observables are unlikely to be detected if $g_\infty - \alpha$ is not very small. The reason is that if ß has a zero it must be of infinite order [2,4]. So, for example, if

$$\beta \sim (g - g_\infty)^2\, e^{\frac{1}{g - g_\infty}} ,$$

for $g \uparrow g_\infty$, $\alpha < g_\infty$, then

$$g(\lambda) \underset{\lambda \to \infty}{\sim} g_\infty - 1/\ln \ln \lambda .$$

In contrast, when $g_\infty = \alpha$, $g(\lambda)$ is already at its asymptotic value, and so the asymptotic behavior of $D_{\mu\nu}$ given by (1) should set in rapidly. Just how rapidly is an open calculation question.

At present fixed-point behavior is almost certainly ruled out [5] at momentum transfers $\lesssim 0.4$ GeV/c by the present agreement between the theoretical value of the anomalous magnetic moment of the muon [6] and its latest experimental value obtained by the CERN Muon Storage Ring Collaboration [7]. For momentum transfers in the range 0.4 - 7.4 GeV/c the onset of fixed-point behavior is restricted [5] by the accuracy achieved by the HEPL-Stanford-Pennsylvania measurements of the absolute cross sections for wide-angle Bhabha scattering at center-of-mass energies of 7.0 and 7.4 GeV [8], muon pair production at 7.4 GeV [9], and the agreement of these cross sections with finite-order perturbation theory. Constraints on possible fixed-point behavior above 7.6 GeV, the maximum possible energy available at SPEAR-II, must await results from PEP and PETRA.

II. Necessary Condition for an UV-Stable Fixed-Point in QED

Obviously it would be very nice to find out if quantum electrodynamics really does have an UV-stable fixed-point and if so, to make quantitative predictions. The existence problem was considerably simplified by Baker and Johnson [10] when they stated a _necessary_ condition for a fixed-point. It may be stated as follows: Consider a mutilated model of electrodynamics with massless electrons [11] and whose amplitudes have no vacuum polarization subgraphs [12-15]. Then the function F_1, defined by one of the model's two-point functions, namely

$$\langle 0| T(j_\mu(x) j_\nu(0))|0\rangle^{\text{one-electron-loop}}$$
$$= \frac{F_1(\alpha)}{\pi^4 x^6}\left(g_{\mu\nu} - 2\,\frac{x_\mu x_\nu}{x^2}\right), \quad (x \neq 0) \quad (2)$$

must have a zero at $\alpha = \alpha_\infty$ on the positive real axis if electrodynamics has an
UV-stable fixed-point:

$$\beta(g_\infty) = 0^\infty \implies F_1(\alpha_\infty) = 0^\infty.$$

The additional requirement that F_1 vanish with an infinite-order zero if ß does
was first obtained by Adler [2]. Other derivations of this result are given in
Refs. [14,16]. Through fourth order[*] [17]

$$F_1(\alpha) = 1 + \frac{3}{4}\left(\frac{\alpha}{\pi}\right) - \frac{3}{32}\left(\frac{\alpha}{\pi}\right)^2 + O\left\{\left(\frac{\alpha}{\pi}\right)^3\right\}.$$

The superscript "one-electron-loop" indicates that the matrix element on the
right-hand side of (2) only receives contributions from photon self-energy
graphs having a single electron loop, by definition. The existence of the
two-point function in (2) in each order of perturbation theory and its structure on the right-hand side are a consequence of the results obtained in Ref.
[12]. From here on α will refer to the square of the coupling constant e of
this model: $\alpha = e^2/4\pi$.

A zero of the Gell-Mann-Low function ψ [11,18] on the positive real axis
at α_0 must be of infinite order [2]:

$$\psi(\alpha_0) = 0^\infty.$$

For definiteness assume that α_0 is the first such zero of ψ. By positivity
$\alpha_0 > \alpha_{phys}$, the fine structure constant [2]. If ψ has an infinite-order zero
at α_0 and $\alpha_\infty = \alpha_{phys}$, then $g_\infty = \alpha_{phys}$ if ß is summed loop-wise [2]. If
$\alpha_\infty > \alpha_{phys}$, then it is possible that $\alpha_\infty = \alpha_0$ [2]. The position of g_∞ on the
positive real axis when $\alpha_\infty = \alpha_0$ cannot be fixed without additional information.
Because F_1 is by definition a vastly simpler function than ß, it is worthwhile
to study massless electrodynamics without vacuum polarization subgraphs in some
detail.

---

[*]   Our F_1 and that of Baker and Johnson [10] are related by F_1 (Baker-
Johnson) $= \frac{\alpha}{3\pi} F_1$.

III. Conformal Invariant QED

Since the above model of electrodynamics has no mass and requires no cut-offs if its charged sectors are restricted to the generalized Landau gauge [12], it is dilatation invariant. We will assume that the Dyson-Wick rotation to a Euclidean metric has been made at the outset. This suggests that it may be possible to enlarge the symmetry group of the model to the full fifteen parameter conformal group. Further investigation shows that this is not possible because the conventional gauge-fixing term in the Lagrangian of massless quantum electrodynamics is not conformal invariant [19,20]. However, it was first observed by Abdellatif [21] that if the gauge-dependent sectors of this model are calculated in a non-local gauge defined by the propagator

$$D_{\mu\nu}(u,v) = \frac{i}{4\pi^2}\left[\left(1-\tfrac{1}{2}G(\alpha)\right)\left(\frac{g_{\mu\nu}}{(u-v)^2} - 2\,\frac{(u-v)_\mu (u-v)_\nu}{(u-v)^4}\right)\right.$$
$$\left. - 2\left(\frac{(u-v)_\mu}{(u-v)^2} - \frac{(u-B)_\mu}{(u-B)^2}\right)\left(\frac{(v-u)_\nu}{(v-u)^2} - \frac{(v-B)_\nu}{(v-B)^2}\right)\right], \tag{3}$$

then they are conformal covariant. This is subject to the proviso that the gauge-labelling vector B transforms as the components of a position four-vector under a special conformal transformation. The additional gauge parameter G in (3) defines the generalized Landau gauge [12]:

$$G(\alpha) = 1 - \frac{3\alpha}{8\pi} + O\left\{\left(\frac{\alpha}{\pi}\right)^2\right\}.$$

Since the propagator (3) can be written in the form

$$D^B_{\mu\nu}(u,v) = D_{\mu\nu}(u-v) + \frac{\partial}{\partial u^\mu}(\cdots) + \frac{\partial}{\partial v^\nu}(\cdots) + \frac{\partial^2}{\partial u^\mu \partial v^\nu}(\cdots),$$

where $D_{\mu\nu}$ is the generalized Landau gauge propagator

$$D_{\mu\nu}(u-v) = \frac{i}{4\pi^2}\left(\frac{g_{\mu\nu}}{(u-v)^2} + \frac{G(\alpha)}{4}\frac{\partial}{\partial u^\mu}\frac{\partial}{\partial v^\nu}\ln(u-v)^2\right), \tag{4}$$

the B-dependent terms in $D^B_{\mu\nu}$ have no effect on physical observables. Further discussion of this non-local gauge may be found in Refs. [19,22-24].

Hereafter we shall refer to Euclidean massless electrodynamics in the gauge defined by (3) and with vacuum polarization subgraphs omitted as conformal in-

variant quantum electrodynamics (CIQED). The main purpose of studying CIQED
is to obtain information about the analyticity properties in α of F_1. The
principal result to date has been obtained by Adler [25]. His analysis indicates that F_1 is unlikely to develop an infinite-order zero as α approaches a
positive singularity from below if the singularity of the photon propagator in
(3) at u = v is cut off as described in [25]. In what follows we will describe
an extension to CIQED of proposals by several authors [26] for the non-perturbative solution of conformal-invariant field theories.

4. CONFORMAL SYMMETRY AND F_1

The power of this enlarged symmetry group is illustrated by the severe
constraints it places on the two- and three-point functions of CIQED. For
example, the unrenormalized, unregularized improper vertex function in CIQED
has the structure [27]

$$\Gamma_\mu(x,y;z) = - \langle 0| T(\Psi(x)\, j_\mu(z)\, \overline{\Psi}(y))|0\rangle^{\text{one-electron-line}}$$

$$= \frac{E_1(\alpha)\, F_1(\alpha)}{4\pi^4}\; \frac{\gamma\cdot(x-z)\,\gamma_\mu\,\gamma\cdot(z-y)}{(x-z)^4 (y-z)^4} \tag{5}$$

$$+ \frac{E_2(\alpha)\, F_1(\alpha)}{4\pi^4}\; \frac{\gamma\cdot(x-y)}{(x-y)^2 (x-z)^2 (y-z)^2} \left(\frac{(z-x)_\mu}{(z-x)^2} - \frac{(z-y)_\mu}{(z-y)^2} \right),$$

provided the gauge parameter B in the photon propagator (3) is set equal to x
or y. The quantities E_1 and E_2 are coefficients appearing in the Wilson expansion of the C = -1, J = 1 part of the product $\psi(x)\overline{\psi}(y)$ of electron fields
when B = x or y:

$$\Psi_i(x)\,\overline{\Psi}_j(y) \underset{x\to y}{=} \frac{1}{4}\, \gamma^\mu_{ij} \left(E_1\, g_{\mu\nu} + E_2 (x-y)_\mu (x-y)_\nu / (x-y)^2 \right) j^\nu(y) + \cdots . \tag{6}$$

Because Z_2 is cutoff-independent in each order of α in CIQED, the expansion
coefficients E_1 and E_2 are non-singular at x = y; they are simply numerical
functions of α. The motivation for setting B = x or y in (5) is to avoid the
introduction of non-trivial conformal scalar form factors in $\Gamma\mu$. Such form
factors can be constructed from harmonic ratios of four space-time points, e.g.,
x, y, z and B. The case B = z also gives useful information as we will see
below. The general case when B $\neq$ x, y, z has been considered by Christ [23]
and Schnitzer [24].

In addition, the unrenormalized, unregularized electron propagator in CIQED is simply the free propagator within a constant factor

$$S(x) = i\langle 0|T(\Psi(x)\overline{\Psi}(y))|0\rangle \quad \text{one-electron-line}$$

$$= \frac{C(\alpha)}{2\pi^2}\frac{\gamma\cdot x}{x^4}\ . \tag{7}$$

Here C is a finite function of α in each order of perturbation theory that has no explicit dependence of B. This observation was first made by Abdellatif [21].

Now notice that the two- and three-point functions of CIQED introduce five numerical functions of α, namely F_1 in (2), G in (3), E_1 and E_2 in (5), and C in (7). Our goal is to find five independent equations relating these functions and thereby obtain F_1. It should be pointed out straight away that if the coefficients of C, E_1, E_2, F_1, and G have no essential singularities in α, then obviously F_1 cannot have an infinite-order zero, and hence quantum electrodynamics cannot have a fixed point. It may be that our inability to find more than four independent equations among these functions is related to this fact.

One equation is immediately obtained from Ward's identity, namely [27]

$$C = (E_1 + E_2)F_1\ . \tag{8}$$

Two more independent equations are obtained by substituting the conformal-covariant Ansätze (5) and (7) for Γ_μ and S into the Schwinger-Dyson equation for S:

$$\frac{1}{i}\gamma^\mu\partial_\mu S(x-y) \quad + \quad \text{⟨diagram⟩} \quad = \delta^4(x-y).$$

They are [28]

$$1/C = 1 + \frac{\alpha}{8\pi} + \frac{\alpha G}{4\pi} \tag{9}$$

$$1/F_1 = \left(1 + \frac{3\alpha}{8\pi}\right)E_1 + \left(1 + \frac{3\alpha}{16\pi}\right)E_2\ . \tag{10}$$

$$- 372 -$$

Finally, a somewhat detailed study [29] of $\Gamma_\mu(x,y,z)$ in the gauge B = z shows that exact conformal symmetry cannot be maintained unless

$$C = G. \tag{11}$$

Although the fifth equation has so far eluded us, (8)-(11) contain a considerable amount of information.

Firstly, when (9) is combined with (11), C and G are determined to all orders in α simply by solving a quadratic equation. Taking the root corresponding to

$$C(0) = G(0) = 1$$

we obtain

$$\frac{1}{C} = \frac{1}{2} + \frac{\alpha}{16\pi} + \frac{1}{2}\left(1 + \frac{5\alpha}{4\pi} + \frac{\alpha^2}{64\pi^2}\right)^{1/2}. \tag{12}$$

Thus, if F_1 has an infinite-order zero at $\alpha = \alpha_\infty > 0$, the electron propagator (7) of CIQED is regular at $\alpha = \alpha_\infty$.

Secondly, (8) and (12) indicate that $E_1 + E_2$ must develop an essential singularity at $\alpha = \alpha_\infty$ if $F_1(\alpha_\infty) = 0^\infty$. Thirdly, because (8)-(10) imply that

$$G = \frac{E_1 + \frac{1}{4}E_2}{E_1 + E_2},$$

$E_1 + \frac{1}{4}E_2$ must also develop an essential singularity at $\alpha = \alpha_\infty$, since G is regular there by (11) and (12). It is known [29] that when the expansion coefficients in (6) are calculated in the gauge defined by (3) with B = ∞ or, equivalently, in the generalized Landau gauge, the values of E_1 and E_2, denoted by E_1^∞ and E_2^∞, satisfy

$$E_1^\infty + \frac{1}{4}E_2^\infty = 1.$$

Therefore, the determination of whether F_1 has an infinite-order zero is now reduced to determining how the two coefficients of j_μ in the short-distance expansion of the product $\psi(x)\bar{\psi}(y)$ of two electron fields transforms in going

from the generalized Landau gauge ($B = \infty$ in (3)) to the gauge in which $B = x$
or y. The analysis is to be carried out in a model of quantum electrodynamics
with massless electrons and no vacuum polarization subgraphs.

References

[1]   Symanzik, K.: Comm. Math. Phys. 18, 227 (1970);

[2]   Adler, S.L.: Phys. Rev. D5, 3021 (1972),
                      D7, 1948 (E) (1973);

[3]   Sirlin, A.: Phys. Rev. D5, 2132 (1972),
        Lautrup, B.: Nucl. Phys. B105, 23 (1976);

[4]   Bernstein, J.: Nucl. Phys. B95, 461 (1975);

[5]   Fry, M.P.: Phys. Rev. D (to be published);

[6]   Calmet J. et al.: Rev. Mod. Phys. 49, 21 (1977);

[7]   Bailey, J. et al.: Phys. Lttrs. 67B, 225 (1977);

[8]   O'Neill, L.H. et al.: Phys. Rev. Lttrs. 37, 395 (1976);

[9]   Hofstadter, R.: in "Proceedings of the 1975 International Symposium
        on Lepton and Photon Interactions at High Energies", edited by
        W.T. Kirk (SLAC, Stanford University, Stanford, Calif., 1975)
        pp. 869-912;

[10]   Baker, M. and K. Johnson: Phys. Rev. D3, 2541 (1971);

[11]   Gell-Mann, M. and F.E. Low: Phys. Rev. 95, 1300 (1954);

[12]   Johnson, K., R. Willey and M. Baker: Phys. Rev. 163, 1699 (1967);

[13]   Baker, M. and K. Johnson: Phys. Rev. 183, 1292 (1969),
                                   D3, 2516 (1971);

[14]   Johnson, K. and M. Baker: Phys. Rev. D8, 1110 (1973);

[15]   Adler, S.L. and W.A. Bardeen: Phys. Rev. D4, 3045 (1971);

[16]   Manoukian, E.B.: Phys. Rev. D12, 3365 (1975);

[17]   Rosner, J.L.: Phys. Rev. Lttrs. 17, 1190 (1966),
                       Ann. of Phys. 44, 11 (1967);

- 374 -

[18]   Wilson, K.: Phys. Rev. $\underline{D3}$, 1818 (1971);

[19]   Englert, F.: Nuovo Cim. $\underline{16A}$, 557 (1973);

[20]   Callan Jr., C.G. and D.J. Gross: Phys. Rev. $\underline{D11}$, 2905 (1975);

[21]   Abdellatif, R.A.: Ph.D. Thesis, 1970, University of Washington,
        Seattle (unpublished);

[22]   Adler, S.L.: Phys. Rev. $\underline{D6}$, 3445 (1972),
                      $\underline{D7}$, 3821 (E) (1973),
        Mintchev, M.C., V.B. Petkova and I.T. Todorov, Scuola Normale
        Superiore Report No. 12/74 (unpublished),
        Ngak-Hua Ng, J.: Ph.D. Thesis, 1974, University of Washington,
        Seattle (unpublished);

[23]   Christ, N.: Phys. Rev. $\underline{D9}$, 946 (1974);

[24]   Schnitzer, H.: Phys. Rev. $\underline{D8}$, 385 (1973);

[25]   Adler, S.L.: Phys. Rev. $\underline{D8}$, 2400 (1973),
                      $\underline{D10}$, 2399 (1974),
                      $\underline{D15}$, 1803 (E) (1977);

[26]   Polyakov, A.M.: JETP Lttrs. $\underline{12}$, 381 (1970),
        Migdal, A.A.: Phys. Lttrs. $\underline{37B}$, 386 (1971),
        Parisi, G. and L. Peliti: Nuovo Cim. Lttrs. $\underline{2}$, 627 (1971),
        Mack, G. and K. Symanzik: Comm. Math. Phys. $\underline{27}$, 247 (1972),
        Mack, G. and I.T. Todorov: Phys. Rev. $\underline{D8}$, 1764 (1973),
        Fradkin, E.S. and M.Ya. Palchik: Nucl. Phys. $\underline{B99}$, 317 (1975);

[27]   Fry, M.P.: Nuovo Cim. $\underline{31A}$, 129 (1976);

[28]   Fry, M.P.: Nucl. Phys. $\underline{B107}$, 535 (1976);

[29]   Fry, M.P.: Nucl. Phys. $\underline{B121}$, 343 (1977).

Quantum Theory of Synchroton Radiation

Heimo G. Latal

Institut für Theoretische Physik, Universität Graz,
A—8010 Graz, Austria

The classical relativistic expression for the synchrotron radiation spectrum has been generalized to include quantum corrections up to second order in $\hbar\omega/E$ [1]. The modifications can be traced to the discrete nature of the emission act and thus arise from conservation of energy and momentum at the vertex. In order to exhibit the relevance of the quantum effects let us first briefly review the salient features of the classical treatment.

We start with the expression for the spectrum in terms of general currents [2,3]

$$I(E, \hbar\omega, H) = \frac{\omega^2}{16\pi^3 \hbar c^3} \int d\Omega \, [J]^2 , \tag{1}$$

where $[J]^2$ represents the invariant magnitude of the Fourier transform of the current sources

$$[J]^2 = |\vec{J}^*(k) \cdot \vec{J}(k)| - |J_o(k)|^2 , \quad J_\sigma(k) = \int d^4x \, e^{-i(k \cdot x)} j_\sigma(x) . \tag{2}$$

In the classical theory we consider the current of an electron of energy E undergoing circular motion perpendicular to the magnetic field H with the relativistic cyclotron frequency $\omega_o = ecH/E$. Then (2) is of the form [3], ($\beta^2 = 1 - (mc^2/E)^2$),

$$[J]_{cl}^2 = \frac{2(2\pi ec)^2}{\omega_o} \left\{ [\beta J_\nu'(\nu\beta \sin\theta)]^2 + [ctg\,\theta\, J_\nu(\nu\beta \sin\theta)]^2 \right\}, \tag{3}$$

where J_ν is a Bessel function and the prime denotes the derivative with respect to its argument. The index ν of the Bessel function, called harmonic index, is the ratio of the emitted frequency to the cyclotron frequency

$$\nu = \frac{\omega}{\omega_o} = \frac{\hbar\omega}{mc^2}\,\frac{E}{mc^2}\,\frac{H}{H_{cr}} \approx 1.7 \times 10^8\, \omega(eV)\,E\,(GeV)/H\,(kG). \tag{4}$$

Here we introduced the quantum mechanical critical field

$$H_{cr} \equiv \frac{m^2 c^3}{e\hbar} = 4.414 \times 10^{13}\, G,$$

in order to facilitate the later comparison with quantum theory. The angular integrations in (1) can be performed by standard methods to yield Schott's classical expression for the synchrotron spectrum [4] ($\alpha \equiv e^2/\hbar c$)

$$I_{cl}(E,\hbar\omega,H) = \alpha\omega\beta\left[2J_{2\nu}'(2\nu\beta) + 2\nu(1-\beta^{-2})\int_0^\beta dx\, J_{2\nu}(2\nu x)\right]. \tag{5}$$

For large ν and $\beta \sim 1$ further simplifications are possible, so that we finally arrive at the well-known result

$$I_{cl}(E,\hbar\omega,H) = \frac{\alpha\sqrt{3}}{2\pi}\,\frac{c}{\lambda_e}\,\frac{H}{H_{cr}}\,\varkappa(\omega/\omega_c), \tag{6a}$$

where λ_e is the electron's Compton wavelength,

$$\varkappa(z) \equiv z \int_{z}^{\infty} dx \, K_{5/3}(x) , \qquad (6b)$$

and

$$\omega_c = \frac{3}{2} \left(\frac{E}{mc^2}\right)^3 \omega_0 = \frac{3}{2} \omega_0 \left(1-\beta^2\right)^{-3/2} . \qquad (6c)$$

From these expressions we see that the character of the spectrum is determined by the mathematical properties of the object $J_\nu(\nu z)$ for $\nu \gg 1$, $z \sim 1$. An essential feature of this Bessel function is that for such parameter values it has a sensitive dependence on z and an insensitive dependence on ν. As a consequence the peak of the spectrum (6), which occurs at a frequency $\omega \approx 0.3\omega_c$, is controlled by the minute offset $1-\beta^2$. It will turn out that these properties of the spectrum are the reason for its possible drastic alteration by quantum effects. It should be noted, however, that this sensitivity can also be exploited by introducing an index of refraction of an ambient medium [5], or in inner bremsstrahlung processes [3].

For the quantum mechanical treatment we have to insert into (2) the usual transition current

$$j_\sigma(x) = i e \, \overline{\Psi}_f(x) \, \gamma_\sigma \, \Psi_i(x) , \qquad (7)$$

and modify $[J]^2$ by summing and/or averaging over all relevant quantum numbers. Here $\Psi_{i,f}$ represent the exact wave functions for an electron in a magnetic field [6], which are principally characterized by an energy quantum number n ("Landau levels")

$$E = mc^2 \left[1 + \left(\frac{p_3}{mc}\right)^2 + 2n \frac{H}{H_{cr}}\right]^{1/2} , \quad n = \frac{H_{cr}}{2H} \left(\frac{E\beta}{mc^2}\right)^2 . \qquad (8)$$

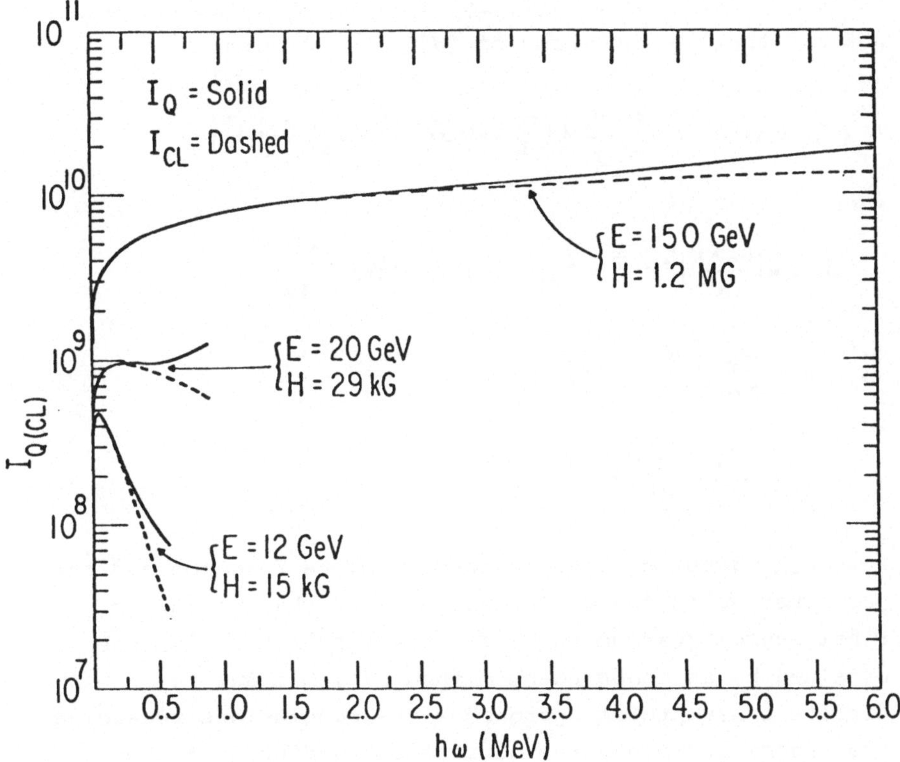

Fig.: Comparison of the classical synchrotron spectrum I_{cl} (eq.(6)) and the spectrum modified by second order quantum corrections I_Q (eq.(5.12a,b) of [1]).

After some tedious but straightforward calculations one obtains the leading terms of $[J]_Q^2$ in the form

$$[J]_Q^2 = 8\pi^2 e^2 c\, \lambdabar_e \frac{E}{mc^2} \frac{H_{cr}}{H} \left\{ \left[\frac{\hbar\omega}{E} \sin\theta\, \mathcal{G}'(x) \right]^2 + \left[ctg\,\theta\, \mathcal{G}(x) \right]^2 \right\} , \tag{9a}$$

where

$$\mathcal{G}(x) \equiv \left[\frac{(n-\lambda)!}{n!} \right]^{1/2} e^{-\frac{x}{2}} x^{\frac{\lambda}{2}} \mathcal{L}_{n-\lambda}^{(\lambda)}(x) \quad , \quad \mathcal{G}'(x) \equiv \frac{\partial \mathcal{G}(x)}{\partial x} \; ; \tag{9b}$$

$$x = \frac{H_{cr}}{2H} \left(\frac{\hbar\omega}{mc^2} \sin\theta \right)^2 \; ; \tag{9c}$$

$$\lambda = \nu \left(1 - \frac{\hbar\omega}{2E} \sin^2\theta \right). \tag{9d}$$

Here $\mathcal{L}_{n-\lambda}^{(\lambda)}$ denotes a generalized Laguerre function and the parameter λ is the difference between the initial and final energy quantum numbers. This result has also been obtained by an independent method [7], but due to different asymptotic approaches the conclusions regarding the second order quantum corrections differ from those of [1].

Whereas in the classical picture the radiation is emitted continuously, in the quantum theory we have discrete photon emission involving transitions between Landau levels with large differences in their characteristic quantum numbers [cf.(9d) and (4)]. In addition, conservation of momentum in the field direction leads to a recoil of the electron which gives rise to the nonlinear dependence of λ on ω in (9d) and thus eventually leads to corrections to the classical spectrum. In fact, in the limit $\hbar\omega/E \to 0$, the classical expression has to be recovered. Such a comparison of (9) with (3) suggests that there must be a

connection between the asymptotic behavior of the Laguerre
and Bessel functions. The basis for this correspondence is
provided by the Erdélyi expansion theorem [8], which in
our case leads to the identification

$$g(x) \approx J_\lambda(\lambda y \sin\theta) \quad , \quad g'(x) \approx \frac{E}{\hbar\omega}\frac{\beta}{\sin\theta} J_\lambda'(\lambda y \sin\theta) ; \tag{10a}$$

with

$$y = \beta\left[1 - \frac{\hbar\omega}{2E}(1-\beta^2\sin^2\theta) + \frac{1}{8}\left(\frac{\hbar\omega}{E}\right)^2\left(1-\frac{1}{3}\sin^2\theta - 4\cos^2\theta\right)\right] , \tag{10b}$$

where in the last expression all terms up to second order
in $\hbar\omega/E$ have been taken into account rigorously. Within
the same approximations it is even possible to factorize
the first and second order contributions in the form

$$J_\lambda(\lambda y \sin\theta) \approx J_{\bar{\nu}}(\bar{\nu} z \sin\theta) ; \tag{11a}$$

where

$$\bar{y} = y\left(1 + \frac{\hbar\omega}{E}\sin^2\theta\right) , \tag{11b}$$

$$z = \beta\left[1 + \frac{1}{8}\left(\frac{\hbar\omega}{E}\right)^2\left(1-\frac{1}{3}\sin^2\theta - 4\cos^2\theta\right)\right]. \tag{11c}$$

A comparison with the classical expression (3) then implies
that quantum mechanical corrections to the spectrum arise
from introducing a double expansion in $\hbar\omega/E$ in the Schott
formula, i.e.

$$\nu \to \nu\left[1 + a_1\frac{\hbar\omega}{E} + a_2\left(\frac{\hbar\omega}{E}\right)^2 + \cdots\right] , \tag{12a}$$

$$\beta \to \beta\left[1 + b_1\frac{\hbar\omega}{E} + b_2\left(\frac{\hbar\omega}{E}\right)^2 + \cdots\right] . \tag{12b}$$

The results of Sokolov [6] and Schwinger [9] lead to the identification [cf.(9d), (10b) for $\Theta \approx \pi/2$] : $a_1 \approx -1/2$, $b_1 \approx -\frac{1}{2}(mc^2/E)^2$. This at first right would indicate that only a_1 contributes significantly to the first order quantum corrections. However the peculiar sensitivity of the Bessel functions in fact make both a_1 and b_1 contribute equally to the net first order quantum shift [cf.(11b) and (4)]:

$$\omega \to \omega \left(1 + \frac{\hbar\omega}{E}\right). \tag{13}$$

The principal second order correction, as a result of these investigations, stems from the coefficient (11c), i.e. $b_2 \approx +1/12$.

From these results one may now judge the relative importance of the corrections: Since the average photon energy is $\langle \hbar\omega \rangle \approx 0.5 \frac{E^2}{mc^2} \frac{H}{H_{cr}}$, we find that the classical description (6) will not be affected by the first order terms unless

$$\frac{E}{mc^2} \frac{H}{H_{cr}} \gtrsim 1 \quad , \quad \text{or} \quad E(GeV) \, H(kG) \gtrsim 10^7.$$

On the other hand, the quantum mechanical spectrum for "soft" photon emission, i.e. for $\hbar\omega < mc^2$, is approximated by [1,3]

$$I_Q(E, \hbar\omega, H) = \frac{\alpha\sqrt{3}}{2\pi} \frac{c}{\lambdabar_e} \frac{H}{H_{cr}} \times \left(\frac{\omega}{\omega_c} \left[1 + \frac{\hbar\omega}{E}\right]\left[1 - \left(\frac{\hbar\omega}{2mc^2}\right)^2\right]^{3/2} \right), \tag{14}$$

and this diverges from (6) at experimentally detectable levels whenever

$$\left(\frac{E}{mc^2}\right)^2 \frac{H}{H_{cr}} \gtrsim 1 \quad , \quad \text{or} \quad E^2(GeV) \, H(kG) \gtrsim 5 \cdot 10^3.$$

Thus we are faced with the surprising feature that the second order terms can be more significant than the first order corrections and may even dominate the classical expression. For higher photon energies the analytic approximations become more complex [1]; the resulting behavior of the spectrum at $\hbar\omega = 2mc^2$ may then be interpreted as reminiscent of a phase transition [3]. In the Figure we show three examples corresponding to conditions which are technically attainable at various accelerator installations; in each case the classical and quantum mechanical spectra are experimentally distinguishable.

References

1. H.G. Latal and T. Erber, to appear in Annals of Physics.
2. W.E. Thirring, "Principles of Quantum Electrodynamics", Academic Press, New York, 1958.
3. T. Erber, D. White, and H.G. Latal, Acta Phys.Austriaca $\underline{44}$, 315 (1976) and $\underline{45}$, 29 (1976).
4. G.A. Schott, Ann. der Physik $\underline{24}$, 635 (1907); "Electro-magnetic Radiation and the Mechanical Reactions arising from It", Cambridge University Press, 1912.
5. J. Schwinger, Wu-yang Tsai, and T. Erber, Ann. Phys. $\underline{96}$, 3o3 (1976);
   T. Erber, D. White, Wu-yang Tsai, and H.G. Latal, Ann. Phys. $\underline{1o2}$, 4o5 (1976).
6. A.A. Sokolov and I.M. Ternov, "Synchrotron Radiation", Academie Verlag-Pergamon Press, New York, 1968.
7. J. Schwinger and Wu-yang Tsai, to appear in Annals of Physics.
8. A. Erdélyi, Golden Jubilee Volume of the Indian Mathematical Society, 235 (1959); Jr. of Math. Phys. $\underline{1}$, 16 (196o).
9. J. Schwinger, Proc. Nat. Acad. Sci. $\underline{4o}$, 132 (1954).

Self-Consistent Quark Bags

J. Rafelski

Cern — Geneva

I. Introduction and Overview

We do not understand the forces that keep the quarks (semi-) permanently bound in hadrons. The known experimental properties of quark bound states -- hadrons -- appear self-contradictory, when confronted with our past experience with nuclear and atomic physics. The basic problem is that experimental information indicates that hadronic properties are well described in a model of weakly interacting (quasi) particles. A possible solution to this dilemma, on which we base our investigations, is a model[1] in which the bound state of a few heavy particles[2] is viewed as a collection of quasi-particles of relatively small mass, each moving in an average "shell" potential independently of the others. It is only when we forcefully destroy this hadronic "bag" that the true nature of the underlying very massive fields becomes more apparent -- but with no free quarks escaping from the region of the interaction. In this picture there are two possibilities to view the absence of asymptotically free quark states: either their free mass is still higher than presently available energies $m_q \gtrsim 10$ GeV, or their decay width is very high[2]. In the latter case it is necessary to consider quarks as quantities with integral quantum numbers.

Since the spectroscopic information available today is very detailed, it would seem that any profound attempt to deal with the binding of quarks and hadronic structure requires inclusion of known symmetries [flavour SU(4), colour SU(3), etc.]. In our approach we will consider the above symmetries as being exact and will focus on extending the qualitative understanding of strongly bound fermion states. We will introduce the internal symmetries only when needed to construct hadrons from quarks. Our aim is to describe several mechanisms for strong binding inherent in the strongly interacting field theories.

The basic role in our investigations will be played by the (quark) fermion field Ψ and its transition amplitude Ψ_0. For example, we have

$$\Psi_0(x) = \langle qq | \Psi(x) | q_0 qq \rangle \qquad (1.1)$$

We will consider several different interactions of both massive and massless quark fields. In the absence of an exact method to treat strongly interacting quantum fields we will consider the strongly bound states in the self-consistent bag approximation[3], discussed in detail further below. In this approximation all quantum fields are replaced by classical fields with some constraints on the fermion fields that arise in view of the anticommutation properties of the quantum fields. This simple prescription must be amended suitably in some cases, in particular when a vector type interaction of the fermion field is considered. This is a particularly interesting case. If the assertion that the interaction considered here is representative of the interacting Yang-Mills fields is correct, we are led to the conclusion that the properties of strongly bound states generated by the coloured vector gluons will have some unusual properties not at all expected from classical potential models. Also some other subtle properties stemming from the quantum character of the fields must not be forgotten.

The motivation for transition from q-number to c-number fields in the domain of strong binding resides in two intuitive notions:

i) The c-number theory arises when the expectation value of the Hamiltonian describing a model is considered in the zero-average momentum frame of trial state $|t\rangle$ formed in the direct product space of gluons and quarks using a localized, coherent gluon state and a quasi-fermion state.

ii) In the cases of interest the interacting c-number fields generate stable, localized solutions which usually are referred to as *solitons*.

The investigations concerning

a) the region of validity of the bag approximation, and

b) the relevance of the c-number solutions in describing strongly bound states in three space dimensions

are as yet inconclusive[4]. It is our basic assumption that the bag approximation is valid whenever

$$M \ll N m_q . \tag{1.2}$$

Here m_q is the lightest "free" field excitation that can be reached from the c-number bound state of mass M (M includes the mass of the gluon field as well). N is the number of constituents. This condition is self-explanatory -- the bound states are "stiff" with respect to vacuum fluctuations if Eq. (1.2) is satisfied.

As the result of the bag approximation we are left with a c-number theory of interacting fields. The properties of soliton solutions[5,6] of interacting fermion fields are interesting in themselves, apart from the point of view presented above. We will consider several general properties of such solutions with the help of virial relations in Section 3. In particular, we will derive a relation between the kinetic energy of the fermions and the potential interaction energy of the gluon field[7].

In Section 4 we present a brief description of the numerical methods and algorithms[5] used to generate examples of the c-number solutions in three-dimensional space that are of interest to us. We will discuss the properties of radially symmetric solutions in Section 5. In particular we consider[3] the *simplest* possible system consisting of a fermion field $\bar{\Psi}$ (quark field) interacting with a scalar meson (gluon) field Φ [see Eq. (2.1)].

Our investigations differ at this stage in two points from those of Bardeen et al. and Vinciarelli[1] and Creutz[8]:

i) We obtain actual solutions for the quark bag by extensive numerical studies in the three-dimensional space rather than in the unphysical one-dimensional space;

ii) choosing (in a special case) the gluon potential $U(\Phi^2) = -\frac{1}{2}\,\mu^2\Phi^2$, we show that in the physical, three-dimensional space the existence of the quark bag does not critically depend on an inherent non-linearity of the gluon field, but can also be viewed as a consequence of the quark-gluon interaction.

In the limit of the vanishing mass of the scalar gluon field we are led to consider in Section 6 the properties of a self-interacting Dirac field. We view the special form of the self-interaction chosen there as representative of many other possible self-interactions as much as the scalar gluon interaction was. Hopefully the results we obtain with these simple Lagrangians remain valid when more complicated interactions are considered.

Finally, in Section 7 we consider the particular case of a vector interaction. In contrast to the scalar interactions considered previously, the vector interaction can bind a c-number field with arbitrary strength, leading to mixtures between particle and antiparticle, single-particle states. The necessary modifications of the theory, as well as particular solutions of a model, are described.

II. Self-Consistent Quark Bags

As indicated in the above discussion, we take the self-consistent solution of the coupled Dirac-meson fields to represent a bound state of strongly interacting fermions. This physical picture corresponds to the idea of a quark trapped in a self-consistent potential well[1,3]. If taken seriously, this remark means that the solutions of the interacting fields contain information about experimentally measurable quantities. A preferred model then can emerge that has the best phenomenological properties. I have found that such a pet theory is that of interacting fermion-scalar gluon fields

$$\mathcal{L}_1 = \bar{\Psi}(\gamma \cdot p - m)\Psi + g\Phi\Psi\Psi + \partial_\mu\Phi\,\partial^\mu\Phi - \big(U(\Phi) - U(o)\big). \tag{2.1}$$

In the first part of this section we will make a connection between the classical theory and quantum field theory, using as an example the Lagrangian (2.1), where we sometimes also write for the gluon potential U the simplest term

$$U(\Phi^2) \rightarrow \tfrac{1}{2}\mu^2\Phi^2. \qquad (2.2)$$

In order to treat the strongly interacting ($g^2 \gg 1$) limit of the quantum field theory, induced by the Lagrangian Eq. (2.1), let us now consider quasi-particles. These are entities carrying, typically, the same internal quantum numbers, for example spin, as the particles, but subject to a quite different interaction from the free modes. The advantage that is gained consists in the modification of the interaction; in particular, quasi-particles interact weakly and we may use the well-proved perturbative methods to treat the residual inter-action. A quasi-fermion state with quantum numbers $|p\rangle$ can be obtained from the quantum operator $\hat{\Psi}(\vec{x},0)$ by projecting with the associated amplitude $\psi_p(\vec{x})$ (a cir-cumflex over a quantity indicates its q-number character)

$$|p\rangle = \int \hat{\Psi}^+(\vec{x},0)\, \psi_p(\vec{x})\, |0\rangle\, d^3x. \qquad (2.3)$$

Here $|0\rangle$ denotes the ground state. The equations of motion for the amplitude ψ_p will be determined *a posteriori*. It is advantageous to introduce the quasi-particle (index p) and antiparticle (index n) operators in the usual manner:

$$\hat{\beta}_p^+ = \int \hat{\Psi}^+(\vec{x},0)\, \psi_p(\vec{x})\, d^3x,$$
$$\hat{\gamma}_n^+ = \int \psi_n^+(\vec{x})\, \hat{\Psi}(\vec{x},0)\, d^3x. \qquad (2.4)$$

The completeness of the basis $\{\psi_p,\psi_n\}$ is reflected in

$$\delta^3(\vec{x}-\vec{x}')\,\delta_{\alpha\beta} = \sum_n \psi_n^\alpha(\vec{x})\,\psi_n^{\beta+}(\vec{x}') + \sum_p \psi_p^\alpha(\vec{x})\,\psi_p^{\beta+}(\vec{x}'). \qquad (2.5)$$

Here (α,β) are the internal structure indices, for example spinor indices, while the sums also stand for possible integrals over continuous parts of the spectra. We now see that the internal structure of the quasi-particles is that of the particles and is a consequence of the (anti-)commutation relations of the fields $\hat{\Psi}$. Any quasi-particle state may be constructed from a product of the quasi-fermion operators acting on the ground state. Thus, for example, the three-quasi-fermion baryon state B is

$$|B\rangle = \sum_{ijk} b_{ijk}\, \hat{\beta}_i^+\hat{\beta}_j^+\hat{\beta}_k^+\, e^{-i\int \hat{\Pi}_\Phi(\vec{x},0)\,\psi_c(\vec{x})d^3x}\, |0\rangle. \qquad (2.6)$$

with suitable coefficients b_{ijk} that generate the symmetry properties of the state. The remaining factor in Eq. (2.6) will be explained in a moment.

The "bag" field for the quarks is generated by the quasi-fermionic source

$$S = \langle B | \hat{\bar{\Psi}} \hat{\Psi} | B \rangle. \tag{2.7}$$

The presence of a strong source, the strength being determined to a large extent by g^2, generates an average gluon "shell" potential φ_c

$$\varphi_c = \langle B | \hat{\phi} | B \rangle. \tag{2.8}$$

To achieve this property, we have introduced the exponential in Eq. (2.6), which makes $|B\rangle$ a coherent state for the meson field, where

$$\pi_\phi = \frac{\delta \mathcal{L}}{\delta \phi} \tag{2.9}$$

and

$$[\hat{\phi}(\vec{x}, 0), \hat{\pi}_\phi(\vec{x}', 0)] = i \delta^3(\vec{x} - \vec{x}'). \tag{2.10}$$

Equipped with the trial state $|B\rangle$ (and similarly a meson state $|M\rangle$) we can minimize the invariant mass that is associated with them, as a function of the unknown functions φ_c and ψ_p. The Lorentz invariant momentum P_μ of our trial state $|B\rangle$ is given by

$$P_\mu = \int \langle B | \hat{T}_\mu^\nu(x) | B \rangle d^4 O_\nu, \tag{2.11}$$

where T_μ^ν is the energy momentum tensor following from the Lagrangian $\mathcal{L}$ under consideration. The hypersurface in the four-dimensional space is denoted by $d^4 O_\nu$. Note that we have chosen, in the above discussion,

$$d^4 O_\nu = n_\nu d^3 x, \tag{2.12}$$

with $n_\nu = (1,0,0,0)$ associated with the laboratory frame of reference. We have written the fully covariant form of P_μ in Eq. (2.11) to show the Lorentz covariance of the bag approach. Furthermore, within the ansatz for $|B\rangle$ we have chosen, we search for solutions with vanishing momentum $\vec{P}$, i.e.

$$\int \langle B | \hat{T}_0^i | B \rangle d^3 x = 0 \tag{2.13}$$

If our solution satisfies condition (2.13), it is called a baryon in zero average momentum frame. By construction of our trial state we obtain the property

$$\langle B | \hat{\pi}_\phi(x) | B \rangle = 0, \tag{2.14}$$

which guarantees the vanishing of the average gluon momentum. Similarly, it is easy to show that the sum of quasi-fermion momenta vanishes. This is in agreement with a picture in which quarks follow closely the motion of the gluon bag and vice versa.

In the zero average momentum frame we therefore have

$$P_0 = M = \int d^3x \, \langle B | \vec{T}_0{}^0 | B \rangle = \int d^3x \, \mathcal{M}_B , \qquad (2.15)$$

where for $\mathcal{L}_1$ defined by Eq. (2.1) we find, using the exact symmetry of all quarks in the bag:

$$\mathcal{M}_B = N \psi_i^+ (\vec{\alpha} \cdot \vec{p} + \beta (m_q - g \varphi_c)) \psi_i$$
$$+ \tfrac{1}{2} | \vec{\nabla} \varphi_c |^2 + U(\varphi_c) - U_0 , \qquad (2.16)$$

where N = 3 is the number of (valence) quarks.

In principle, the expectation value (2.15) should include the zero-point energy of the fermion field:

$$E_0 = -\tfrac{1}{2} \left(\sum_p \epsilon_p [\varphi_c] - \sum_n \epsilon_n [\varphi_c] \right) . \qquad (2.17)$$

Here ϵ_p and ϵ_n are the positive and negative frequency solutions of the Dirac equation: E_0 is a (divergent) implicit function of φ_c. It is possible to include in the functional $U(\varphi_c)$ the necessary counterterms -- the renormalized zero-point energy then is still a function of φ_c and should, in principle, be included in Eq. (2.16). However, such a contribution can be absorbed in the function $U(\varphi_c)$. Since we have, in principle, no idea about the form of U, we need not consider this point in further detail for the moment.

A result similar to Eq. (2.16) is also obtained for the mesons:

$$\mathcal{M}_M \equiv \langle M | \mathcal{H} | M \rangle = \psi_i^+ [\vec{\alpha} \cdot \vec{p} + \beta (m_q - g \varphi_c)] \psi_i$$
$$- (i \rightleftarrows) + \tfrac{1}{2} | \vec{\nabla} \varphi_c |^2 + U(\varphi_c) - U_0 . \qquad (2.18)$$

The usual (-) sign for antiquasi-fermion amplitude ψ_j in Eq. (2.18) is necessary to change the sign of the negative frequency mode associated with the solution of its Dirac equation. For a scalar potential φ_c the spectrum of the Dirac equation is symmetric between particles and antiparticles, that is

$$\epsilon_p = -\epsilon_{ap} \qquad (2.19a)$$

and the wave functions transform like

$$\psi_p = \gamma_5 \psi_{ap}^* \qquad (2.19b)$$

Therefore the invariant mass for all hadrons can be written as a function of the number of bound quarks and antiquarks in the bag only:

$$M_N = \int d^3x \; \mathcal{M}_N = \int d^3x \left[N \left\{ \psi_0^\dagger (\vec{\alpha}\cdot\vec{p} + \beta(m_q - g\varphi_c)) \psi_0 \right\} + \frac{1}{2}(\vec{\nabla}\varphi_c)^2 + (U(\varphi_c) - U_0) \right] = M_{N,f} + M_g . \tag{2.20}$$

The index "0" on the spinor wave function indicates that we must choose the lowest energy particle solution (which in the case of scalar interaction is the lowest positive energy eigenvalue solution of the Dirac equation) in order to obtain the smallest value of M_N. The equations of motion satisfied by ψ_0 and φ_c follow now from the variational condition:

$$\delta\left(M_N - \epsilon_0 N \int d^3x \; \psi_0^\dagger \psi_0\right) = 0, \tag{2.21}$$

where we have introduced the usual Lagrange multiplier in order to normalize the solutions:

$$\int \psi_0^\dagger \psi_0 \, d^3x = 1 . \tag{2.22}$$

We find from Eq. (2.21) the self-consistent equations of motion

$$\left[\vec{\alpha}\cdot\vec{p} + \beta(m_q - g\varphi_c)\right] \psi_0 = \epsilon_0 \psi_0 , \tag{2.23}$$

$$-\Delta \, g\varphi_c + \frac{\partial U}{\partial \varphi} = g^2 N \, \bar{\psi}_0 \psi_0 . \tag{2.24}$$

N is equal to three for baryons and two for mesons. When solving the above equations, we consider all dimensional quantities in units of m_q. If we obtain the solutions as a function of g^2, then it is possible to consider the case $N = 1$ only, and the relevant parameter is

$$g'^2 = g^2 N . \tag{2.25}$$

We note that M_N can be written as

$$M_N = N \left(\epsilon_0 \left[g'^2\right] + \frac{1}{g'^2} \int \left(\frac{1}{2}(\vec{\nabla}g\varphi_c)^2 + \frac{1}{2}\mu^2(g\varphi_c)^2\right) d^3x \right) . \tag{2.26}$$

Thus M_N/N is only a function of g'^2. We thus have seen that self-consistent solutions of classical interacting fields are closely related to the quark-bag solutions of N quarks. In the quark-bag approach the main approximation has been the following: only the diagonal part of the Hamiltonian matrix of the interacting fields in a trial state has been considered. The off-diagonal parts remain to be diagonalized and the renormalization program carried out.

It is instructive to notice that the choice of φ_c that is self-consistent in the sense of Eqs. (2.23) and (2.24) makes the remaining interaction off-diagonal.

The full Hamiltonian of q-number fields is (up to the necessary symmetrization)

$$\hat{H} = \hat{H}_D + \hat{H}_\phi + \hat{H}_I \tag{2.27}$$

with

$$\hat{H}_D = \int d^3x \; \hat{\Psi}^+ (\vec{\alpha}\cdot\vec{p} + \beta m)\hat{\Psi} \tag{2.28a}$$

$$\hat{H}_\phi = \int d^3x \; \tfrac{1}{2}\left(\hat{\pi}^2 + (\vec{\nabla}\hat{\phi})^2 + (U(\hat{\phi}) - U_0)^2\right) \tag{2.28b}$$

$$\hat{H}_I = \int d^3x \; (-g\hat{\phi})\,\hat{\bar{\Psi}}\hat{\Psi} \; . \tag{2.28c}$$

Now let us add and subtract a term in $\hat{H}$

$$\hat{H} = (\hat{H}_D + \hat{H}_{ex}) + \hat{H}_\phi + (\hat{H}_I - \hat{H}_{ex}), \tag{2.29}$$

where

$$\hat{H}_{ex} = \int d^3x \; (-g\varphi_c)\,\hat{\bar{\Psi}}\hat{\Psi}. \tag{2.30}$$

The approach presented above corresponds to such a choice of a trial state that the *diagonal* matrix elements of the new interaction

$$\hat{H}_I' = \hat{H}_I - \hat{H}_{ex} \tag{2.31}$$

vanish, while the Hamiltonian

$$\hat{H}_o = (\hat{H}_D + \hat{H}_{ex}) + \hat{H}_\phi \tag{2.32}$$

is diagonal.

An argument in favour of the bag approximation, applicable only in the limit of strong coupling, involves the large separation in energy between the mass of the trial state, M_N, and the free fermion mass, m. I believe that the second-order contributions to M from the off-diagonal matrix elements will be of the order of

$$\delta M \lesssim \left[\int \psi_i^+ \, g\varphi_c \, \psi_i \, d^3x\right]^2 \Big/ \left[Nm_q - M_N\right]$$

$$\approx \left(\frac{M_N}{Nm_q}\right) M_N \; . \tag{2.33}$$

If this is the case, then the ratio

$$\delta = M/m_q \tag{2.34}$$

is the small parameter that determines the validity of the bag approximation.

Finally, we consider the adiabaticity of the fermionic motion. When a quasi-fermion is excited inside the bag, we have to know if the bag should be adjusted instantly, that is the excited state considered as the source on the right-hand

side of Eq. (2.23). I will now argue that the opposite should be expected: the
bag does not have time to adjust to the internal excitation.

While the transition time Δt_{ij} from the quasiparticle state i to the quasi-
particle state j is

$$\Delta t_{ij} \approx \pi / (\epsilon_j - \epsilon_i)^{-1},\tag{2.35}$$

the simultaneous adjustment of the gluon bag involves the time

$$\Delta t_\varphi \approx \pi / (M_j - M_i)^{-1} > \Delta t_{ij}.\tag{2.36}$$

The latter relation follows from the observation that the adjustment of the bag
improves the parameters of the trial state and so requires

$$M_j < N\epsilon_j + (M_i - N\epsilon_i),\tag{2.37}$$

[except in the (undiscovered) case that the lowest particle eigenenergy ϵ is not
associated with the lowest mass M]. Thus the bag approximation implies presumably
a Born-Oppenheimer type of adiabaticity -- the motion of the fermions is faster
than the rate of change of the self-consistent potential.

III. The Virial Approach to Classical Field Equations

Before actually discussing the solutions of particular models, let us see
how much we can learn about the general properties of self-consistent solutions
from the equations of motion[7]. Basic to our considerations will be a (virial)
relation describing the kinetic energy of the fermion field ψ in some prescribed
potential V(x), where V is a matrix in the spinor space and can, in principle,
consist of all possible couplings (S, V, P, A, T). Let H_D be the Dirac operator

$$H_D = \vec{\alpha} \cdot \vec{p} + \beta m + V.\tag{3.1}$$

Then we have

$$[\vec{x} \cdot \vec{p}, H_D] = i\vec{\alpha} \cdot \vec{p} - i\vec{x}(\vec{\nabla} V).\tag{3.2}$$

Taking the expectation value of Eq. (3.1) between localized eigenfunctions ψ_i of
H_D, we find the *virial* equation for the Dirac field:

$$\int d^3x \ \psi_i^\dagger \vec{\alpha} \cdot \vec{p} \psi_i = \int d^3x \ \psi_i^\dagger \vec{x} \cdot (\vec{\nabla} V) \psi_i.\tag{3.3}$$

This is a very useful relation which is valid only when ψ_i is an eigenfunction of
H_D. We record that all of the ψ_i considered above are localized discrete eigenstates in configuration space. We now apply Eq. (3.3) to the Lagrangian field
theory describing a scalar meson field in interaction with the Dirac field: the
Lagrangian has been given in Eq. (2.1) and the equations of motion by Eqs. (2.23)

and (2.24) with N = 1 and $\varphi_c = 0$. Then Eq. (3.3) for the kinetic energy of the Dirac field becomes

$$\int d^3x \; \psi_i^\dagger \, \vec{\alpha} \cdot \vec{p} \, \psi_i = \int d^3x \; (-\vec{x} \cdot \vec{\nabla} \varphi_c) g \, \bar{\psi}_i \psi_i \; . \tag{3.4}$$

We find after some manipulations that

$$\int d^nx \; \psi_i^\dagger \vec{\alpha} \cdot \vec{p} \, \psi_i = \int d^nx \left[\frac{n-2}{2} |\vec{\nabla}\varphi_c|^2 + n\,(U - U_0) \right] + ST, \tag{3.5}$$

with a vanishing surface term for localized solutions, with

$$n \equiv \vec{\nabla} \cdot \vec{x} \tag{3.6}$$

as the dimensionality of the space. Normally n = 3, but to dramatize the uniqueness of the three-dimensional, physical space, let us consider the number of dimensions as a parameter.

The Hamiltonian associated with the Lagrangian (2.1) is, with a time-independent φ_c:

$$H^S = \int d^nx \left[\psi_0^\dagger \, (\vec{\alpha} \cdot \vec{p} + \beta m) \, \psi_0 \right.$$
$$\left. - g\,\varphi_c \bar{\psi}_0 \psi_0 + \tfrac{1}{2} |\vec{\nabla}\varphi_c|^2 + U(\varphi_c) - U_0 \right]. \tag{3.7}$$

We may use the above result, Eq. (3.5), together with the equations of motion (2.23) and (2.24) to obtain the following expression for the Hamiltonian:

$$E^S_{eff} = \int d^nx \left\{ \frac{n-3}{2} |\vec{\nabla}\varphi_c|^2 + (n+1)\left[U(\varphi_c) - U_0 \right] + \left(\frac{m}{g} - \varphi_c \right) \frac{\partial U}{\partial \varphi_c} \right\} . \tag{3.8}$$

As is well known, the above equation (3.8) cannot be used as a basis of a variational principle. For known φ_c, which minimizes Eq. (3.7), it gives the proper value of H^S. Therefore we may view E^S_{eff} as an expression defining the total energy.

A striking feature of Eq. (3.8) is the fact that for n < 3 (n is the number of space dimensions) the effective kinetic energy of the scalar field becomes negative definite. It just vanishes for n = 3. Thus the energy content of E^S_{eff} in a normal number of space dimensions is only implicitly dependent on the derivatives of the field φ_c. The negative definite kinetic energy for n < 3 signals a possible instability of the Hamiltonian; the energy of the solution could be reduced by a small non-continuous variation of φ_c.

Another notable feature is that any φ_c^4 proportional term in U cancels out in three-dimensional space (in n-dimensional space φ_c^{n+1} cancels). This is an important feature, since φ_c^4 is commonly held responsible for the stability of a theory with spontaneously broken symmetry[1,4]. In the case of a one-dimensional world ($n = 1$), where the φ_c^2 contribution vanishes, this is obviously a necessary term to stabilize the theory. It would therefore seem that a φ^6 plays the role of the ψ^4 term in three space dimensions, as compared with one-dimensional models.

Very often it is the field

$$\chi = m/g - \varphi_c \tag{3.9}$$

that is introduced, instead of φ_c. The advantage is that there is no explicit mass term of the fermion field in the Hamiltonian. Our result may be easily adapted and we find for the Hamiltonian

$$H^{S'} = \int d^n x \left[\Psi_k \, \vec{\alpha} \cdot \vec{p} \, \Psi_k + g \, \chi \, \bar{\Psi}_k \Psi_k + V(\chi) + \tfrac{1}{2} (\vec{\nabla}\chi)^2 \right], \tag{3.10}$$

the result

$$E_{eff}^{S'} = \int d^n x \left[\tfrac{n-3}{2} (\vec{\nabla}\chi)^2 + (n+1) V(\chi) - \chi \, \frac{\partial V}{\partial \chi} \right], \tag{3.11}$$

where

$$V(\chi) = U(\varphi_c) - U_0 . \tag{3.12}$$

The field χ may be considered the effective mass of the Dirac field ψ. The mass of the *free* Dirac field is now

$$m = g \, \chi \quad (x \to \infty), \tag{3.13}$$

which is dependent on the specific form of $V(\chi)$. Further, we note that since the integrability of the Hamiltonian H^S requires $U(\varphi_c = 0) = U_0$, we find $V(m/g) = U_0$.

In the case of the σ model the function V is considered to be

$$V = H (\chi^2 - f^2)^2 . \tag{3.14}$$

In three-dimensional space we obtain with this form of V:

$$E_{eff}^{'(3)} = \int d^3 x \, 4 H f^2 (f^2 - \chi^2) . \tag{3.15}$$

If χ_s is allowed to vary only between $+f$ and $-f$ then the last expression is explicitly positive definite. Thus the form of the solution is essential for the determination of a lower bound on E_{eff}. However, an upper limit for E_{eff} may be obtained setting $\chi^2 = 0$ in Eq. (3.15) and taking the volume of the solution for the integral:

$$E_{eff}^{'(3)} \lesssim 4\pi/3 \, R^3 \, 4 H f^4 . \tag{3.16}$$

We now consider the case of the self-interacting Dirac field, where H assumes the form, for scalar-type self-interaction[6])

$$H^P = \int d^n x \left[\Psi_0^+ (\vec{\alpha}\cdot\vec{p} + \beta m_q) \Psi_0 - \frac{G}{2} (\bar{\Psi}_0 \Psi_0)^2 \right]. \tag{3.17}$$

In view of the equations of motion,

$$\left[\vec{\alpha}\cdot\vec{p} + \beta (m_q - G\bar{\Psi}_0\Psi_0) \right] \Psi_0 = \epsilon_0 \Psi_0 , \tag{3.18}$$

the virial relation reads as

$$T_0 = \int d^n x \, \Psi_0^+ \vec{\alpha}\cdot\vec{p} \, \Psi_0 = -\frac{G}{2} \int d^n x \left[(\vec{x}\cdot\vec{\nabla})(\bar{\Psi}_0\Psi_0)^2 \right]. \tag{3.19}$$

Upon partial integration of the right-hand side, we obtain, up to a vanishing surface term,

$$T_0 = -n V_0 , \tag{3.20}$$

where

$$V_0 = -\int d^n x \, \frac{G}{2} (\bar{\Psi}_0 \Psi_0)^2. \tag{3.21}$$

The energy for the self-interacting Dirac field can be written now as

$$E^P_{eff} = (n-1)(-V_0) + m S_0 . \tag{3.22}$$

S is the scalar integral

$$S_0 = \int d^n x \, (\bar{\Psi}_0 \Psi_0). \tag{3.23}$$

This result, Eq. (3.22), corrects a superficial impression that H^P, Eq. (3.17), is unbound, since it has the structure $x^2 - x^4$. This means that the kinetic energy of the Dirac field more than offsets the attractive self-interaction V_0. Since $-V_0$ is always positive, we find that for all dimensions the positivity of the solution depends on the sign of the scalar integral. Further, we note that the eigenfrequency can be written as

$$\epsilon_0 = (n-2)(-V_0) + m S_0 . \tag{3.24}$$

A similar derivation can be carried out in the case of a Lorentz vector field A_μ of mass μ_v in interaction with the Dirac field. For the Lagrangian

$$\mathcal{L}_{A'} = -\frac{1}{2} \partial_\mu A^\nu \partial^\mu A_\nu + \frac{1}{2} \mu_v^2 A_\mu A^\mu$$
$$- g_v A_\mu \bar{\Psi} \gamma^\mu \Psi + \Psi(\gamma\cdot p - m)\Psi, \tag{3.25}$$

we find

$$E^{A'}_{eff} = -\int d^n x \left[\frac{n-3}{2} (\nabla\cdot A_\mu)^2 + \frac{n-1}{2} \mu_v^2 A^2 - m \bar{\Psi}_0 \Psi_0 \right], \tag{3.26}$$

which is negative definite, considering the longitudinal component A_0 only. We note that the longitudinal part of the vector-type interaction is repulsive in the particle-particle channel and attractive in the particle-antiparticle channel.

The situation changes when the sign in the part of the Lagrangian corresponding to the free vector field is changed[9]):

$$\mathcal{L}_A = \tfrac{1}{2}\partial_\mu A^\nu \partial^\mu A_\nu - \tfrac{1}{2}\mu_v^2 A^2 - g_v A_\mu \bar{\Psi}\gamma^\mu \Psi + \bar{\Psi}(\gamma \cdot p - m)\Psi. \tag{3.27}$$

We then find

$$E_{eff}^A = \int d^n x \left[\frac{n-3}{2}(\nabla A_\mu)^2 + \frac{n-1}{2}\mu_v^2 A^2 + m\,\bar{\Psi}_k \Psi_k \right], \tag{3.28}$$

which is positive definite, constrained to the longitudinal part of A_μ, provided that the scalar integral $\int d^n x\,\bar{\Psi}_0 \psi_0 > 0$. The above-described change in sign accomplishes at the same time a change in the "polarity" of the vector interaction -- the longitudinal part is now attractive in the particle-particle channel, while the particle-antiparticle channel becomes repulsive. The perturbative quantum field theory, if based on Eq. (3.27) would suffer from the well-known difficulties associated with the possible need for negative metric particles (ghosts) to guarantee a spectrum bounded below. Therefore, such modifications are usually not considered seriously. Such an example considered in the frame of classical field theory can serve as an educational example in order to gain experience with "attractive" vector-type fields encountered in non-Abelian meson theories (quantum chromodynamics). Returning for a moment to Eq. (3.26), it should be mentioned that it is in principle possible to find a solution in which the space-vector part of A_μ dominates -- thus allowing a stable solution, even with the conventional choice for the sign of the vector field action. We will return to the discussion of the vector field further below.

In principle, the solutions of the coupled non-linear equations of motion need not be actual stable minima of the action -- they allow the first variation of the action to vanish, but no information is available about the second variations. Let us consider here again, as an example, the case of an externally prescribed potential V_{ex}. Then the Hamiltonian is

$$H = \int d^3 x\,\Psi^\dagger \left(\vec{\alpha}\cdot\vec{p} + \beta m + V_{ex} \right)\Psi \tag{3.29}$$

and the equation of motion follows from equating to zero the first variation of $(H - \varepsilon N)$ with respect to ψ. Here N is the norm of the field ψ, while ε is the Lagrange multiplier that ensures the normalizability of the solution. We find, as usual, the eigenvalue equation

$$\left(\vec{\alpha}\cdot\vec{p} + \beta m + V_{ex} \right)\Psi_k = \varepsilon_k \Psi_k. \tag{3.30}$$

As is well known, the set of eigensolutions $\{\psi_k\}$ is complete. Suppose we take a trial function

$$\psi_t = \sum a_k^t \psi_k \tag{3.31}$$

in an attempt to minimize the Hamiltonian, Eq. (3.29). Using the expansion (3.31) with the complete basis set $\{\psi_k\}$ generated by Eq. (3.30), we find

$$(H - \epsilon_t N) = \sum_k |a_k^t|^2 (\epsilon_k - \epsilon_t) . \tag{3.32}$$

The spectrum of the Dirac equation is unbounded below as is the expression (3.32).

From this example we recognize the need to constrain the number of allowed degrees of freedom of the Dirac field. We must exclude from our considerations the possibility of a transition that a classical Dirac particle can undergo into a (classically) unoccupied state of arbitrary large negative frequency. In quantum field theory such a state corresponds to an antiparticle state of positive energy. We have seen in Section 2 that this constraint arises naturally in the self-consistent bag approximation.

Thus the solutions under investigation here must be obtained by a method that allows *a priori* rejection of all unwanted modes of the Dirac equation. In a particular application, this means that the Dirac equation must always be solved exactly for some prescribed potential and an eigenstate ψ_k obtained.

In the case of the interacting Dirac-scalar fields described in Section 2, we actually minimize the action constrained by this consideration, provided that we solve Eq. (2.23) exactly for a given φ_c. Then we may consider the Hamiltonian to be given by

$$H^s = \epsilon_o[\varphi_c] + \int d^3x \left[\tfrac{1}{2}(\nabla\varphi_c) + U(\varphi_c) - U_o\right], \tag{3.33}$$

where ϵ_0 is given implicitly by the solution of the eigenvalue problem:

$$\left[\vec{\alpha}\cdot\vec{p} + \beta(m - g\varphi_c)\right]\psi_o = \epsilon_o \psi_o . \tag{3.34}$$

The above equation must always be solved exactly in order to obtain the Dirac eigenvalue ϵ_0 as a functional depending on φ_c. Using the apparent relation

$$\frac{\delta\epsilon_o}{\delta\varphi_c(x)} = -g \,\overline{\psi}_o(x)\,\psi_o(x), \tag{3.35}$$

we recover the conventional equation of motion for the field $\varphi_c(x)$ that we now write in the form

$$-\Delta\varphi_c + \frac{\partial U}{\partial\varphi_c} = -\frac{\delta\epsilon_o}{\delta\varphi_c(x)} . \tag{3.36}$$

The question which we consider next is: are the solutions, constrained to the lowest positive frequency (particle) solutions of the Dirac field, stable?

A method employed in the study of classical field theories by Goldstone and Jackiw[10] is the linear transformation

$$x' = a x \tag{3.37}$$

The virial relation (3.4) is then obtained from

$$\frac{\partial}{\partial a} \left[H(ax) / N(ax) \right]\Big|_{a=1} = 0 , \tag{3.38}$$

where N is the norm of the Dirac field, while the stability of the solutions should be tested by computing:

$$T = \frac{\partial^2}{\partial a^2} \left[H(ax) / N(ax) \right]\Big|_{a=1} . \tag{3.39}$$

Naively, T should be positive definite in order to have a minimum against the elongations, Eq. (3.37). However, should we obtain a negative value for T, it may in our case only indicate that a number of negative frequency modes have been included in the trial wave function. With a complete set ψ_k, cf. Eq. (3.30), we may write

$$\psi_o (ax) = \sum_k b_k (a) \psi_k (x) \tag{3.40}$$

and therefore we obtain for the Hamiltonian H^S in Eq. (3.33):

$$H^S[ax] = a^3 \sum_k |b_k(a)|^2 \epsilon_k \left[\varphi(ax) \right] + E_\varphi (ax), \tag{3.41}$$

where E_φ is the energy content of the scalar field. We record that the second variation of Eq. (3.41) includes

$$\frac{\partial^2}{\partial a^2} H^S[ax]\Big|_{a=1} = \sum_k \left| \frac{\partial b_k}{\partial a} \Big|_{a=1} \right|^2 \epsilon_k \left[\varphi(x) \right] + \ldots . \tag{3.42}$$

Since the spectrum is not positive definite, there is no way to determine the sign of the expression (3.42). With

$$\frac{\partial b_k}{\partial a}\Big|_{a=1} = \int \psi_k^\dagger \, \vec{x} \cdot \vec{\nabla} \, \psi_o \, d^3x , \tag{3.43}$$

we recognize that the term (3.42) does not vanish.

An explicit example of a stable theory with non-positive T is found considering the self-interacting Dirac field. We find for the Hamiltonian (3.17):

$$H^P(ax) / N(ax) = \int d^3x \left\{ \psi^\dagger \left[\frac{1}{a} \vec{\alpha} \cdot \vec{p} + \beta m \right] \psi \right.$$

$$\left. - \frac{1}{2a^3} (\bar{\psi}\psi)^2 \right\} / \int d^3x \, \psi^\dagger \psi , \tag{3.44}$$

wherefrom we find for the second derivative

$$\frac{\partial^2}{\partial a^2} \left(H(ax)/N(ax) \right)\Big|_{a=1} = -\frac{6}{2}\frac{G}{2}\int d^3x \,(\bar\Psi\Psi)^2, \qquad (3.45)$$

a result that is explicitly negative definite. This would make us worry about
the stability of the self-interacting field, had we not convinced ourselves that
such a result can arise as a consequence of the indefinite character of the Dirac
equation spectrum and has nothing to do with the stability of the solutions.

IV. Numerical Methods

Our aim in this section is to illustrate the numerical methods involved in
minimization of the mass (or Hamiltonian) M_N in Eq. (2.20) as a function of the
fields ψ_0 and φ_c, with the constraint that ψ_0 be the lowest frequency eigenmode
of the quasi-fermion field.

From the considerations of the last section, we recognize the need to constrain the number of allowed degrees of freedom of the Dirac field, as arose
naturally in the bag approximation. We must exclude from our considerations the
possibility of a transition that a classical Dirac particle can undergo into a
(classically) empty state of arbitrary large negative frequency. In a particular
application this means that the Dirac equation must always be solved exactly for
some prescribed potential, and an eigenstate ψ_0 obtained. We must thus consider
the Hamiltonian to be given implicitly by Eq. (3.33), where ε_0 is given implicitly
by the solution of Eq. (2.23) with the eigenvalue condition

$$\lim_{|\vec{x}|\to\infty} \Psi_0(\vec{x}) = 0. \qquad (4.1)$$

Any trial value of φ_c in Eq. (3.33) will give an upper limit for H^s.

But how do we actually obtain a self-consistent solution for φ_c with ψ_0 constrained as described above to the lowest particle solution? It is very inviting
to proceed with an iterative algorithm in which, for a given spherical $\varphi_{c,n-1}$ and
fixed g and μ/m, a solution ψ_0 of Eq. (2.23) is obtained as described for spherical potentials further below. From Eq. (2.24) we then determine a new function
$\varphi_{c,n}$. In practical calculations it turns out that the radius of convergence of
such an iteration decreases rapidly as a function of g^2, making it virtually impossible to obtain a solution for $g^2 \gtrsim 2$, a region in which we are mainly interested.

I have developed[5] an algorithm that has so far shown an unrestricted radius
of convergence when applied to equations of the type (2.23) and (2.24). Let us
introduce for the purpose of clarity two additional functions

$$\phi_c = g\,\psi_c \tag{4.2}$$

$$f = \psi_c / g \tag{4.3}$$

Equations (2.23) and (2.24) can now be written in the generalized and self-explanatory fashion

$$D[\phi]\,\psi = 0, \tag{4.4}$$

$$Kf = \rho_s[\phi], \tag{4.5}$$

$$\phi_c = g^2 f. \tag{4.6}$$

Here D and K are linear operators, while the source ρ_s is $\bar{\psi}_0\psi_0$. Next we note that, using an arbitrary linear functional L, we can write Eq. (4.6) in the form

$$g^2 = L\phi / Lf. \tag{4.7}$$

We will return to the discussion of the actual form of L further below.

Our algorithm is: given a function ϕ_{n-1} the scalar density is found solving Eq. (4.4). Then we obtain from Eq. (4.5)

$$f_n = K^{-1}\rho_s[\phi_{n-1}] \tag{4.8}$$

and Eq. (4.7) gives us in turn

$$g_n^2 = L\phi_{n-1} / Lf_n. \tag{4.9}$$

The algorithm loop is closed by the use of Eq. (4.6)

$$\phi_n = g_n^2 f_n. \tag{4.10}$$

Note that we have iterated the coupling constant g^2. For a given starting value ϕ_0 a convergent point

$$g^2 = \lim_{n \to \infty} g_n^2 \tag{4.11}$$

may be found that is dependent on the actual form of the initial function ϕ_0. I have found empirically that

$$\phi_c = \lim_{n \to \infty} \phi_n \tag{4.12}$$

is uniquely determined by g^2 and that there is only one g^2 for each given ϕ_0 in the case of Eqs. (2.23), (2.24), and other similar theories discussed further below.

Now, as to the choice of the operator L, the difference between ϕ_{n-1} and $g_n^2 f_n = \phi_n$ needs to be small, in the sense that we are searching for the minimum of the integral

$$I_h = \int w(r) \left[\phi_{n-1} - g_n^2 f_n \right]^2 dr \qquad (4.13)$$

as a function of g_n^2. Here $w(r)$ is a weighting function that may be dependent on ϕ_{n-1}. This prescription gives us

$$L\phi = \int w(r) \, \phi \, dr. \qquad (4.14)$$

The choice $w(r) = 1$ is usually sufficient to have a convergent algorithm. However, $w(r) = \rho_s[\phi_{n-1}]$ gives a quicker convergence in most cases. I would like to conjecture that the algorithm described above should be convergent in most similar cases of physical interest, provided that a suitable choice for the operator L has been made.

Although Eqs. (2.23) and (2.24) look very simple, it is extremely difficult to obtain a self-consistent solution. In particular, we have been able only to analyse in detail the spherically symmetric case. With the usual ansatz for the ground-state wave function

$$\psi_0(x) = \frac{1}{r(4\pi)^{1/2}} \begin{pmatrix} u\, \xi_s \\ i\, \dfrac{\vec{\sigma}\cdot\vec{x}}{|x|}\, v\, \xi_s \end{pmatrix}, \qquad (4.15)$$

where ξ_s are the usual spin functions, we find the *radial* equations following from Eqs. (2.23) and (2.24)

$$u' - r^{-1}u - (m - g\varphi_c + \epsilon_0)\, v = 0 \qquad (4.16)$$

$$-v' - r^{-1}v + (m - g\varphi_c - \epsilon_0)\, u = 0 \qquad (4.17)$$

$$-r^{-1}(rg\varphi_c)'' + \mu^2 g\varphi_c = Ng^2(u^2 - v^2)r^{-2}. \qquad (4.18)$$

Both spin states have the same equations of motion. Thus not only SU(3) (flavour) but also SU(6) (flavour + spin) are exact symmetries at this point. Furthermore, since N enters as a multiplicative factor, the solution is only a functional of (Ng^2), as noted in Eq. (2.26). A solution of Eqs. (4.16) to (4.18) will be obtained by an iterative procedure.

V. Properties of Bound States of Scalar Self-Consistent Bags

Properties of bound states obtained within several models will be described here. In particular, we are interested in the geometric properties of the quasi-fermion amplitudes ψ_0 and mass densities $\mathcal{M}$. Furthermore, we consider also g^2 and μ/m systematics of the observable matrix elements of ψ_0, such as the charge radius, the magnetic moment of the proton, and the axial coupling constant of the semi-leptonic charged current.

It is my feeling that although the average properties (involving matrix elements) of our variational solutions, belonging to some specific models, may be closer to reality than should be expected, a more detailed description is required for comparison of the theoretical and experimental form factors. In particular, we should recall that the amplitudes ψ_0 are, strictly speaking, transition matrix elements between N and N $\pm$ 1 fermion states, and not probability amplitudes, as in the case of weakly interacting fields. Also, we have considered here only "valence" quarks -- the sea quarks do also contribute significantly, in particular to the inelastic form factors.

In Fig. 1 we show some of the typical solutions associated with the Lagrangian (2.1) of the interacting fermion-scalar gluon fields[3] for $U(\phi^2) = \tfrac{1}{2}\mu^2\phi^2$ and N = 3. The self-consistent quasi-particle (bound quark) mass

$$m^* = m_q - g\,\varphi_c \tag{5.1}$$

is plotted for several values of the gluon mass μ/m_q and the coupling constant g^2 as a function of $r = |\vec{x}|$. For $\mu = 0.04m_q$ and $\mu = 0.02m_q$ we also show the vector

$$\rho_v = \psi_0^+ \psi_0 \tag{5.2}$$

and scalar

$$\rho_s = \bar{\psi}_0 \psi_0 \tag{5.3}$$

densities. We note that, in Fig. 1, ρ_s is enhanced by a factor of 10 in relation to ρ_v. For μ of the order of the bare quark mass m, $m^*(r)$ has a pronounced minimum. For coupling constants smaller than those considered here, i.e. for $g^2 \lesssim 10$, we have also found solutions *that were evenly distributed over the volume*. The transition from the volume to surface solution as a function of g^2 and μ/m is smooth. Qualitatively this behaviour can be easily understood in view of the particular nature of the scalar coupling. Disregarding the spin effects, the spectrum equivalent Schrödinger potential is given by

$$V_{eff} \sim \frac{(m - g\varphi_c)^2}{2m} \tag{5.4}$$

We notice that if $g\varphi_c(0) > m$, then a minimum of V_{eff} will occur at $m = g\varphi_c(R)$. Depending on the precise form of $g\varphi_c$, this minimum may be broad enough to support a wave function localized around R. Alternatively, only a minor perturbation of the volume solution may be found.

In Fig. 2 the mass density $\mathcal{M}$ is shown as a function of r, as defined by Eq. (2.16) for N = 1. We note, in particular, the existence of "surface" solutions for which $\mathcal{M}$ peaks at a certain distance R for large values of the coupling constant, while there are also volume solutions for smaller coupling constants.

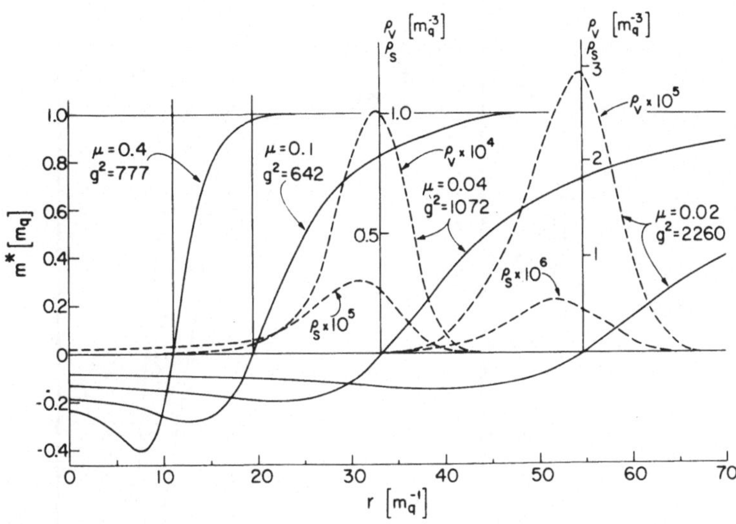

Fig. 1

The effective quasi-particle mass $m^* = m_q - g\phi_c$ (in units of m_q) as a function of r (in units of m_q^{-1}) for μ/m_q = 0.02, 0.1, 0.4 and g^2 = 2260, 1072, 642, 777, correspondingly. The dashed curves are the vector and scalar densities for $\mu = 0.4m_q$, $g^2 = 642$ and $\mu = 0.02m_q$, $g^2 = 2260$. The vector density is normalized to one.

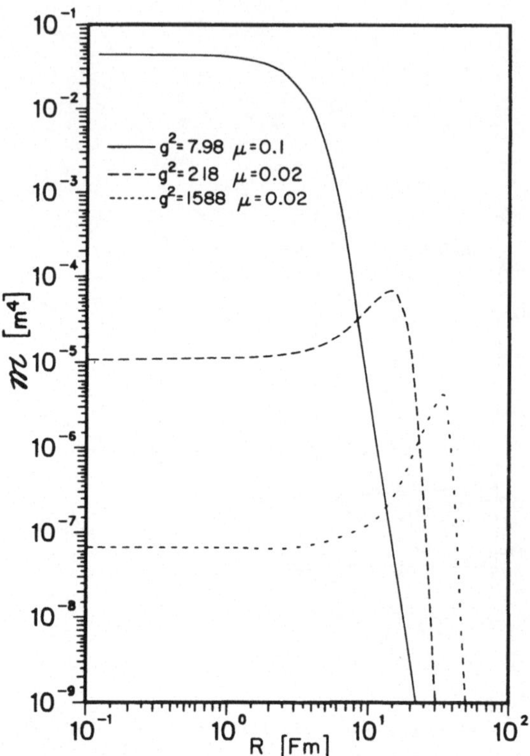

Fig. 2

The mass density of the soliton solution as a function of r; several values of the characteristic parameters g^2 and μ in a double-logarithmic plot.

In Fig. 3 the masses[3] of the bound states M_N are given as functions of g^2 for the gluon masses of 0.4, 0.1, 0.02 (m_q). We see that smaller gluon masses give considerably smaller bag masses. As g^2 rises we would find a point g_c^2 such that

$$2 M_2 < \mu \, , \quad g^2 > g_c^2 \, (\mu/m_q) \, . \tag{5.5}$$

For $g^2 > g_c^2$ it can be argued that no free gluons can exist owing to their strong coupling to the hadronic states. As g^2 is increased with a fixed gluon mass μ, M_N decreases monotonically.

However, weakly bound exotic states are found as shown in Fig. 3. These states are conglomerates of the allowed hadronic states. This lack of saturation is particular for the scalar interaction. When other (repulsive) interactions are included, such states may easily become unbound.

There is an easy way of understanding the functional form of the curves shown in Fig. 3. The mass consists of two terms, cf. Eq. (2.20),

$$M_N = M_f + M_g \, , \tag{5.6}$$

where M_g is the energy contained in the gluon field φ_c. For *large* g^2 the eigen-frequency of the Dirac equation solution is

$$\epsilon_0 = R^{-1} \, , \tag{5.7}$$

where R is the radius of the bag. This relation is confirmed by results shown in Fig. 4, where the product $\epsilon_0 \cdot R$ is shown as a function of g^2, for large values of g^2. Let us assume that

$$M_g = \gamma \, g^2 R^\beta \, , \tag{5.8}$$

where β may be considered as a slowly varying function of R. The constants γ and β depend also upon μ/m. Thus we have

$$M_N = \frac{N}{R} + g^2 \gamma R^\beta = \frac{N}{R} \left(1 + \frac{g^2}{N} \gamma R^{\beta+1} \right) . \tag{5.9}$$

Upon minimization at fixed g^2 and μ/m with respect to R

$$\frac{\partial M_N}{\partial R} = 0 = -\frac{N}{R^2} + g^2 \gamma \beta R^{\beta-1} \, . \tag{5.10}$$

We obtain

$$\frac{g^2}{N} \gamma R^{\beta+1} = \beta^{-1} \, , \tag{5.11}$$

and for Eq. (5.9)

$$M_N = M_f \left(1 + \beta^{-1} \right) . \tag{5.12}$$

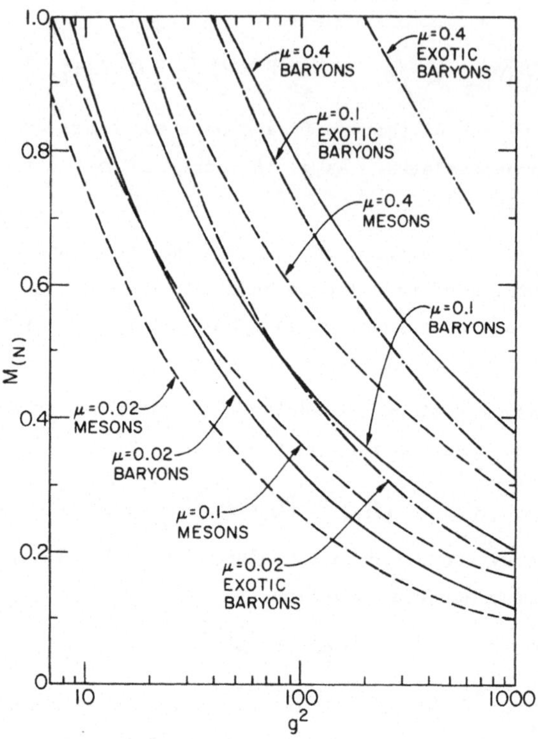

Fig. 3

The total mass M_N in units of the quark mass of several quark states; baryons (N = 3), mesons (N = 2), and exotic baryons (N = 6) as a function of the coupling constant g^2 for several values of the gluon mass μ/m_q = 0.02, 0.1, 0.4.

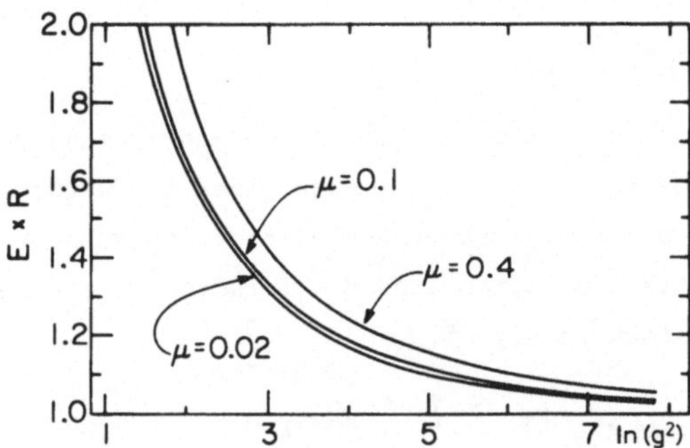

Fig. 4 :   The product of the lowest positive Dirac eigenvalue with r.m.s. radius of the amplitude ψ_0 as a function of $\ln g^2$.

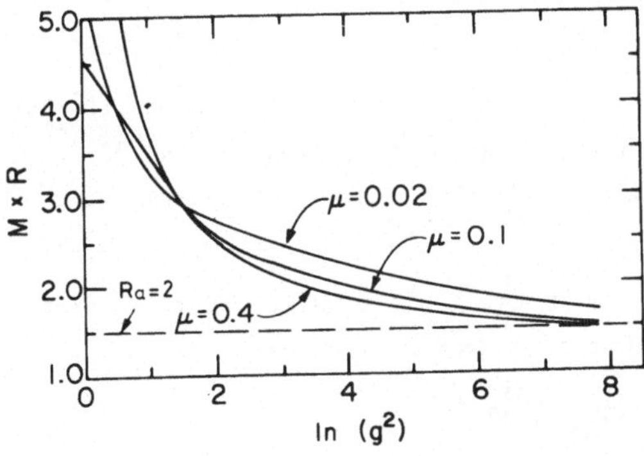

Fig. 5 :   The product of the mass of the quark bag with the r.m.s. radius as a function of $\ln g^2$.

Defining

$$R_a = M_f / M_g, \qquad (5.13)$$

we find

$$R_a = \beta \qquad (5.14)$$

Thus the ratio of the fermionic to the gluonic part of the hadron mass has a geometrical meaning, since the value of β, Eq. (5.8), determines the nature of the solution. For $\beta = 3$ we have a volume bag with the glue distributed over the whole hadron. For $\beta = 2$ we have a surface solution, and $\beta = 1$ indicates the possibility of a string-like solution. As shown in Fig. 5,

$$P = M_N R/N \simeq 1 + 1/R_a \qquad (5.15)$$

tends to the value 1.5 for large g^2, consistent with $R_a \lesssim 2$. We recall that for the SLAC bag[1] $R_a = 2$ (surface bag) and for the MIT bag[11] $R_a = 3$ (volume bag).

We can use relation (5.11) to eliminate R completely from Eq. (5.9). Then, with R_a as the second variable apart from g^2 we find

$$M_N \sim N^{R_a/(R_a+1)} \, (g^2)^{-(R_a+1)^{-1}}, \qquad (5.16)$$

where the dependence on N follows from Eq. (2.26) which can also be written as

$$M_N(g^2) = N M_1(N g^2). \qquad (5.17)$$

In passing we note that the lack of saturation of the scalar interaction is apparent in Eq. (5.16). The energy per quark M_N/N is proportional to $N^{(R_a+1)^{-1}}$ and is therefore decreasing with an increasing number of quarks in the bag (of course, only colour singlets are considered).

We consider the limit $g^2 \to \infty$, $m_q \to \infty$ in Eq. (5.16), while μ remains finite. If there is a relation between the values of m_q and g^2 such as

$$g^2 \sim \left(\frac{m_q}{\mu}\right)^{R_a+1}, \qquad (5.18)$$

then M_N stays constant in the limit $(m_q, g^2) \to \infty$ and we have permanent confinement in the sense that, though the free quark mass m_q is infinite, the mass of the bound state remains finite. It is apparent that the relation (5.18) between the parameters of the theory is rather arbitrary. The above remark is meant only to alert the reader to the fact that the ratio of the bound state mass M to the free quark mass m_q can be made arbitrarily small.

We now proceed to calculate other hadronic properties. We take wave functions with exact SU(6) symmetry of spin and flavour, and an extra internal quantum number, colour, in which the baryonic wave functions are antisymmetric, thus following the concepts set forth in the SLAC bag paper[1]). The baryon multiplet $\underline{56}$ has 56 states counting the spin states. Since the absolute value of the free quark mass m_q is not known, we choose to consider only quantities that are scale-independent. The simplest ones are the products of the r.m.s. radius of the baryons and mesons with their masses; N·P, Eq. (5.15), which are shown in Figs. 6a and 6b. The experimental number to compare with for baryons is, most likely, the product of the proton charge radius, 0.8 fm, with the average mass of the $\underline{56}$ multiplet, $M_{56} = 1280$ MeV, which is $5.2\hbar c$. We see that this lies well within values spanned by our calculations. We note that 5.2 implies for $N = 3$ and $\varepsilon_0 \sim 1/R$ that $R_a \sim 1.7$. Thus it would seem that hadrons are surface-like, or $\varepsilon_0 \not\sim 1/R$.

We can also calculate the absolute value of the magnetic moment of the proton, using as the basic unit $e\hbar/2M_{56}c$, that is the Bohr magneton *in units of the computed mass* (rather than the experimental value). Therefore the unknown quark mass m_q cancels out from our result. For comparison with the experiment we must scale up the experimental value of μ_p to account for the larger multiplet mass than that of the proton[3]); we obtain for comparison with Fig. 6c a value

$$\mu_p^{exp} = 2.79 \times 1280/938 = 3.8.$$

We see that the self-consistent bag results are of the right magnitude. We consider also the axial coupling constant g_A of the neutron decay process. The experimental value is $g_A^{exp} = 1.25$. The results shown in Fig. 6d are in the range $0.6 < g_A < 1$.

Finally, in Fig. 6e the ratio R_a as computed from the solutions is shown. We find $1 < R_a \lesssim 2$, indicating that the self-consistent quark bag is mostly surface-like. The small values of R_a for *small* g^2 cannot be as clearly associated with β, Eq. (5.8), since relation (5.7) is not satisfied in this range of g^2, see Fig. 5.

As the final point on the subject, let us note that there is a simple relation between g_A, μ_p, and M_N, that can be derived using the Dirac equation (2.23) and which can be cast into the form

$$\mu_p = (0.5 + 0.3\,g_A) \cdot N \cdot (1 + R_a^{-1})$$
$$= 2.625\,(1 + R_a^{-1}). \tag{5.19}$$

In the second line of Eq. (5.19) we have used $g_A = 1.25$, $N = 3$. Thus if we were able to find a $U(\phi)$ such that μ_p would also have the desired value 3.8, R_a would necessarily turn out to be 2.2. This means that in a "perfect" scalar bag about 70% of the mass would be carried by the quarks. We further note that the bag could be volume-like. Since g_A must be 1.25, we know that a solution with $\varepsilon_0 \langle r^2 \rangle^{1/2} > 1$ is needed. In this case the value $R_a = 2.23$ does not imply a surface-like solution. In view of the above remarks, it is quite possible that a theory based on scalar interaction with very good phenomenological properties could be found.

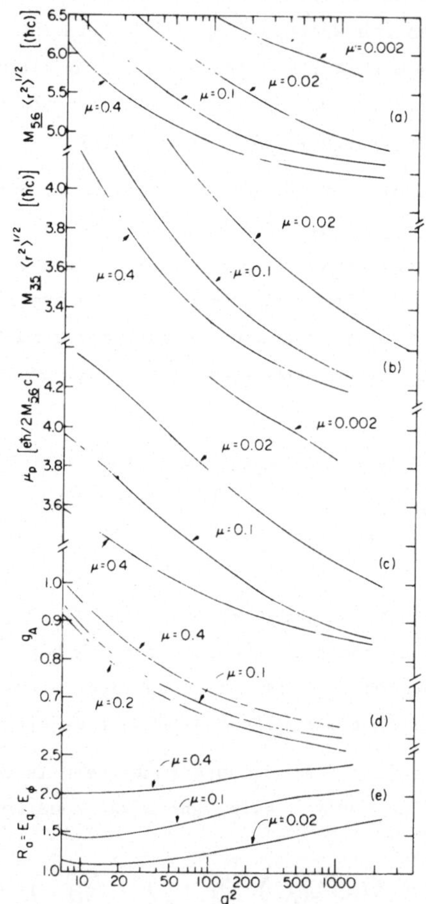

Fig. 6

The structure constants of SU(6) hadrons as a function of the coupling constant g^2 for several values of the gluon mass, μ/m_q = 0.02, 0.1, 0.4. a) The product of the mass of the baryon with its size. b) The same as in (a) for mesons. c) The magnetic moment of protons. d) The axial-coupling constant g_A. e) The ratio R_a of the energy (momentum) carried by the quarks to that carried by the neutral glue.

VI. Properties of Solutions of a Self-Coupled Dirac Field

We have argued that localized, stable solutions of classical field equations are the basic ingredients in the description of extended objects in quantum field theory. We now turn to the discussion of the self-coupled Dirac field $\bar{\psi}$ in *three-dimensional* space governed by the action

$$L = \int d^4x \left[\bar{\psi}\, \gamma \cdot p\, \psi - m_q \bar{\psi}\psi + \frac{G}{2}(\bar{\psi}\psi)^2 \right], \tag{6.1}$$

where m_q is the mass of the free ($G = 0$) fermion field. G, like the weak coupling constant, has the dimension of inverse mass squared.

The quantum field theory associated with Eq. (6.1) is known to lead to a non-renormalizable perturbation expansion in the vicinity of $G = 0$. I believe that the difficulties associated with the perturbation expansion do not provide a compelling argument for rejection of the Lagrangian (6.1). In particular it is quite possible that the properties of the theory change with increasing strength of the coupling G in a non-analytic fashion. I find[6] that for $G > G_{min} = 4.47/(4\pi m_q^2)$ localized solutions of the classical field equation arise.

Another way of seeing the self-interaction invokes the infinite gluon mass limit of a scalar meson field discussed in the last section. We have considered the properties of the Hamiltonian (3.17) associated with the self-coupled field already in Section 3. We therefore turn now to the discussion of the properties of the solutions obtained by the methods introduced in Section 4, with the modification that the equation defining a field ϕ in terms of $\bar{\psi}_0\psi_0$ is very simple now and reads

$$\phi = G\, \bar{\psi}_0 \psi_0 . \tag{6.2}$$

From this point on the iteration procedure can proceed as described previously. Use of numerical continuation of the solutions as a function of the coupling constant increased the convergence speed of the iteration. The solutions ψ_0 that will be discussed below are normalized to unity. We note a similar scale invariance as already noted with meson fields, Eq. (2.26); given an arbitrary norm

$$Q = \int d^3x \, \psi_0^+ \psi_0 , \tag{6.3}$$

we can introduce

$$\psi_0' = Q^{1/2} \psi_0 . \tag{6.4}$$

Then ψ_0' is a solution of the problem with the parameter

$$G' = G/Q , \tag{6.5}$$

that leads to the total energy

$$H'(G/Q) = H(G)\, Q .$$

(6.6)

The dimensionless parameter of the model is

$$\beta = m_q^2\, G/4\pi .$$

(6.7)

The lowest mass solution associated with lowest positive frequency solution has been computed. The energy (H/m_q) and the frequency (ε_0/m_q) are shown in Fig. 7a. We recognize that both functions fall monotonically with increasing β. We also find a minimum value of G for which the mass of the bound state is smaller than m. It is $G_{min} = 4.47/(4\pi m_q^2)$, as mentioned previously. Interestingly enough, the qualitative behaviour of $H(Gm_q^2)$ changes when H is plotted in units of $G^{-1/2}$, as shown in Fig. 7b. We find

$$(H/G^{-1/2}) = (4\pi\beta)^{1/2}\, (H/m_q),$$

(6.8)

which monotonically rises. The lines in Fig. 7a may be understood also as obtained for fixed m with changing G, while in Fig. 7b, G is fixed and m_q varies.

In Fig. 8 the forms of the solutions for some representative values of $\beta = 22, 475, 2125$ are shown. The effective mass m^* as a function of r, is shown in Fig. 8a. With increasing coupling strength the size of the solution increases significantly, while the value of m^* at origin decreases. Notably, it remains always negative. In Figs. 8b, c, and d, the vector ρ_v and scalar ρ_s densities belonging to the same solutions are shown. As β increases, the vector density becomes more and more localized around the point where $m^* = 0$. The scalar density remains distributed over the whole volume of the solution. In that respect the self-interacting field solutions differ from the solutions involving a gluon field. There we have seen that the scalar density, like the vector density, is localized at the surface of the solutions. The volume character of ρ_s in the present case is necessary to allow the relatively constant effective mass m^* over the space occupied by the solution.

In view of the results presented in Figs. 7 and 8, the self-interacting fermion field discussed here may also be taken as a prototype of a self-consistent quark bag *without gluons*. All we need to do in order to generate the bag solutions from the soliton solutions given above is to perform the transformation described in Eqs. (6.3) to (6.5) with Q = N. N is the number of quarks: N = 3 for baryons, N = 2 for mesons (the scalar self-coupling also does not distinguish between quarks and antiquarks). A self-interacting field provides a natural explanation of the phenomenon of quark confinement in that the free particle of mass m_q cannot be considered as a quantum of the fermion field for $G > G_{min}$. Also, the absence of the gluons in the quark bag is satisfactory.

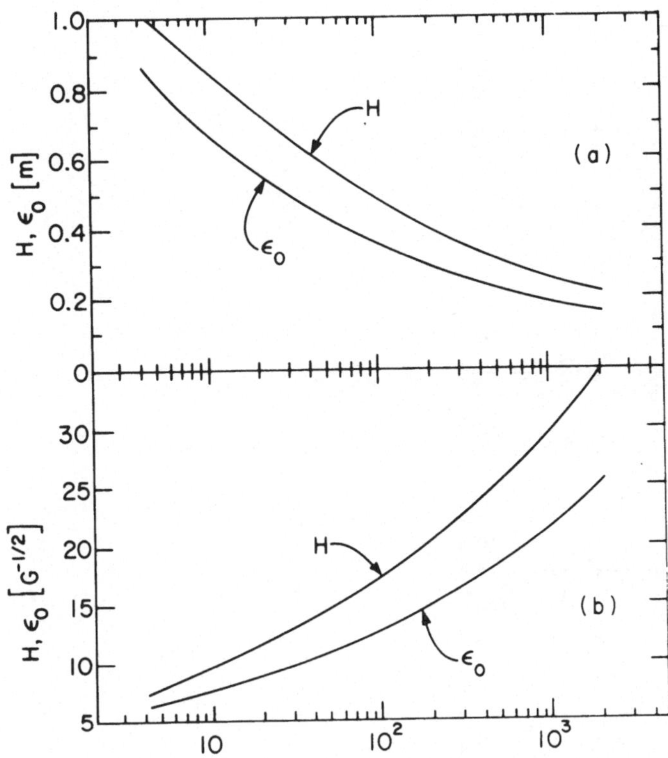

Fig. 7 : The energy H and the eigenfrequency ϵ_0 as a function of the dimensionless coupling constant β: a) in units of m_q; b) in units of $G^{-1/2}$.

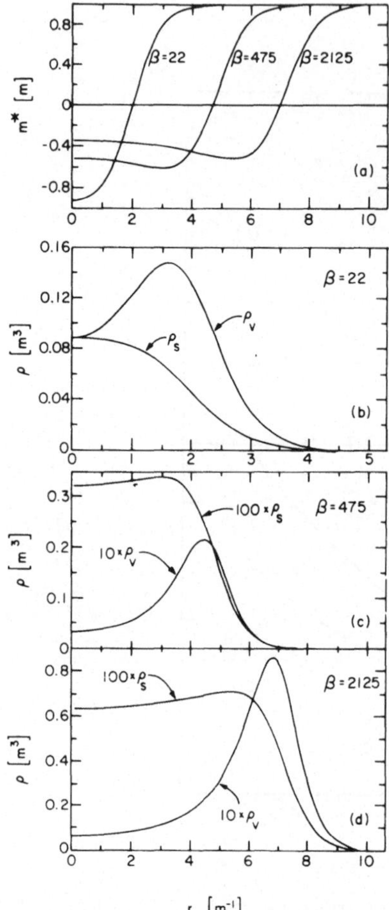

Fig. 8 :   The soliton solutions of self-coupled Dirac fields for β = 22, 475, 2125.   a) The effective mass m^*;   b), c), d) the scalar and vector densities.

VII. Supercritical Fields in Quantum Field Theory

In this section we shall consider the supercritical binding ($\epsilon_0 < -m$) in the frame of the bag approach. Since some of the theoretical details concerning the bag approximation differ slightly from our previous considerations, we will briefly rederive the self-consistent c-number equations, with particular emphasis on the new case $\epsilon_0 < -m$. A simple and possibly relevant example of supercritical binding arises when considering strongly bound states in the (classical) field theory of *attractive* massive vector mesons interacting with fermion fields.

We consider a fermion field coupled to a vector meson field with the Hamiltonian[9] derived from a slight generalization of Eq. (3.25) (with $A \cdot A \equiv A^2 = A_0^2 - \vec{A} \cdot \vec{A}$)

$$H = \tfrac{1}{2} \left\{ \int d^3x \, \tfrac{1}{2} \left[\hat{\Psi}^+ (\vec{\alpha} \cdot \vec{p} + \beta m - g_v(\vec{\alpha} \cdot \hat{\vec{A}} - A^0)) \, \hat{\Psi} \right] \right.$$
$$\left. - \tfrac{1}{2} \int d^3x \left[\hat{\pi}_\mu^2 + (\nabla \hat{A}_\mu)^2 + W(\hat{A} \cdot \hat{A}) - W(0) \right] + h.c. \right\}, \tag{7.1}$$

where $W(A^2)$ is an arbitrary potential energy of the vector field A_μ, which leads to the equations of motion

$$(\vec{\alpha} \cdot \vec{p} + \beta m - g_v(\vec{\alpha} \cdot \hat{\vec{A}} - A_0)) \, \hat{\Psi} = i \frac{\partial \hat{\vec{\Psi}}}{\partial t} \tag{7.2}$$

$$\hat{\pi}_\mu - \Delta A_\mu + \hat{A}_\mu \frac{\partial W}{\partial A^2} = \frac{g_v}{2} \left[\hat{\bar{\Psi}}, \gamma_\mu \Psi \right] . \tag{7.3}$$

We also have the auxiliary condition

$$\langle ps | \, \partial_\mu \hat{A}^\mu \, | ps \rangle = 0 \tag{7.4}$$

for any physical state $|ps\rangle$.

For strong interactions it is most convenient to choose as the basis for the expansion of the Fermi field ψ the complete set of solutions of the Dirac equation with an average field A_c (p stands for both discrete and continuous indices):

$$(\vec{\alpha} \cdot (\vec{p} - g_v \vec{A}_c) + \beta m + g_v A_c^0) \Psi_p = \epsilon_p \Psi_p. \tag{7.5}$$

The field operator is expanded in the quasi-particle Fock space as

$$\Psi(\vec{x}, 0) = \int_m^\infty b_\epsilon \Psi_\epsilon(\vec{x}) \, d\epsilon + \sum_{m > \epsilon_p \geq -m} b_p \Psi_p(\vec{x}) + \int_{-\infty}^{-m} d_\epsilon^+ \Psi_\epsilon(\vec{x}) \, d\epsilon, \tag{7.6}$$

whenever the field A_c^μ is undercritical. A different expansion is appropriate, when one of the bound states of Eq. (7.6) is embedded in the negative energy continuum *and* the total energy remains positive. We explain this remark further below. Then we have

$$\Psi(\vec{x},0) = \int_m^\infty b_\epsilon \Psi_\epsilon \, d\epsilon + \sum_{m > \epsilon_p > -m} b_p \Psi_p + b_0 \int_{-\infty}^{-m} a(\epsilon) \Psi_\epsilon \, d\epsilon$$

$$+ \int_{-\infty}^{-m} d\epsilon'^\dagger \left[\int_{-\infty}^{-m} h_{\epsilon'}(\epsilon) \Psi_\epsilon \, d\epsilon' \right], \tag{7.7}$$

where $a(\epsilon)$ and $h_{\epsilon'}(\epsilon)$ are suitable functions that describe a particle resonance
in a negative energy continuum[12]. If more than one resonance is embedded in the
continuum, a generalization of Eq. (7.7) is in order.

It is important to stress here the difference between the self-consistent
field problem and the case of an external field. The neutral vacuum, charac-
terized by $b_p|0\rangle = d_n|0\rangle = 0$, may be a *stable,* zero-energy state even should super-
critical binding described by (7.7) occur, since the total energy of the excited
state $b_0^\dagger|0\rangle$ may be positive upon consideration of the meson energy included in
Eq. (7.1), in contradistinction to the situation with *external* potentials. To
put it in another way, when fermions move in a *self-consistent* (shell) potential,
then the total energy of the bound state is given by the sum of the Dirac eigen-
energies and of the energy associated with the shell potential, sometimes also
called "correlation energy". The stability condition of the conventional ground
state characterized by $b_p|0\rangle = d_n|0\rangle = 0$ is

$$\epsilon_0 + E_{meson} > 0 \tag{7.8}$$

and not

$$\epsilon_0 > -m \tag{7.9}$$

as would be the case if external fields were considered. It must be expected that
the critical potential strength at which $\epsilon_0 = -m$ can be exceeded by the *self-
consistent* potential and a resonance be embedded in the negative energy continuum
of the Dirac equation. In view of Eq. (7.8) we recognize that even if the self-
consistent field is supercritical, the neutral vacuum may remain stable. This
last remark can only be appreciated after a thorough discussion of the case of an
external potential[12].

For the sake of argument, let us assume now that we have found a theory such
that Eq. (7.8) is violated. We then would find that a charged ground state $b_0^\dagger|0\rangle$
has lower energy than the neutral state $|0\rangle$ -- then the latter state is the false
vacuum, which under certain circumstances could undergo a transition into the true
vacuum. Fortunately, we will find below that it is difficult to violate condition
(7.8). I will show here that Eq. (7.8) is satisfied in the model discussed, although $\epsilon_0 < -m$, beyond a certain coupling strength. This means that despite the
resonance of the Dirac equation we do not have a resonance of the Hamiltonian.

Our choice of the mean field A_c^μ was directed by our desire to eliminate from the ground state most contributions from the virtual quasi-particle excitations.

Therefore $A_c^\mu = \langle ps|A^\mu|ps\rangle$ and we obtain

$$-\Delta A_c^\mu + A_c^\mu \left(\frac{\partial W}{\partial A^2}\right)_{A_c^2} = g_v \langle ps|\tfrac{1}{2}[\hat{\bar{\psi}}, \gamma^\mu \hat{\psi}]|ps\rangle. \tag{7.10}$$

We consider the state $b_0^\dagger|0\rangle = |p\rangle$ as a trial state. Then Eq. (7.10) becomes (neglecting virtual vacuum polarization charge density):

$$-\Delta A_c^\mu + A_c^\mu \left(-\frac{\partial W}{\partial A^2}\right)_{A_c^2} = g_v \psi_0 \gamma^\mu \psi_0. \tag{7.11}$$

Here ψ_0 is the coefficient of b_0 in Eq. (7.7) or the lowest energy particle wave function in Eq. (7.6).

At this point, the interaction must be modified[9] to allow bound fermion states, bound only by "electric" forces. That is to say the sign on the right-hand side of Eq. (7.11) has to be changed, in order to obtain localized solutions with $\vec{A}_c = 0$ [see Eqs. (3.25) to (3.28)]

$$-\Delta A_c^\mu + A_c^\mu \left(\frac{\partial W}{\partial A^2}\right)_{A_c^2} = -g_v \psi_0 \gamma^\mu \psi_0. \tag{7.12}$$

This change is also motivated by the consideration of possible internal symmetry of the fields of interest -- the sign may be understood in analogy with the different possible signs of the z-component of the isospin. This change also means that the energy of the classical bound state becomes

$$E_v = E_{Dirac} + E_{meson} = \int d^3x\, \psi_0^\dagger (\vec{\alpha}\cdot\vec{p} + \beta m - g_v(\vec{\alpha}\cdot\vec{A} - A^0))\psi_0$$
$$+ \tfrac{1}{2}\int d^3x\left[(\nabla A_c^\mu)^2 + W(A_c\cdot A_c) - W(0)\right] = \int d^3x\, \mathcal{H}, \tag{7.13}$$

where we find

$$E_{Dirac} = \epsilon_0 \tag{7.14}$$

In view of the arbitrary (but motivated) changes involved in writing (7.13), what follows should be understood only as illustrating a type of behaviour may be characteristic to the Yang-Mills fields.

Let us devote the rest of our discussion to the special case

$$W = \mu^2 A^2, \tag{7.15}$$

where μ is the vector meson mass. We recall from Eq. (3.28) that the total energy is given by

$$E_{eff}^A = \int d^3x\, (\mu_v^2 A^2 + m\bar{\psi}_0 \psi_0). \tag{7.16}$$

The energy E_{eff}^{A} is positive definite if the scalar integral $\int d^3x \, \bar{\psi}_0 \psi_0$ is positive. No lower bound for the scalar integral is known when vector fields are present, but in all examples studied numerically it has been found to be positive definite and very small for large g_v^2.

Equations (7.5) and (7.12) have been solved numerically with $W = \mu_v^2 A^2$ for the lowest energy particle state such that the Dirac wave functions (either eigenstates or resonances) are normalized to one. In Fig. 9 we show the energy of the bound fermion, E_{Dirac} [Eq. (7.14)], and the total energy E_v [Eq. (7.13)], as a function of the coupling constant g_v for vector meson masses $\mu = 0.2$ and $0.4m$. The energy is measured as usual in units of the bare fermion mass m. In this example, we see that the total energy is always positive for coupling constants $g_v^2/4\pi \gtrsim 4$, while the Dirac eigenvalues E_{Dirac} take large negative values. The scalar integral is a negligible contribution to E_v, Eq. (7.13), for strong coupling. Exact agreement of numerical results with those expected from the virial theorem [Eq. (7.16)] (as long as $E_{Dirac} > -m$) has been found. This is also a good check of the computer codes involved in solving the non-linear equations (7.5) and (7.12).

The main conclusion that can be drawn from this exercise is the recognition that in this simple model *the total energy of the strongly bound state lies above that of the usual vacuum,* i.e. $E_v > 0$ even when the self-consistent potential is supercritical, i.e. $\epsilon_0 < -m$. Therefore the usual vacuum is the stable ground state of the theory. Further we note that essential in obtaining a proper solution of Eqs. (7.5) and (7.12) is a proper treatment of resonant states in the antiparticle continuum when they occur. The proper description of the system involves the amplitude:

$$\psi_0 \cong \int_{-\infty}^{-m} a(E) \, \psi_E \, dE. \tag{7.17}$$

Without the proper understanding of the resonant spectrum a solution would seem discontinuous at the critical points, since we would take the lowest *discrete* state in place of the resonance.

Another way of seeing the difference between the presently considered case and the external field problem is to consider the singularity structure of the time transform of the Green's function

$$\tilde{S}(\vec{x}, \vec{x}'; \epsilon) = \int d(t-t') \, e^{-i\epsilon(t-t')} \, S_p(x, x'), \tag{7.18}$$

where p indicates the dependence of S on the path chosen in the ϵ plane when inverting Eq. (7.18). Two such paths are shown in Fig. 10 implied by Feynman boundary conditions. The integration path C' of Fig. 10 should be taken to obtain the

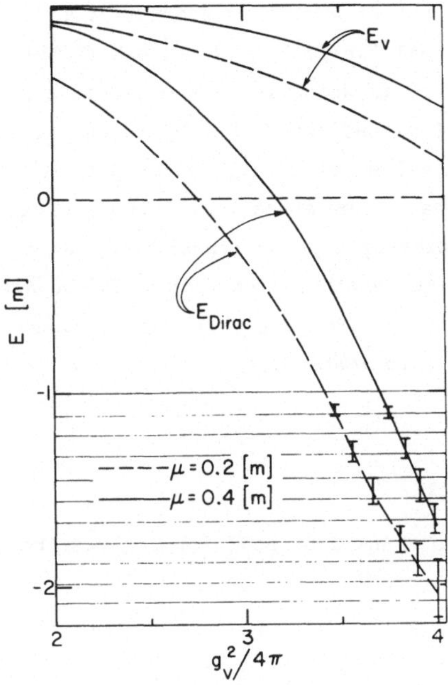

Fig. 9

The Dirac eigenvalue and total energies of interacting Dirac-vector meson fields.

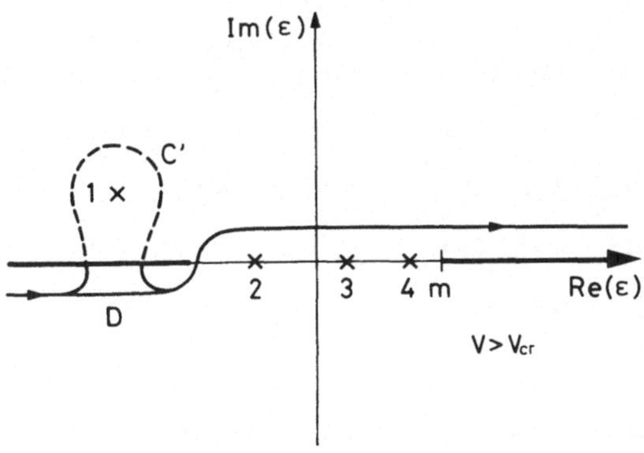

Fig. 10

The contours and the singularities of the Green's function in over-critical potentials.

Green's function in the case of the interacting fields, and not the path D used in the case of external fields[12]. We note here that the pole in the complex ε plane shown in Fig. 10 is actually on the second ε-sheet. A quite different result would follow if we had found that E_v in Eq. (7.13) is actually a negative quantity at certain values of g_v^2. Then the path D would have to be taken and we would expect that the vacuum state is the charged vacuum. We would then face the problem of a macroscopic phase transition. At every point in space the vacuum would be charged. Thus another procedure would be needed to determine the new ground state, since it is impossible to describe such a global change of the world in the single-mode approximation.

Acknowledgement

I would like to thank J.S. Bell for reading the manuscript and his helpful comments.

References

1) W.A. Bardeen, M.S. Chanowitz, S.D. Drell, M. Weinstein and T.-M. Yan, Phys. Rev. D 11, 1094 (1975).
    P. Vinciarelli, Nuovo Cimento Letters 4, 905 (1972) and Nuclear Phys. B89, 463 (1975).

2) J.C. Pati and A. Salam, Phys. Rev. D 10, 275 (1974).
    Q. Shafi, Phys. Letters 69B, 473 (1977).

3) J. Rafelski, Phys. Rev. D 14, 2358 (1976).

4) D.E.L. Pottinger and R.J. Rivers, Nuclear Phys. 117, 189 (1976).
    R. Friedberg and T.D. Lee, Phys. Rev. D 15, 1694 (1977).

5) J. Rafelski, Nuovo Cimento Letters 17, 575 (1976).

6) J. Rafelski, Phys. Letters 66B, 262 (1977).

7) J. Rafelski, Phys. Rev. D 16, 1890 (1977).

8) M. Creutz, Phys. Rev. D 12, 3126 (1975).

9) J. Rafelski and B. Mueller, Phys. Rev. D 14, 3532 (1976).

10) J. Goldstone and R. Jackiw, Phys. Rev. D 11, 1486 (1975).

11) A. Chodos, R.L. Jaffe, K. Johnson, C.B. Thorn and V.F. Weisskopf, Phys. Rev. D 9, 3471 (1974).

12) J. Rafelski, L.P. Fulcher and A. Klein, Fermions and bosons interacting with arbitrarily strong external fields, to be published in Phys. Reports.